Visual Science

Proceedings of the
1968 International Symposium

International Symposium on Visual Science

Planning Committee Members

H. Ward Ewalt, Jr., Chairman
Pittsburgh, Pennsylvania

Anthony J. Adams
Division of Optometry
Indiana University
Bloomington, Indiana

Garland Clay
Ardmore, Oklahoma

Henry B. Peters
School of Optometry
Medical Center
University of Alabama
Birmingham, Alabama

Robert C. Phillips
New Brighton, Pennsylvania

John R. Pierce
School of Optometry
Medical Center
University of Alabama
Birmingham, Alabama

Bernard W. Shannon
Mauston, Wisconsin

Melvin D. Wolfberg
Selinsgrove, Pennsylvania

Visual Science

*Proceedings of the
1968 International Symposium*

Edited by

John R. Pierce

John R. Levene

INDIANA UNIVERSITY PRESS

BLOOMINGTON AND LONDON

Contents

v

Part IV.   Clinical and Applied Aspects of Visual Science

List of Contributors

S. HOWARD BARTLEY, Department of Psychology, Michigan State University, East Lansing, Michigan

IRVIN M. BORISH, Division of Optometry, Indiana University, Bloomington, Indiana

JOHN LOTT BROWN, Kansas State University, Manhattan, Kansas

JOHN H. CARTER, Pennsylvania College of Optometry, Philadelphia, Pennsylvania

RAYMOND CROUZY, National Museum of Natural History, Paris, France

RUSSELL De VALOIS, School of Optometry, University of California, Berkeley, California

JAY M. ENOCH, Department of Ophthalmology, Washington University, Saint Louis, Missouri

RICHARD L. FEINBERG, National Institute of Neurological Diseases and Blindness, National Institutes of Health, Bethesda, Maryland

WILLIAM FEINBLOOM, New York, New York

ADRIANA FIORENTINI, Instituto Nazionale di Ottica, Florence, Italy

NATHAN FLAX, Optometric Center of New York, New York

GLENN A. FRY, College of Optometry, Ohio State University, Columbus, Ohio

CLARENCE H. GRAHAM, Department of Psychology, Columbia University, New York, New York

SHERMAN L. GUTH, Department of Psychology, Indiana University, Bloomington, Indiana

RICHARD HELD, Department of Psychology, Massachusetts Institute of Technology, Cambridge, Massachusetts

DANIEL KAHNEMAN, Department of Psychology, Hebrew University, Jerusalem, Israel

YVES LE GRAND, Laboratory of Applied Physics, National Museum of Natural History, Paris, France

WILLIAM M. LYLE, University of Waterloo, Waterloo, Ontario

LAWRENCE W. MACDONALD, Boston, Massachusetts

ELWIN MARG, School of Optometry, University of California, Berkeley, California

HELEN M. PAULSON, United States Naval Submarine Base, Groton, Connecticut

ANTOINETTE PIRIE, Nuffield Laboratory of Ophthalmology, University of Oxford, England

DONALD G. PITTS, Wright-Patterson Air Force Base, Ohio

HERBERT A. W. SCHOBER, Institute for Medical Optics, University of Munich, Germany

GEORGE WALD, Harvard University, Cambridge, Massachusetts

W. D. WRIGHT, Imperial College of Science and Technology, London, England

Foreword

By definition, visual science is an interdisciplinary science overlapping the broad areas encompassed by anatomy, psychology, physiology, physiological optics, physics, and chemistry. While well aware of this overlapping, we have arbitrarily attempted to arrange the papers under four main areas, ranging from basic science issues to their practical and clinical applications.

John R. Pierce
John R. Levene

Indiana University, 1969

Preface

The International Symposium on Visual Science, April 2–4, 1968, co-sponsored by the American Optometric Association and Indiana University, brought together the most disciplinarily diversified group of visual scientists ever assembled under the single theme of visual science, and with representation from many parts of the world. The program participants were deliberately selected from among nominees of numerous organizations and individuals with the objective of bringing into full view the contribution potentials of the many disciplines toward the solutions of the variety of visual science problems ordinarily attacked only within the communicational confines of single disciplines.

In addition to the papers given by renowned researchers, the conference included an unusual number of exhibits of rare books, current books, research techniques, demonstration experiments, and audiovisual teaching aids related to visual science. An incidental by-product was the preparation and distribution of five hundred copies of a very recently compiled *Directory of Visual Science Personnel,* a list of 1,350 names and addresses put together by Professor John R. Levene. A first supplement to this *Directory* is already in progress. Another by-product was the more or less spontaneous assemblage of a small, highly representative, internationally diverse group of about twenty, individually invited by the conference chairman, H. Ward Ewalt, Jr., O.D., to try to adopt a suitable working definition of visual science. This the group accomplished, and, in addition, they laid the groundwork for the establishment of an internationally representative organization of visual science personnel. The three-day conference, completely free of business sessions and with only one organized social function, a dinner without speeches, provided maximum opportunity for the almost four hundred registrants to participate in the broad exchange of information and views which this conference was designed to encourage.

That the spirit of interdisciplinary exchange successfully pervaded the conference was clearly documented by the spoken and written comments of appreciation expressed by a great many of the participants during and after the Conference, several of whom even expressed a bit of grateful surprise that visual science could be thought of as a broad interdisciplinary body of knowledge.

Particularly significant to me in my professional affiliation was the fact that the initiative for this conference was taken by the American Optometric Association. This reflects a gratifying sense of responsibility on the part of optometry in giving recognition to the basic science, or sciences, the "materia optica," on which it must depend for its own vitality.

Undoubtedly the most obvious deduction that can be made even from a cursory review of the contents of this volume is that the field of visual science, however tightly defined, is a vast one requiring a broad range of talent and sophistication. Hopefully this collection of contributions may be one more milestone in the development of greater interdisciplinary exchange of information and experience in this fascinating field of study.

                                                        H. W. Hofstetter

Introduction

Glenn A. Fry

The Status of
Physiological Optics

When the invitation was extended to me to participate in this program, I was offered the position as lead-off man for this particular session with the hope that I would try to outline what physiological optics is and how it relates to the total scheme of visual science. I am pleased to try to do this because I feel that important steps can be taken to improve the status of physiological optics and that of visual science in general, and I want to solicit your help.

We can begin by discussing the relation of physiological optics to optics and physics.

Ronchi of the Institute of Optics in Florence has written a book called *Optics, the Science of Vision*. He has defended this point of view first of all on the simple ground that the word optics comes from a Greek word meaning eye, and, secondly, that the early history of optics justifies it. It turns out, however, that over the years the word optics has come to mean something entirely different from the science of vision. It now means the science of light. In the early days the eye was the only device which a physicist could use to detect the presence of light. It would be very difficult under these conditions to dissociate the science of light from the science of vision. There finally came a period in which all the classical treatments of optics referred to optics as having three divisions: physical optics, geometrical optics, and physiological optics. Physical optics and geometrical optics cover the science of light, and physiological optics the science of vision. This subdivision was definitely formalized when Helmholtz wrote his great treatise on physiological optics, which established physiological optics as a unique science.

It is too bad that physics did not designate the science of light as *photics,* and in fact it is sometimes called this. This would have allowed those who are interested in the eye to use the term *optics* to designate the science of vision. It has actually turned out that the word optics has been taken over by physicists to designate the science of light. As a result, people who are interested in the science of vision can no longer lay exclusive claim to the use of the word optics.

What has happened to the word optics is also happening to the word optical. We used to refer to spectacle lenses as optical aids. On the one hand, this could mean that such an aid is another kind of optical device akin to a camera lens, and on this account it can be called an optical device. On the other hand, the expression optical aid could imply that the device is an aid to the human flesh and blood "optikos." We have settled this issue by calling the spectacle lens an ophthalmic aid or an optical aid for the eye.

Another example is optical illusion. We are already switching over to calling some of these illusions visual illusions. Mirror images and mirages are still optical illusions, but the moon illusion and the Müller-Lyer illusions are visual illusions.

One wonders what will happen to the *optic nerve.* Must it become another *ophthalmic nerve?*

Let us turn now to consider whether *physiological optics* can still be used to designate the *science of vision.* Although physiological optics as defined by Helmholtz and Southall included the whole of visual science, a physicist today will not regard anything having to do with vision as optics, or physiological optics, unless it specifically involves the image-forming mechanism of the eye or the stimulation of photoreceptors by light.

It is because of this restricted use of the concept of physiological optics that there is a trend away from the use of the term physiological optics to designate visual science. Just to illustrate the trend, the Optical Society of America for many years has offered programs and symposia on physiological optics. Recently this Society has set up special interest groups. Such groups include groups interested in such things as the laser on the one hand, and vision on the other. The vision group was quite insistent on having itself referred to as the "vision group" rather than the "physiological optics group."

There is at the present time a trend to move everything in the direction of the use of the term *visual science*. In the National Institutes of Health, for example, we have recently segregated the activity in the field of vision under what we call the Visual Science Study Section. There is a move on foot to establish a new institute within the National Institutes of Health which will be referred to as the Eye Institute. I hope that some day this will become a reality, but now it is only in the planning stage. This is cited simply as an indication of the trend away from the use of the term physiological optics and toward the use of the eye and vision.

A conference on graduate training in physiological optics, which was organized by Mrs. Norma Miller several years ago, wound up with a list of recommendations, one of the most important of which was that the science of vision be called visual science and not physiological optics.

The split between physics and vision is more real than a mere quibbling of words. This is illustrated by what has been happening to schools of optometry.

The fact that the early programs in optometry at the universities began as programs in departments of physics and have subsequently become dissociated from the departments of physics can be interpreted as just another indication of visual science cutting its apron strings from physics. The programs at Columbia University, The Ohio State University, and the University of California at Berkeley all originated as programs in departments of physics and gradually emerged as separate schools and colleges. We are very grateful for this tradition and are especially pleased to have had some eminent physicists associated with the programs in optometry: Southall at Columbia, Sheard and Smith at Ohio State, and Minor at California to mention several of them. The new programs in optometry, however, such as the one at Indiana and the new one at Alabama which is now emerging, are developing separately from physics.

I think that as long as the Optical Society exists, it will promote papers and symposia which have to do with the eye as an image-forming mechanism and with the use of the eye in conjunction with other kinds of optical devices, such as the telescope and microscope, and also with the use of the eye as a device for detecting the presence of radiation as in colorimetry and photometry. But the Optical Society of America can no longer afford to be concerned with the whole of visual science.

Because the Optical Society can no longer afford to be concerned with the whole of visual science, those of us in visual science have to look around for a different kind of home base than an organization like the Optical Society of America.

At the end of the Rochester Conference on Graduate Training in Visual Science a committee was set up to explore the establishment of a

Society of Visual Science. Dr. Sylvester Guth was made chairman of the committee. Dr. Guth at that time met with a certain amount of resistance to the establishment of a new organization. There are several reasons why we should now review the needs for a new organization. One of the most significant things is that the newly formed Association of University Professors of Ophthalmology and the Association of Schools and Colleges of Optometry have been trying to develop rapport and mutual understanding and are actively exploring the development of a new Society of Visual Science in which optometrists, ophthalmologists, and all others interested can participate. It is not intended that this would impair the effectiveness of the American Academy of Optometry or the American Academy of Ophthalmology and Otolaryngology or the Association for Research in Ophthalmology. No arbitrary line need be drawn between basic visual science and clinical studies, but the kinds of papers presented at the meetings of these different kinds of organizations might make it possible for a given researcher or practitioner to choose from among the several meetings those that best fit his interest.

Meetings of a visual science society would de-emphasize clinical problems and eye disease, and it is to be hoped that this shift in emphasis toward the eye and vision would attract to these meetings anatomists, physiologists, biologists, zoologists, biophysicists, chemists, and physicists who are concerned about the science of vision.

Up to now I have said little about the role of psychology, physiology, anatomy, and biophysics and biochemistry to visual science. The dependence of visual science on these basic disciplines is obvious. Let us approach this in terms of what happens on a university or college campus. On a campus which has no department of ophthalmology or school of optometry, those interested in visual science must find their homes in the several departments of physics, psychology, physiology, and so on. There is no reason why such people could not participate in the affairs of a society of visual science and at the same time maintain loyalty and responsibility to their basic disciplines. Even on a campus where a college of optometry exists, the present pattern is not to set up departments of physics, psychology, or physiology within the college of optometry, but to require the students in optometry to take courses in these disciplines in the appropriate departments in other colleges of the university.

I propose, however, that whenever there does exist a school of optometry on a campus, it should constitute a real home or center for visual science. It should stimulate interdepartmental activity and interest in visual science.

I want to congratulate Indiana University on its establishment of a Division of Optometry. I want to express my pride in its growth and in its wonderful new building and facilities. It has the potential for becoming an outstanding center for visual science.

In order to provide rapport and communication between vision researchers in different departments, we have established an Institute for Research in Vision at The Ohio State University, which serves the double role of coordinating vision research activity on the campus and involving the university in applied problems in industrial vision, illuminating engineering, etc.

A somewhat similar interdepartmental center for Visual Science has been established at the University of Rochester, and this is of special interest because no school of optometry is involved.

Although it appears that what has been called physiological optics must now be called visual science, one can get a very good idea of what is covered by visual science by looking at the physiological optics courses offered in a college of optometry at the graduate and undergraduate levels.

At The Ohio State University, for example, we have been very careful to differentiate between courses in physiological optics on the one hand and courses in clinical optometry and diseases of the eye on the other. These cover the image-forming mechanisms of the eye, the motility of the eye and the motor adjustments of the eye, the sensory mechanisms of vision and the perception of color, direction, distance, time, etc.

It may be noted also that we include under physiological optics such things as circulation, pressure phosphenes, intraocular pressure, movements of the eyelids, and secretion of tears by the lacrymal gland.

We must include within the scope of visual science the different branches of applied visual science. The applications of visual science are almost too numerous to mention, but it is possible to formulate a simple concept of what is meant by applied visual science.

Let us start with the concept of a visual environment. A visual environment is occupied by a seeing person. By contrast, we can think of environments occupied by blind people; the contrast is terrific. In a visual environment we need a source of light, which is symbolized by the light bulb. We need illuminating engineers to develop bigger and better sources of light. The light has to illuminate objects. Here we run into some specialized problems in getting the light to the object through the atmosphere, and this involves meteorology. We also have the same kind of problems in getting the light from the object to the eye. In some situations we use fiber optics to transport light from the source to the object and from the object to the eye. The ophthalmoscope is another example of a special way of dealing with this problem. We have to get light into and out of the eye in order to see the fundus. There are a large number of problems involved in the design of objects to make them visible. The design of visual displays constitutes a special branch of applied and visual science, and the people who are working in this field are usually psychologists or engineers.

The spectacle lens represents the area in which the optometrist operates. His field is not limited to the design of eye wear, although that is an important aspect of his job. He is concerned also with image formation in the eye and the transfer of this information to the cortex, and, in addition, he has to be concerned with the motor mechanisms controlling fixation, accommodation, and convergence.

The illuminating engineer is primarily concerned with getting light from the source to the object; he has to pay special attention to the light from the source that is directly falling on the eye and producing glare. The architect and the interior designer have to be concerned with the distribution of objects and background in the entire field of view and the general lighting of the total environment.

Once the information fed in through the pupil of the eye is relayed to the brain, it has to be interpreted. This process is one of concern to psychologists, human engineers, educators, communication scientists, artists, philosophers, linguists, semanticists, epistomologists, and aestheticians.

Psychologists and human engineers are interested in the use of the eyes in performing tasks, and this includes memory, learning, information processing, motivation, and fatigue.

The training of an optometry student includes a certain amount of emphasis on applied visual science. He is taught the basic principles of both human engineering and illuminating engineering. An optometrist has to visualize his patient's problems at his work and at his play. We are teaching optometrists how to cope with this problem not only in the examination room, but also as consultants in industry, in schools and transportation, and elsewhere. We are trying to provide them with the necessary know-how to do a good job as consultants in these specialized areas.

We should not try to replace human engineers with visual scientists. The human engineer has to be concerned with sensory inputs other than vision and the interrelations between the different kinds of inputs. He should feel free to seek help from visual scientists on his visual problems, and he must know in what ways a visual scientist can help him. On the other hand, the visual scientist must understand the human engineer's problems in order to help him.

The eye is the only type of sense organ involved in illuminating engineering, and if the visual scientist wants to become an illuminating engineer, he is not handicapped by his specialization in vison and can become an illuminating engineer if he has, in addition, the proper background in engineering. There are many reasons, however, why a visual science society cannot replace an illuminating engineering society.

It has been suggested that instead of setting up a visual science society, we should set up a combined society for vision and hearing.

Disciplines like psychology, physiology, neurophysiology, and biophysics are equally concerned with vision and hearing. Human engineering is concerned with both. On the other hand, optometry programs and hearing programs on university campuses have been independent of each other. In our colleges of medicine we have separate departments of ophthalmology and otolaryngology. We have separate acoustical and optical societies which are concerned with the physical aspects of light and sound. The Armed Forces Committee on Vision is paralleled by a Committee on Hearing and Bioacoustics, but up to now the two have been independent.

The Institute for Neurological Diseases and Blindness started off with a single study section which reviewed proposals for research in both vision and hearing, but several years ago it became necessary to set up a separate section for Visual Science.

Although the American Academy of Ophthalmology and Otolaryngology and eye-ear-nose-throat clinics and journals still exist, the trend is toward a split between these two fields. Graduates of colleges of medicine must now elect to prepare for one or the other of these two specialties.

The Armed Forces Committee on Vision in August of 1968 is sponsoring a joint meeting between the Committee on Vision and the Committee on Hearing and Bioacoustics. This kind of meeting will help determine whether the interests of both groups can be best served by a unified society. Communication between the groups could be achieved by occasional joint symposia or by depending on meetings and societies in biophysics, psychology, physiology, or neurophysiology, which cover both areas.

One can summarize what has been said above in four simple statements:

1. The names optics and physiological optics can no longer be used to designate visual science. Visual science should be called visual science.

2. The scope of visual science can be defined by relating it to optometry and ophthalmology, illuminating engineering, human engineering, and the basic disciplines. The content of visual science is spelled out in the undergraduate and graduate courses in physiological optics offered by a college of optometry.

3. There is a need for the development of a society of visual science

at the national or international level. In this connection it is not the aim to discourage the Optical Society from continuing to deal with problems which involve both vision and optics. A new society would, however, relieve the Optical Society of the responsibility of providing a home for the whole of visual science.

4. The new society should not be a combined society for vision and hearing at the outset, but the possibility of a future merger should be kept in mind.

In England and on the continent of Europe I suspect that the problem is somewhat different from that in North America. I am grateful, therefore, that we have with us a number of distinguished visitors from across the Atlantic who can look at our problem objectively and advise us. It seems to me that nearly everywhere in Europe visual science is still regarded a part of optics, and thus the problem of communication between disciplines is somewhat different.

VISUAL SCIENCE

Proceedings of the 1968 International Symposium

Part I

Physiological Aspects

George Wald

Molecular Basis
of Visual Excitation

I have often had cause to feel that my hands are cleverer than my head. That is a crude way of characterizing the dialectics of experimentation. When it is going well, it is like a quiet conversation with Nature. One asks a question and gets an answer; then one asks the next question, and gets the next answer. An experiment is a device to make Nature speak intelligibly. After that one has only to listen.

Backgrounds

As a graduate student at Columbia University, I was introduced to vision by Selig Hecht in a particularly provocative way. Hecht was one of the great measurers of human vision, like Aubert, König, and Abney before him. But he was not content merely to measure. He wanted to understand what lay behind the measurements, what was going on at the molecular level in vision.

There a door was opened for him, while still a graduate student at Harvard, by the great Swedish physical chemist Svante Arrhenius. Hecht has told me of the excitement with which he read Arrhenius's new book *Quantitative Laws in Biological Chemistry* (Arrhenius, 1915). It offered the hope of translating accurate measurements on whole organisms into the simple kinetics and thermodynamics of chemical reactions in solution.

This article is the lecture George Wald delivered at Stockholm, Sweden, December 12, 1967, when he received a Nobel Prize in physiology. It is published here with the permission of the Nobel Foundation, is included in the complete volume of *Les Prix Nobel en 1968,* in the series *Nobel Lectures* (in English) published by the Elsevier Publishing Company, Amsterdam and New York, and appeared in *Science* Volume 162, Number 3850, October 11, 1968, pp. 230-239.

In this vein Hecht applied his measurements and those of earlier workers to constructing a general conceptual model for the photoreceptor process. A photosensitive pigment, S, was dissociated by light into products, P + A, one of them responsible for excitation. In turn P + A, or a variant, P + B, recombined to regenerate S. In continuous light these opposed reactions achieved a pseudo-equilibrium, a photo-stationary state, that underlay the steady states of vision in constant illumination (Hecht, 1931, 1934, 1937–38, 1938).

I left Hecht's laboratory with a great desire to lay hands on the molecules for which these were symbols. That brought me to Otto Warburg in Dahlem, where I found vitamin A in the retina (Wald, 1933, 1934–35), There were good reasons to look for it there, as I found out later, and that is the way I wrote my paper. Dietary night blindness, a condition already known in ancient Egypt, had been shown in Denmark during World War I to be a symptom of vitamin A deficiency (Blegvad, 1924), and Fridericia and Holm (1925) and Tansley (1931, 1933) had shown that vitamin A deficient rats synthesize less rhodopsin than normal animals do. But vitamins were still deeply mysterious, and at that time one hardly expected them to participate directly in physiological processes. I think this was the first instance of so direct a connection, though Warburg and Christian (1932, 1933) were already analyzing the first yellow enzymes, and shortly their chromophore riboflavin would prove to be vitamin B_2 (Kuhn, György, and Wagner-Jauregg, 1933; Theorell, 1934, 1935).

After that, things happened quickly. I went to Karrer in Zurich, who with Morf and Schopp (1931) had the year before established the structure of vitamin A, to complete its identification in the retina. Then I went on to Meyerhof in Heidelberg, to do something else; but with a shipment of frogs that had gone astray, I found retinene, an intermediate in the bleaching of rhodopsin, on the way to vitamin A (Wald, 1934, 1935–36A). Years later Ball, Goodwin, and Morton (1946, 1948) showed in Liverpool that retinene is vitamin A aldehyde. At Morton's suggestion the names of all these molecules have recently been changed, in honor of the retina, still the only place where their function is understood. Vitamin A is now retinol, retinene is retinal (Fig. 1); there is also retinoic acid.

That early *Wanderjahr* in the laboratories of three Nobel laureates—Warburg, Karrer, Meyerhof—opened a new life for me; the life with molecules. From then on it has been a constant going back and forth between organisms and their molecules—extracting the molecules from the organisms to find what they are and how they behave, returning to the organisms to find in their responses and behavior the greatly amplified expression of those molecules.

A basic characteristic of the scientific enterprise is its continuity. It is an organic growth, to which each worker in his time brings what he can; like Chartres or Hagia Sofia, to which over the centuries a buttress was added here, a tower there. Hecht's work was most intimately bound up with that of men who had worked generations before him: Hermann Aubert (1865) in Breslau, Arthur König (1930) in Berlin, Abney (1895) in England. Now I entered into such a relationship with Willy Kühne of Heidelberg. Kühne had taken up rhodopsin immediately upon Franz Boll's discovery of it in 1877 (Boll, 1877), and in two extraordinary years he and his co-worker Ewald learned almost everything known about it for another half-century (Kühne, 1878; Ewald and Kühne, 1878; Kühne, 1879; Voit, 1900). It was largely on the basis of Kühne's observations

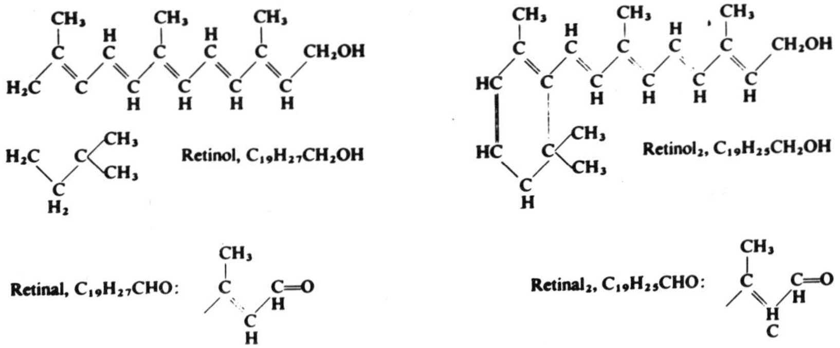

Figure 1. Structures of retinol₁ (left) and retinol₂ (right) (vitamins A₁ and A₂) and their aldehydes, retinal₁ and retinal₂ (formerly retinenes 1 and 2).

that I could conclude that rhodopsin is a protein, a carotenoid-protein such as Kuhn and Lederer (1933) had just shown the blue pigment of lobster shells to be, that, in the retina, under the influence of light, engages in a cycle of reactions with retinal and vitamin A (Wald, 1934, 1935A, 1935–36A, 1935–36B).

I owe other such debts to past workers in far-off places. Köttgen and Abelsdorff (1896) had found the visual pigment from eight species of fish to have difference spectra displaced considerably toward the red from the rhodopsins of frogs, owls, and mammals. Trying to check this observation at Woods Hole, I was surprised to find the same rhodopsin-retinal-vitamin A cycle in fishes there as in frogs (Wald, 1935B, 1936–37). It turned out that Köttgen and Abelsdorff had worked entirely with fresh-water fishes. On turning to them, I found another visual pigment, por-phyropsin, engaged in a cycle parallel with that of rhodopsin, but in which new carotenoids replace retinal and vitamin A (Wald, 1937A, 1938–39A). On the basis of these observations, it was suggested that the substance that replaces vitamin A in the visual system of freshwater fishes be called vitamin A_2 (Edisbury, Morton, and Simpkins, 1937). In what follows I shall call it retinol$_2$, and its aldehyde retinal$_2$. These substances differ from their analogues in the rhodopsin cycle only in possessing an added double bond in the ring (Fig. 1) (Morton, Salah, and Stubbs, 1947; Farrar, Hamlet, Henbest, and Jones, 1952).

Shortly afterward it emerged that such familiar euryhaline and hence potentially migratory fishes as salmon, trout, and the freshwater eel possess mixtures of rhodopsin and porphyropsin, in which the system commonly associated with the spawning environment tends to predominate. Other such euryhaline fishes as the white perch and alewife possess this system virtually alone (Wald, 1938–39B, 1941–42). The bullfrog has porphyrop-sin as a tadpole, and changes to rhodopsin at metamorphosis (Wald, 1945–46). The sea lamprey *Petromyzon marinus* possesses mainly rho-dopsin on its downstream migration to the sea, but on going upstream as a sexually mature adult to spawn has changed to porphyropsin (Wald, 1956–57). Denton and Warren having shown that the rhodopsins of deep-sea fishes have spectra displaced to shorter wavelengths than those of surface forms (that is, to λ_{max} about 480 mμ) (Denton and Warren, 1956; Munz, 1957; Wald, Brown, and Brown, 1957), it developed that the European eel at sexual maturity, preparatory to migrating to the Sargasso Sea to spawn, transfers from its previous mixture of rhodopsin and porphyropsin to deep-sea rhodopsin (Carlisle and Denton, 1957; Brown and Brown, 1958). The chemical pattern of visual systems main-tains close relationships with the evolution, development, and way of life of these and other animals (Wald, 1958, 1960A, 1960B). I am glad to say that the pursuit of molecules has not taken me out of biology, but led me more deeply into it.

Molecular Architecture

Let me now leave this history and say where it has brought us.

All the visual pigments we know are built upon a common plan. All of them consist of retinal bound as chromophore to a type of protein, called an opsin, found in the outer segments of vertebrate rods and cones and the analogous rhabdomeres of invertebrate eyes. In vertebrates, the two retinals, 1 and 2, join with two great families of opsins, those of the rods and those of the cones, to form the four major pigments of vertebrate vision (Wald, 1937B; Wald, Brown, and Smith, 1953, 1954–55) (see chart 1).

Chart 1

					Usual λ_{max} (mμ)
			+ Rod Opsin	light → Rhodopsin	500
Retinol$_1$	DPN+ / DPN — H	Retinal$_1$			
			+ Cone Opsin	light → Iodopsin	562
(Alcohol dehydrogenase)					
			+ Rod Opsin	Porphyropsin	522
Retinol$_2$	DPN+ / DPN — H	Retinal$_2$		light	
			+ Cone Opsin	light → Cyanopsin	620

The retinols are oxidized to the corresponding retinals by alcohol dehydrogenases. The first such enzymes we examined, those of frogs and fishes, employed DPN as coenzyme (Wald and Hubbard, 1948–49; Wald, 1949, 1950). Other systems have since been found that prefer TPN as coenzyme (Futterman, 1963), and recent work, as now seems inevitable, is multiplying the numbers of such enzymes, some of which may act preferentially on retinol among the alcohols (hence retinol dehydrogenases) (Koen and Shaw, 1966).

Under physiological conditions the equilibrium between retinol and retinal lies far over toward reduction, so far that retinol is oxidized to retinal only to the degree that the latter is trapped out of the system. In vitro, that can be done with hydroxylamine, which condenses spontaneously with retinal to form retinal oxime:

$$C_{19}H_{27}HC = O + H_2NOH \rightarrow C_{19}H_{27}HC = NOH + H_2O.$$

In the retina, opsin performs this function, trapping retinal as fast as it appears, to form the visual pigments (Wald and Brown, 1950; Wald and Hubbard, 1950; Hubbard and Wald, 1951). Thus it is opsin that regulates how much retinol is oxidized, and visual pigment synthesized.

Along with this parallelism of structure, the visual pigments exhibit an extraordinary parallelism of chemical behavior. The reactions one finds with any of them are usually found, with minor variations, in all the others. The opsins, like other proteins, are species-specific, and often are multiple within a species, as in man. With the different opsins go differences in absorption spectrum, stability, the kinetics of bleaching and regeneration, and other properties. Yet all the visual systems we know represent variations on a central theme. The vertebrates play this out to the end. The invertebrates tend to cut it short at various levels. The action of light on the known invertebrate pigments ends with the production of retinal, in most instances still attached to opsin. Also all the known visual pigments of invertebrates have retinal$_1$ as chromophore (Wald, 1960A).

Some years ago Collins (1953) and Morton and Pitt (1955) provided good evidence that in rhodopsin, retinal is bound to opsin in Schiff base linkage, by the condensation of the aldehyde group of opsin:

$$C_{19}H_{27}HC = O + H_2N\text{-opsin} \rightarrow C_{19}H_{27}HC = N\text{-opsin} + H_2O.$$

Bownds (1967) in our laboratory has recently identified this amino group in cattle opsin as the ϵ-NH$_2$ group of lysine. He has also analyzed the neighboring covalently bound amino acids; together with lysine they constitute a decapeptide segment of the composition: ala$_3$ phe$_3$ thr pro ile ϵ-N-retinyl lysine. Figure 2 shows a model of this segment. Cattle rhodopsin as usually prepared has a molecular weight of about 40,000 (Hubbard and Wald, 1952–53; Hubbard, Gregerman, and Wald, 1952–53). If the molecule is spherical, its diameter is about 40 Å. The chromophore, though its molecular weight is only 282, is about 20 Å long. It looms surprisingly large therefore in the structure of rhodopsin.

To make visual pigments demands not only retinals 1 or 2, but the right shapes of these molecules (Hubbard and Wald, 1952–53). The retinals possess 4 carbon-to-carbon double bonds in the side chain, each of which might potentially exist in either *cis* or *trans* configuration (Fig. 3). The most stable and prevalent form is the all-*trans*. The first double bond at C-7 is always *trans*, since hindrance between the methyl groups on C-1 and C-9 prevents the 7-*cis* linkage from forming. The 9- and 13-*cis* forms are common; but the 11-*cis* linkage was recognized to be highly improbable, since it too involves a large overlap, between the methyl group on C-13 and the hydrogen on C-10. A *cis* linkage always represents

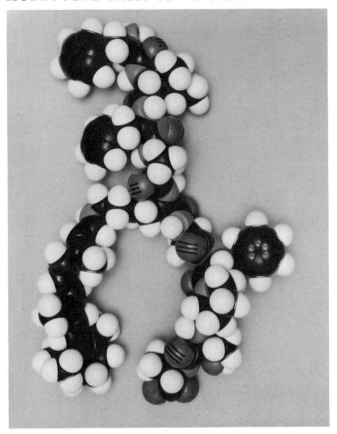

Figure 2.   Model of the chromophoric site of cattle rhodopsin, according to Bownds (1967). The 11-*cis* retinyl chromophore is attached to opsin in Shiff base linkage through an epsilon-amino group of lysine. The covalently bound amino acids in this neighborhood form a decapeptide fragment of the composition *ala₃ phe₃ thr pro ile* ε-*N*-retinyl lysine. Their sequence is not yet known, and is arbitrary in this model.

a bend in the chain; but because of this steric hindrance the 11-*cis* molecule is not only bent but twisted at the *cis* linkage. This departure from planarity, by interfering with resonance, was expected to make the molecule so unstable that one hardly expected to find it (Hubbard and Wald, 1952–53; Hubbard, Gregerman, and Wald, 1952–53; Oroshnik, Brown, Hubbard, and Wald, 1956; Brown and Wald, 1956).

Nevertheless it has turned out that 11-*cis* retinal, once prepared, is reasonably stable provided it is kept dark; and all the visual pigments we know that have been analyzed to this degree, possess as chromophore 11-*cis* retinal₁ of retinal₂ (Brown and Smith, 1953, 1955, 1957).

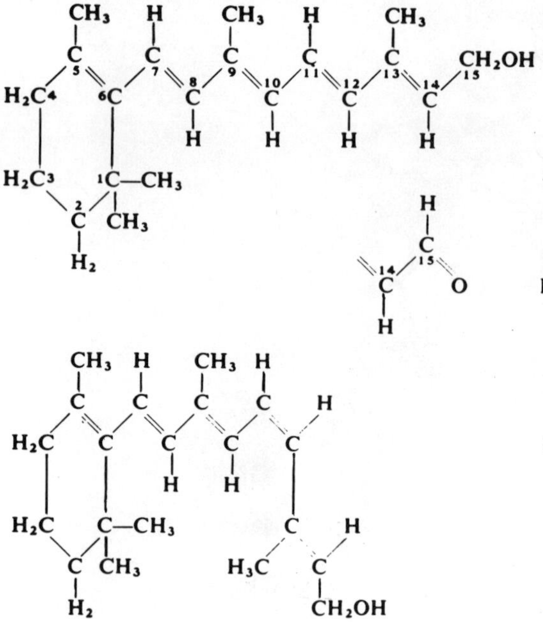

All-*trans*-Retinol (Vitamin A)

Retinal

11-*cis* Retinol

Figure 3. Structures of the all-*trans* 9-*cis,* and 11-*cis* isomers of retinol and retinal.

When, however, a visual pigment is bleached by light, the retinal that emerges is all-*trans*. It must be reisomerized to the 11-*cis* configuration before it can take part again in regenerating the visual pigment. Hence a cycle of *cis-trans* isomerization is an intrinsic part of every visual system we know (Fig. 4).

The 9-*cis* isomer, which is closest in shape to 11-*cis* retinal, also combines with the opsins to yield light-sensitive pigments that behave much as do the visual pigments (Hubbard and Wald, 1952–53; Hubbard, Gregerman, and Wald, 1952–53). We call them the iso-pigments: isorhodopsin, isoporphyropsin (Brown and Smith, 1953, 1955, 1957), and so on. None of them has yet been identified under physiological conditions in a retina. So far as we yet know, the iso-pigments are to be regarded as artifacts.

As prelude to what is about to be said, it should be understood that when any single geometrical isomer of retinal in solution is exposed to light, it rapidly isomerizes to a steady-state mixture of all the possible isomers, in proportions that depend upon the wavelength of the light, and even more upon the polarity of the solvent (Hubbard and Wald, 1952–53; Hubbard, Gregerman, and Wald, 1952–53; Brown and Wald, 1956). In

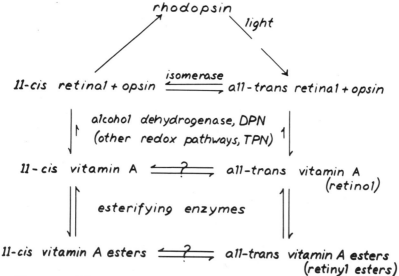

Figure 4. Diagram of the rhodopsin system, showing the isomerization cycle. The bleaching of rhodopsin by light ends in a mixture of opsin
and all-*trans* retinal. The latter must be isomerized to 11-*cis* before it
can regenerate rhodopsin. While that is happening, much of it is reduced to all-*trans* vitamin A, most of which in turn is esterified (Wald,
1935A, 1935-36B; Krinsky, 1958). These products must be isomerized
to or exchanged for their 11-*cis* configuration before engaging in the
resynthesis of visual pigments.

such a homopolar solvent as hexane, about 95 percent of the final, steady-
state mixture is all-*trans*; whereas in such a polar solvent as ethanol, about
50 percent is distributed among *cis* configurations, and a surprisingly
large fraction, about 25 to 30 percent, ends as 11-*cis* retinal. Though
thermodynamically improbable, this is one of the most favored configurations of retinal (Brown and Wald, 1956; Hubbard, 1955–56).

A few years ago Hubbard and Kropf showed that the only action of
light in vision is to isomerize the chromophore of a visual pigment from
the 11-*cis* to the all-*trans* configuration (Hubbard and Kropf, 1958; Kropf
and Hubbard, 1958). Everything else that happens—chemically, physio-
logically, indeed psychologically—represents "dark" consequences of this
one light reaction.

This photochemical step can be isolated by bringing a visual pigment
to very low temperatures. For example, if rhodopsin in a 1:1 mixture of
water and glycerol is brought to liquid nitrogen temperature (about
—190°C), at which the solvent vitrifies, its absorption spectrum narrows,

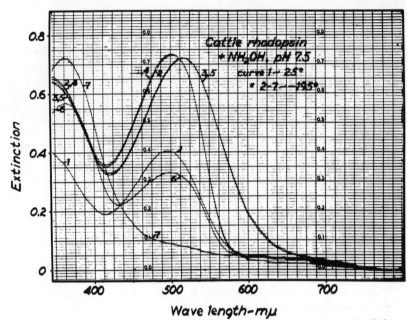

Figure 5.   Interconversion of rhodopsin and prelumirhodopsin by light
at liquid nitrogen temperature. Cattle rhodopsin in 1 : 1 glycerol-water
(1) is cooled to −195°C (2). On long irradiation with blue light
(440 mμ) the spectrum moves to (3). The light has isomerized the
11-*cis* chromophore of rhodopsin to a steady-state mixture of rhodopsin
and prelumirhodopsin (all-*trans*). Irradiation with orange light (600
mμ) re-isomerizes all the all-*trans* chromophore back to 11-*cis*: it is
now again all rhodopsin (4). Reirradiation with blue light brings it
back again to the steady-state mixture (5). On warming in the dark to
25°C, the prelumirhodopsin in this mixture bleaches to all-*trans* retinal
and opsin, the retinal condensing with hydroxylamine to yield retinal
oxime (λ_{max} 367 mμ); the rhodopsin in the mixture remains unchanged.
This product was recooled to −195°C and its spectrum recorded (6).
Finally it was warmed again and bleached completely to all-*trans*
retinal oxime and opsin (7). [From Yoshizawa and Wald (1963)]

rises and shifts slightly toward the red (Fig. 5) (Yoshizawa and Wald,
1963). Exhaustive irradiation with blue light shifts the spectrum further
toward the red, ending with the production of a steady-state mixture, in
which part of the chromophore is still 11-*cis*, hence still rhodopsin, and
the rest has been isomerized to all-*trans* to form the photoproduct, prelumirhodopsin (λ_{max} 543mμ). At this point irradiation with orange light
drives the spectrum back to its original position: the orange light, by
re-isomerizing the all-*trans* chromophore to 11-*cis*, has reconverted all

the prelumirhodopsin back to rhodopsin. Long irradiation with green light can drive the spectrum to still shorter wavelengths, by isomerizing the 11-*cis* chromophore to 9-*cis*; the pigment is now isorhodopsin. Irradiation of the isorhodopsin with blue light again yields the same steady-state mixture of rhodopsin and prelumirhodopsin as before. In this way one can go back and forth without loss as often as one likes, among rhodopsin (11-*cis*), prelumirhodopsin (all-*trans*) and isorhodopsin (9-*cis*). At liquid nitrogen temperature this is a perfectly reversible system.

Comparable changes occur in squid rhodopsin (Yoshizawa and Wald, 1964), chicken iodopsin (Yoshizawa and Wald, 1967), and as I have just learned from Yoshizawa (1967), now back in Osaka, in carp porphyropsin. It is interesting that in all these cases the prelumi photoproduct, though the first step in bleaching, is a more intense pigment than the visual pigment itself, its spectrum both shifted toward the red and considerably taller than that of the visual pigment.

Intermediates of Bleaching

As already said, the irradiation of rhodopsin with blue light at liquid nitrogen temperature produces a steady-state mixture of rhodopsin and prelumirhodopsin. On gradual warming in the dark, the latter goes at specific critical temperatures through a progression of intermediate stages —lumirhodopsin, metarhodopsin I, metarhodopsin II—representing stepwise changes in the conformation of opsin (Fig. 6). Finally the Schiff base linkage hydrolyzes to yield all-*trans* retinal and opsin (Matthews, Hubbard, Brown, and Wald, 1963–64). In the course of these transformations new groups of opsin are exposed: two sulfhydryl (—SH) groups per molecule (Wald and Brown, 1953–54, 1951–52), and one proton-binding group with pK about 6.6, perhaps imidazole (Radding and Wald, 1955–56).

Literal bleaching in the sense of loss of color occurs mainly between metarhodopsins I and II. Visual excitation must have occurred by the time meta II is formed (Fig. 7). This stage is reached within about 1 msec at mammalian body temperature (Hagins, 1956). All subsequent changes are much too slow to be involved in excitation.

Up to metarhodopsin II, the all-*trans* chromophore has remained attached to opsin at the same site. So long as that is so, the absorption of a photon, by isomerizing the all-*trans* chromophore to 11-*cis*, immediately regenerates rhodopsin (Fig. 8).

We have been in the habit of saying that light bleaches visual pigments. What it does however is to isomerize the chromophore. The end of the

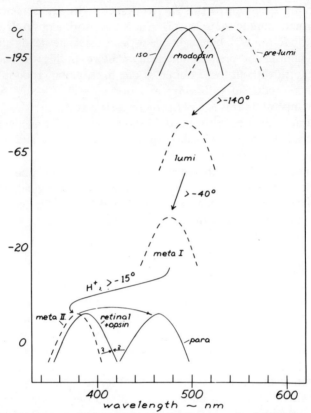

Figure 6.   Intermediates in the bleaching of cattle rhodopsin as they appear by irradiating at liquid nitrogen temperature and then gradually warming in the dark. Irradiation at −195°C yields steady-state mixtures of rhodopsin (11-*cis*) with prelumirhodopsin (all-*trans*) and isorhodopsin (9-*cis*) in proportions that depend on the wavelength of irradiation. On warming in the dark, prelumirhodopsin goes at certain critical temperatures over a succession of all-*trans* intermediates to a final mixture of all-*trans* retinal and opsin.

process, if it is allowed to go to completion, is a steady-state mixture of isomers of the chromophore, in proportions that depend upon the wavelength of irradiation and the relative quantum efficiencies of the photoreactions. A first photon absorbed by rhodopsin can isomerize its 11-*cis* chromophore to all-*trans*, the initial step in bleaching. The absorption of a second photon, however, by any of the all-*trans* intermediates of bleaching can re-isomerize the chromophore to 11-*cis*, regenerating rhodopsin; or to 9-*cis*, forming isorhodopsin. Light not only bleaches visual pigments,

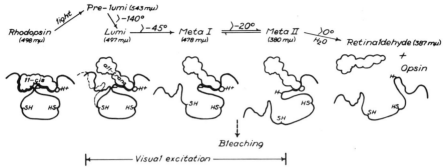

Figure 7.   Stages in the bleaching of rhodopsin. The chromophore of rhodopsin, 11-*cis* retinal, fits closely a section of the opsin structure. The only action of light is to isomerize retinal from the 11-*cis* to the all-*trans* configuration (prelumirhodopsin). Then the structure of opsin opens progressively (lumi- and the metarhodopsins), ending in the hydrolysis of retinal from opsin. Bleaching occurs in going from meta-rhodopsin I to II; and visual excitation must have occurred by this stage. The opening of opsin exposes new chemical groups, including two —SH groups and one H^+-binding group. The absorption maxima shown are for prelumirhodopsin at $-190°C$, lumirhodopsin at $-65°C$, and the other pigments at room temperature.

but can regenerate them or the iso-pigments, depending upon the circumstances.

The final states of bleaching still present problems. Under conditions that promote a surge of metarhodopsin II—notably the exposure of a rhodopsin solution or dark-adapted retina to a short, intense irradiation and recording its consequences in the dark—one finds it flowing in part back into the form of a more highly colored product, so that the absorption in the visible region, having fallen (metarhodopsin II), rises again (Figs. 6, 8). I saw this happening long ago in solutions of frog rhodopsin (Wald, 1937–38), and we have recently followed this change in cattle rhodopsin solutions and frog retinas (Matthews, Hubbard, Brown, and Wald, 1963–64). This seems also to be what Hagins (1956) observed after flashing excised rabbit eyes (his product C). This product (λ_{max} about $465m\mu$) resembles metarhodopsin I in spectrum and has sometimes been confused with it. It may be called pararhodopsin. It is formed from metarhodopsin II in the dark, and more rapidly in the light. Light also drives it back to metarhodopsin II, so that depending upon the wavelength of irradiation one can push this intermediate back and forth between these two states. In the dark, pararhodopsin decays to retinal and opsin (Matthews, Hubbard, Brown, and Wald, 1963–64). Whether it is an essential

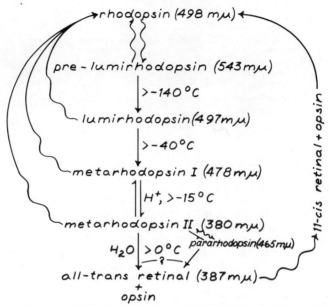

Figure 8. Intermediates in the bleaching and regeneration of rhodopsin. Wavy arrows represent photoreactions, straight arrows thermal
("dark") reactions. The interrelationships of pararhodopsin with the
final products of bleaching still present problems.

intermediate or a bypass in the bleaching of rhodopsin, and whether on
irradiation it can directly regenerate rhodopsin all require further analysis
(Abrahamson and Ostroy, 1967; Bridges, 1967).

I must confess to having become somewhat bored with such minutiae
of the final stages of bleaching, which come much too late to have anything to do with visual excitation. They have suddenly taken on a new
interest, however, owing to an astonishing development. Having had
nothing to do with this, I can praise it freely. For generations past the
chemistry of vision and its electrophysiology have traveled separate paths.
Suddenly many of the details of the bleaching and regeneration of visual
pigments are emerging in a new class of electrical responses.

The arrangements are familiar by which one measures an electroretinogram (ERG) (Fig. 9). An active electrode on the cornea and an
indifferent electrode elsewhere on the eye or on the body connect through
an amplifier to an oscilloscope. On exposing the eye to a flash of light
there is a silent period lasting at least 1.5 msec even in a mammal, and
much longer at lower temperatures; then a biphasic fluctuation of potential, the characteristic a- and b-waves of the ERG. One would suppose

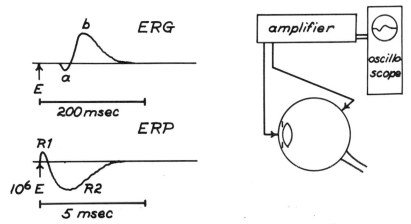

Figure 9.   Diagram to show the essential hookup for observing an electroretinogram (ERG) or early receptor potential (ERP) in a dark adapted vertebrate eye. Each of these responses is a biphasic fluctuation of potential, involving cornea-positive (upward) and cornea-negative (downward) components. Unlike the ERG, the ERP has no measurable latency. For both types of response to be comparable in amplitude, the flash that stimulates the ERP must be of the order of 1 million times more intense.

that the latent interval before the response represents the time needed to bleach the visual pigment to the critical stage, and for secondary events, including the large buildup of amplification, that lead to the response.

About three years ago K. T. Brown and Murakami (1964) found a new electrical response that fills in this interval—the early receptor potential (ERP) (Fig. 9). It has no measurable latency. Ordinarily it takes a flash about one million times as intense as would produce a moderate ERG to evoke an ERP of about the same amplitude. The ERP also is biphasic, consisting of a rapid cornea-positive wave (R1) followed by a slow cornea-negative wave (R2). In the cold (5°C and below), R1 appears alone (Review articles, *Cold Spring Harbor Symp. Quant. Biol.,* 1965).

It is becoming increasingly clear that the ERP has its source in the action of light on the visual pigments themselves. One can get an ERP in retinas cooled to −30°C, heated to 48°C, bathed in glycerol solutions, or fixed in glutaraldehyde. All it requires is intact—and oriented—rhodopsin. In the membranes that compose the rod outer segments of rhodopsin is almost perfectly oriented (Wald, Brown, and Gibbons, 1963). If the orientation is destroyed by heating, the ERP goes with it (Cone and Brown, 1967).

A recent experiment demonstrates the immediacy of these relationships. A flash of light acting on the rhodopsin of a dark-adapted retina having evoked the usual ERP, another flash acting on an intermediate stage of bleaching (probably para- and meta-rhodopsin II) so as to photoregenerate rhodopsin produces a biphasic ERP of reversed polarity (Fig. 10)

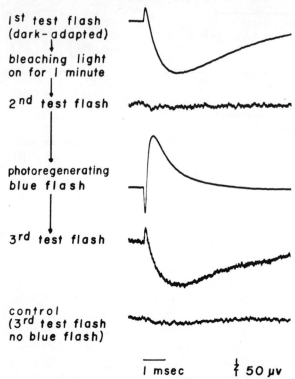

Ist test flash
(dark-adapted)

bleaching light
on for I minute

2nd test flash

photoregenerating
blue flash

3rd test flash

control
(3rd test flash
no blue flash)

I msec   50 μv

Figure 10. Photoregeneration of the early receptor potential (ERP) in the eye of an albino rat. Both the test flash and the bleaching light were composed of long wavelengths primarily absorbed by rhodopsin. The blue photoregenerating flash contained wavelengths absorbed by longer-lived intermediates of bleaching (apparently mainly metarhodopsin II and pararhodopsin). The control trace was obtained from a second eye subjected to the same bleaching light and test flashes, but without the interpolated blue flash, 27°C. The first test flash yields a normal ERP. The bleaching light having removed rhodopsin, a second test flash yields nothing, even though the amplification is increased (see gain index at right). A blue flash, photoregenerating rhodopsin, yields an ERP of reversed polarity. The third test flash, yielding again a normal ERP (high gain!) shows that the blue flash did regenerate rhodopsin. The control flash shows that without the interpolated blue flash, no response is obtained (high gain). [From R. A. Cone (1967)]

(Cone, 1967). That is, the isomerization of the rhodopsin chromophore from 11-*cis* to all-*trans* having triggered reactions that produce the normal ERP, the reverse isomerization from all-*trans* to 11-*cis* induces similar changes of potential reversed in sign.

Such experiments with rat and squid retinas have begun to identify individual components of the ERP with the photoreactions of specific intermediates in the bleaching of rhodopsin (Cone, 1967; Hagins and McGaughy, 1967, 1968). It is hard to see how intramolecular changes of this kind, presumably involving charge displacements, can generate changes of potential between the front and back of the eye; but there is no doubt that they do, and we can only hope eventually to understand them. There is as yet no evidence that the ERP generated by rhodopsin itself is part of the mechanism of excitation; obviously the action of flashes of light on intermediates of bleaching, though they generate various forms of ERP, do not excite. At the least, the ERP offers a new and powerful tool for studying the reactions of the visual pigments *in situ*.

Color Vision Pigments

For many centuries man was the only object of visual investigation. Recently he has again become for many purposes the experimental animal of choice. In certain ways experimenting with man offers unique advantages. With other animals one can pursue a biophysics, but only with man also a psychophysics. Moreover human genetics is by now probably better understood than that of any other animal, and with man one does not have to seek out mutant forms—their curiosity and anxiety bring them in.

A few years ago Paul Brown (1961) designed and built a recording microspectrophotometer in which one could measure the difference spectra of visual pigments in small fragments of retina. With this one could record the difference spectrum of visual pigments in the rod-free area of the human fovea. Just as the spectrum of human rhodopsin agrees with the spectral sensitivity of human rod vision (Wald and Brown, 1958), so the spectrum of foveal pigments accounts for the spectral sensitivity of human cone vision (Brown and Wald, 1963).

The human cones, however, are the receptors of color vision. Since Thomas Young it has been recognized that normal human color vision is trivariant; it involves the interplay of three independent variables. A prevalent thought for many years past has been that it involves three types of cone, each with its own individual pigment.

By irradiating human and monkey foveas with deep red light in the microspectrophotometer, we were able to bleach the red-sensitive pigment alone, and measure its difference spectrum (λ_{max} about 565mμ). When red light caused no further change, yellow light induced a renewed bleaching that yielded the difference spectrum of the green-sensitive pigment (λ_{max} about 535mμ) (Brown and Wald, 1963). Presumably there was also a blue-sensitive pigment that could not be measured adequately with these arrangements.

Shortly afterward it proved possible to measure in the microspectrophotometer the difference spectra of the visual pigments in single parafoveal rods and cones of human and monkey retinas. Such measurements were made simultaneously and independently in our laboratory and by Marks, Dobelle and MacNichol (1964) at Johns Hopkins University (Brown and Wald, 1964). They showed that primate retinas possess, in

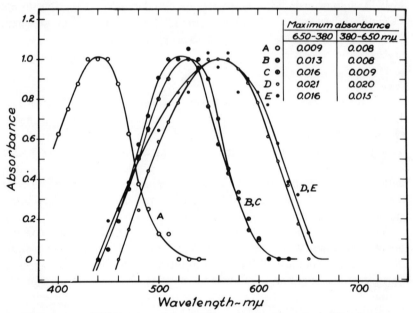

Figure 11. Difference spectra of the visual pigments in five human parafoveal cones: one blue-sensitive, two green-sensitive, and two red-sensitive cones. Each of these difference spectra was obtained by recording in the microspectrophotometer the "dark" and then the "bleached" spectra from 650 to 380 mμ and again in the reverse direction. Then the "bleached" spectra were subtracted from the "dark" spectra, and both curves averaged. The absorbances at λ_{max}, shown at the upper right, indicate the amounts by which these preparations bleached in the course of the two recordings. The second recording, from 380 to 650 mμ, always displays a somewhat smaller absorbance, owing to bleaching. [From Wald and Brown (1965)]

addition to rods with their rhodopsin, three kinds of cone, blue-, green-, and red-sensitive, each containing predominantly or exclusively one of three color-vision pigments, with λ_{max} at about 435, 540, and 565mμ (Fig. 11).

The red- and green-sensitive pigments, after being bleached by light, are regenerated by adding 11-*cis* retinal in the dark (Brown and Wald, 1963). The same seems to be true of the blue-sensitive pigment. Since all the visual pigments of the primate retina, rod, and cone apparently have the same chromophore, 11-*cis* retinal, they must differ in their opsins.

The iodopsins, with λ_{max} near 560mμ, and cyanopsins, with λ_{max} near 629mμ, are widely distributed among animals. Iodopsin is the major, perhaps the only cone pigment in the chicken, pigeon (Wald, Brown, and Smith, 1954–55), cat, snake, and frog (Granit, 1942A, 1942B, 1943A, 1943B); just as cyanopsin seems to be the cone pigment of such retinol$_2$ animals as the tench and tortoise (Granit, 1941A, 1941B), a freshwater turtle and frog tadpoles (Liebman and Entine, 1967). In the few vertebrates whose color vision systems have been analyzed, these pigments are apparently the red-sensitive components. So in man and the monkey, the red-sensitive pigment of color vision is apparently iodopsin (Brown and Wald, 1963), just as it is cyanopsin in the goldfish (Marks, 1965) and carp (Tomita, Kaneko, Murakami, and Pautler, 1967). Recently, pairs of visual pigments resembling in spectrum rhodopsin and iodopsin have been found in two species of crayfish (Wald, 1967A), and here again the "iodopsins" seem to be the red-sensitive components in systems of color discrimination (Wald, 1967–68).

A simple psychophysical procedure has recently been developed for measuring the spectral sensitivities of the three groups of cones in human subjects (Wald, 1964). This is a highly simplified extension of the procedures by which W. S. Stiles (1949, 1959), first measured such sensitivity curves. In my application of this procedure, the eye is continuously adapted to bright colored lights, each of which so lowers the sensitivities of two of the three color-vision mechanisms that the measurements report primarily the properties of the third system. So, for example, exposure of the eye to a brilliant yellow light so lowers the sensitivities of the green- and red-sensitive systems that measurements of the spectral sensitivity are dominated by the blue-sensitive system. Similarly exposure to wave bands in the blue and red, hence purple light, isolates the green-sensitive system; and exposure to bright blue light isolates the red-sensitive system. The sensitivity curves measured in this way come out much as do the difference spectra of the pigments measured in the microspectrophotometer (Fig. 12).

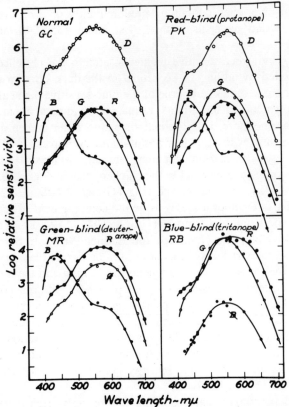

Figure 12. Measurements of spectral sensitivity with a standardized selective adaptation procedure in a normal subject (trichromat), and typical subjects representing each of the three main classes of color blindness (dichromats). In the normal eye, this procedure displays the spectral sensitivity of the dark-adapted fovea (*D*), and those of the blue-, green-, and red-sensitive systems (*B, G, R*), as measured respectively on intense yellow, purple, or blue backgrounds. Each of the color-blinds reveals the operation of only two of the three color-vision pigments; in each case the attempt to measure the third pigment reveals only one of the other mechanisms at a lower level of sensitivity. The crossed out symbols *B, G,* and *R* represent such unsuccessful attempts to measure the missing systems in the color-blind eyes.

Color Blindness

With this simple procedure one can inquire into mechanisms of color blindness (Wald, 1966). Just as normal human color vision is trivariant (trichromatic), that of the usual congenital color-blind is divariant (dichromatic). Theoretically the reduction from three variables to two could occur in many ways, and at any level in the visual pathways. By now however I have examined a reasonably large number of dichromats with this procedure; and each of them apparently lacks one of the three color mechanisms, the other two remaining normal and fully functional (Fig. 12). Depending upon which component is lacking, the three major classes of dichromat can be characterized as blue-, green-, or red-blind.

Every schoolboy learns that color blindness is caused by a sex-linked recessive mutation. That is true however only of red- and green-blindness. About 1 percent of men are red-blind, about 2 percent green-blind, whereas both conditions are very rare in women. Blue-blindness also is very rare, and not sex-linked, affecting only about 1 in 20,000 persons, about 40 percent of whom are women (Wright, 1952).

There is another, more widespread congenital color defect, red- or green-anomaly, closely related to red- and green-blindness. Persons with this condition have three-color vision, but make abnormal color matches. Examined by my procedure, they yield the same type of result as red- or green-blinds, showing that one of their three-color mechanisms is abnormally insensitive, so much so that my procedure does not detect it; in addition, to account for their abnormal color matches this aberrant system must be displaced in spectrum. In one instance, that of a blue-blind subject who was also green-anomalous, the displaced sensitivity curve of the green-receptor could be measured (Wald, 1966). It lay about midway between the normal green- and red-sensitive curves, a position that accounts well for the abnormal color matches.

One of the triumphs of modern biology is to have shown that the usual business of a gene is to specify the amino acid sequence of a protein. For many years geneticists have been unable to decide whether red- and green-blindness involve one or two gene loci on the X chromosome. I think the demonstration that two different proteins, two opsins, are needed to form the red- and green-sensitive pigments settles this issue (Brown and Wald, 1963).

Normal human vision requires the synthesis of four different opsins: one in the rods, to make rhodopsin, and three in the cones, for the color-vision pigments. Each of these must be specified by a different gene. It seems reasonable to assume that two such genes, lying close together in the X chromosome, specify the opsins of the normal red- and green-sensitive pigments (Fig. 13) (Wald, 1966). Mutations in these genes that result in the failure to form either pigment probably account for red- or green-blindness. Other mutations at the same loci, resulting in the formation of abnormally small amounts of visual pigments with displaced spectra, may account for red- and green-anomaly. All these conditions

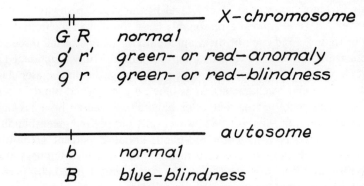

Figure 13. Diagram showing a hypothetical arrangement of genes that specify the three opsins which, with 11-*cis* retinal, form the blue-, green-, and red-sensitive pigments (*B, G, R*) of normal human color vision. Two such genes on the X chromosome are assumed to determine *G* and *R*. Mutations in these genes that result in the production of less effective pigments, displaced in spectrum, are responsible for color-anomaly; and other mutations that result in the failure to form functional visual pigments results in color blindness. The normal condition is dominant to red- or green-anomaly, which in turn is dominant to red- or green-blindness. The location of the gene for the opsin of the normal blue-sensitive pigment is not known; but the mutation responsible for blue-blindness seems clearly to be autosomal, since this condition is almost equally distributed between males and females; and it may be dominant to the normal condition.

breed true, the normal condition being dominant to color anomaly, which in turn is dominant to color blindness as expected. The mutation responsible for blue-blindness must be autosomal; and there is some evidence that it is inherited as an irregular dominant, though too few blue-blind genealogies are yet available to characterize the genetics reliably (Kalmus, 1955).

Color Blindness in the Normal Retina

A recent investigation, taking off from earlier observations by König (König and Köttgen, 1894) and Willmer and Wright (Willmer, 1944; Willmer and Wright, 1945), introduces an altogether different aspect of color blindness. A small, central patch of the normal fovea, subtending a visual angle of about 7 to 8 minutes and hence hardly larger than the fixation area, is blue-blind in the sense of lacking functional blue-sensitive cones (Fig. 14) (Wald, 1967B). This is a matter of retinal topography, not of size of field, for blue-sensitive cones are well represented in a field of

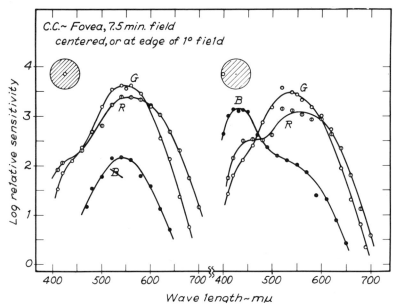

Figure 14. Blue-blindness of the fixation area of the normal fovea. Spectral sensitivities of the blue-, green-, and red-sensitive cones (*B, G, R*), measured in a 7.5-min field fixated either centrally or 7/16° from the fixation point. At the center of the fovea, though *G* and *R* are well represented, all attempts to measure *B* result only in finding *G* or *R* at lower levels of sensitivity. As soon as the field is moved away from the fixation area, *B* appears; simultaneously *G* and *R* decline somewhat in sensitivity, owing probably to smaller densities of cones. [From Wald (1967B)]

this size fixated elsewhere in the fovea or in the peripheral retina. Blue-blindness, though by far the rarest form of congenital color blindness, appears to be the usual condition of the fixation area of the fovea.

This intrusion of color blindness at the center of the normal fovea recalls the old and often repeated observation that other, more peripheral areas of the normal retina are red- or green-blind; and that areas still further out are totally color-blind (Fig. 15). It is difficult to make good color-vision experiments in such outlying areas, and much remains to be done. Nevertheless it now seems possible that all the classic forms of color blindness, including total color blindness, are represented in the normal retina.

What we regard as normal trichromatic vision appears now to be only the special property of a broad annulus of retina stretching from the blue-

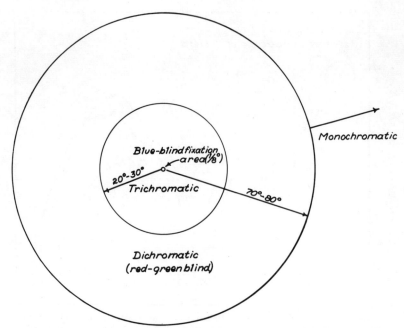

Figure 15. Approximate distribution of color function over the normal human retina. The fixation area is blue-blind, in the sense of lacking functional blue-sensitive cones. From there to about 20° to 30° from the fixation point is trichromatic. Beyond this range, to perhaps 70° to 80° out, the retina behaves as though red- or green-blind; and still further out as totally color-blind (monochromatic). The nature of peripheral color blindness and its mechanisms are still to be explored; yet it seems possible that all the classic forms of color blind-ness are represented in various zones of the normal retina.

blind fixation area at the center to about 20° to 30° visual angle out (Fig. 15). Most of the normal retina is color-blind. The mechanisms of color blindness in the peripheral retina—whether the lack of certain classes of cone or the reduction or fusion of nerve channels from the cones —must still be explored.

However this comes out, it raises an altogether different kind of problem from that of congenital color blindness: not what pigments the organisms can make, but how it distributes those pigments and the cells that contain them. That takes us from molecular genetics to questions of embryogeny and cell differentiation. The human retina, with its complex topography and radial zonation, may be a particularly fortunate place to study such problems.

Afterword

Looking back over this account I am struck with the thought that some of the most significant aspects of the photoreceptor process come from its being laid out in two dimensions: on the molecular level, in two-dimensional arrays of oriented molecules, the membranes that comprise the photoreceptor organelles; and on the cellular level, in the single layer of receptor cells that composes the retinal mosaic. In these arrangements each molecule of visual pigment and each receptor cell can report on its own. The absorption of a single photon by any molecule of rhodopsin among the many millions that a dark-adapted rod contains can excite it (Hecht, Shlaer, and Pirenne, 1941–42); and the early receptor potential signals in detail the synchronized reactions of populations of the visual pigment molecules. Similarly, quite apart from the skillful electrophysiological procedures that measure directly the responses of single retinal units, we have no great difficulty in sampling, through their differences in spectral absorption and sensitivity, the properties and behavior of each receptor type over the surface of such highly variegated retinas as that of man.

Looking back also I feel deep regret to have left out of this account so much that is important: the early and fundamental studies on the photochemistry of rhodopsin by Lythgoe, Dartnall, and Goodeve, for example; Tansley's digitonin to extract visual pigments; Dartnall's realization that when plotted on a frequency scale almost all visual pigments have nearly the same shape of spectrum; Denton's first direct spectrophotometry of retinas; Rushton's and Weale's pioneering ophthalmoscopic measurements of human color vision pigments, and much more.

I am glad, however, to see that each part of this account ends in mid-career, which of course is just as I would want it.

References

Abney, W. de W. (1895). *Color Vision*. Sampson Low, Marston, London.

Abrahamson, E. W. and S. E. Ostroy (1967). Discussions. *Progr. Biophys. Mol. Biol. 17*, 179.

Arrhenius, S. (1915). *Quantitative Laws in Biological Chemistry*. G. Bell, London.

Aubert, H. (1865). *Physiologie der Netzhaut*. E. Morgenstern, Breslau.

Ball, S., T. W. Goodwin, and R. A. Morton (1946). *Biochem. J. 40,* 59.

Ball, S., T. W. Goodwin, and R. A. Morton (1948). *Bichem. J. 42,* 516.

Blegvad, O. (1924). *Amer. J. Ophthal. 7* (series 3), 89.

Boll, F. (1877). *Arch. Anat. Physiol. Physiol. Abt. 1877,* 4.

Bownds, D. (1967). *Nature 216,* 1178.

Bridges, C. D. B. (1967). Discussion. *Comprehensive Biochemistry 27,* 1, M. Florkin and E. H. Stotz, eds., Elsevier, Amsterdam.

Brown, K. T. and M. Murakami (1964). *Nature 201.* 626.

Brown, P. K. (1961). *J. Opt. Soc. Amer. 51,* 1000.

Brown, P. K. and P. S. Brown (1958). *Comparative Biochemistry 1,* 311, G. Wald, ed., Academic Press, New York, 1960.

Brown, P. K. and P. H. Smith (1953). *Fed. Proc. 12,* 606, G. Wald.

Brown, P. K. and P. H. Smith (1955). *Amer. J. Ophthal. 40,* 18, G. Wald.

Brown, P. K. and P. H. Smith (1957). *Mod. Probl. Ophthal. 1,* 173, G. Wald.

Brown, P. K. and G. Wald (1956). *J. Biol. Chem. 222,* 865.

Brown, P. K. and G. Wald (1963). *Nature 200,* 37.

Brown, P. K. and G. Wald (1964). *Science 144,* 45.

Carlisle, D. B. and E. J. Denton (1957). *J. Physiol. 139,* 8.

Collins, F. D. (1953). *Nature 171,* 469.

Cone, R. A. (1967). *Science 155,* 1128.

Cone, R. A. and P. K. Brown (1967). *Science 156,* 536.

Denton, E. J. and F. J. Warren (1956). *Nature 178,* 1059.

Edisbury, J. R., R. A. Morton, and G. W. Simpkins (1937). *Nature 140,* 234.

Ewald, A. and W. Kühne (1878). Untersuchunger über den Schpurpur, I-IV, *Untersuch. Physiol. Inst. Heidelberg 1.*

Farrar, K. R., J. C. Hamlet, H. B. Henbest, and E. R. H. Jones (1952). *J. Chem. Soc. 1952,* 2657.

Fridericia, L. S. and E. Holm (1925). *Amer. J. Physiol. 73,* 63.

Futterman, S. (1963). *J. Biol. Chem. 238,* 1145.

Granit, R. (1941A). *Acta Physiol. Scand. 1,* 386.

Granit, R. (1941B). *Acta Physiol. Scand. 2,* 334.

Granit, R. (1942A). *Acta Physiol. Scand. 3,* 137.

Granit, R. (1942B). *Acta Physiol. Scand. 4,* 118.

Granit, R. (1943A). *Acta Physiol. Scand. 5,* 108.

Granit, R. (1943B). *Acta Physiol. Scand. 5,* 219.

Hagins, W. A. (1956). *Nature 177,* 989.

Hagins, W. A. and R. E. McGaughy (1967). *Science 157,* 813.

Hagins, W. A. and R. E. McGaughy (1968). *Science 159,* 213.

Hecht, S. (1931). Die physikalische Chemie und die Physiologie des Sehaktes. *Erg. Physiol. 32,* 243-290.

Hecht, S. (1937-38). The Nature of the Visual Process. *Harvey Lecture Ser. 33,* 35.

Hecht, S. (1938). *Le Base Chimique et Structurale de la Vision.* Hermann et Cie, Paris, pp. 1-100.

Hecht, S., S. Shlaer, and M. H. Pirenne (1941-42). *J. Gen. Physiol. 25,* 819.

Hubbard, R. (1955-56). *J. Gen. Physiol. 39,* 935.

Hubbard, R. and A. Kropf (1958). *Proc. Nat. Acad. Sci. U.S. 44,* 130.

Hubbard, R. and G. Wald (1951). *Proc. Nat. Acad. Sci. U.S. 37,* 69.

Hubbard, R. and G. Wald (1952–53). *J. Gen. Physiol. 36,* 269.

Hubbard, R., R. I. Gregerman, and G. Wald (1952–53). *J. Gen. Physiol. 36,* 415.

Kalmus, H. (1955). *Ann. Hum. Genet. 20,* 39.

Karrer, P., R. Morf, and K. Schopp (1931). *Helv. Chim. Acta 14,* 1431.

Koen, A. L. and C. R. Shaw (1966). *Biochim. Biophys. Acta 128,* 48.

König, A. (1930). *Gesammelte Abhandlunger zur Physiologischen Optik.* Barth, Leipzig.

Köttgen, E. and G. Abelsdorff (1896). *Z Psych. Physiol. Sinnesorg. 12,* 161.

Krinsky, N. I. (1958). Retinal esterification of vitamin A. *J. Biol. Chem. 232,* 881.

Kropf, A. and R. Hubbard (1958). *Ann. N.Y. Acad. Sci. 74,* 266.

Kuhn, R. and E. Lederer (1933). *Ber. Chem. Ges. 66,* 488.

Kuhn, R., P. György, and T. Wagner-Jauregg (1933). *Chem. Ber. 66,* 317.

Kühne, W. (1878). *On the Photochemistry of the Retina and on Visual Purple,* Michael Foster, ed., Macmillan, London.

Kühne, W. (1879). Chemische Vorgance in der Netzhaut. *Handbuch der Physiologie,* L. Hermann, ed., Volume 3, Part 1, p. 312, Vogel, Leipzig.

Liebman, P. and G. Entine (1967). *Nature 216,* 501.

Marks, W. B. (1965). *J. Physiol. 178,* 14.

Marks, W. B., W. H. Dobelle, and E. F. MacNichol (1964). *Science 143,* 1181.

Matthews, R. G., R. Hubbard, P. K. Brown, and G. Wald (1963–64). *J. Gen. Physiol. 47,* 215.

Morton, R. A., and G. A. J. Pitt (1955). *Biochem. J. 59,* 128.

Morton, R. A., M. K. Salah, and A. L. Stubbs (1947). *Nature 159,* 744.

Munz, F. W. (1957). *Science 125,* 1142.

Oroshnik, W., P. K. Brown, R. Hubbard, and G. Wald (1956). *Proc. Nat. Acad. Sci. U.S. 42,* 578.

Radding, C. M. and G. Wald (1955–56). *J. Gen. Physiol. 39,* 909.

Review Articles (1965). *Cold Spring Harbor Symp. Quant. Biol. 30,* 457-504.

Smith, T. G. and J. E. Brown (1966). *Nature 212,* 1217.

Stiles, W. S. (1949). *Ned. Tijdschr. Natuurk. 15,* 125.

Stiles, W. S. (1959). *Proc. Nat. Acad. Sci. U.S. 45,* 100.

Tansley, K. (1931). *J. Physiol. London 71,* 442.

Tansley, K. (1933). *Proc. Roy. Soc. London Ser. B 114,* 79.

Theorell, H. (1934). *Biochem. Z. 275,* 37, 344.

Theorell, H. (1935). *Biochem. Z. 278,* 263.

Tomita, T., A. Kaneko, M. Murakami, and E. L. Pautler (1967). *Vis. Res. 7,* 519.

Voit, C. (1900). Nachruf auf Willy Küehne. *Z. Biol.* (Munich) *40,* i-viii.

Wald, G. (1933). *Nature 132,* 316.

Wald, G. (1934). *Nature 134,* 65.

Wald, G. (1934–35). *J. Gen. Physiol. 18,* 905.

Wald, G. (1935A). *Nature 136,* 832.

Wald, G. (1935B). *Nature 136,* 913.

Wald, G. (1935–36A). *J. Gen. Physiol. 19*, 351.

Wald, G. (1935–36B). *J. Gen. Physiol. 19*, 781.

Wald, G. (1936–37). *J. Gen. Physiol. 20*, 45.

Wald, G. (1937A). *Nature 139*, 1017.

Wald, G. (1937B). *Nature 140*, 545.

Wald, G. (1937–38). *J. Gen. Physiol. 21*, 795.

Wald, G. (1938–39A). *J. Gen. Physiol. 22*, 775.

Wald, G. (1938–39B). *J. Gen. Physiol. 22*, 391.

Wald, G. (1941–42). *J. Gen. Physiol. 25*, 235.

Wald, G. (1945–46). *Harvey Lect. Ser. 41*, 117.

Wald, G. (1949). *Science 109*, 482.

Wald, G. (1950). *Biochim. Biophys. Acta 4*, 215.

Wald, G. (1956–57). *J. Gen. Physiol. 40*, 901.

Wald, G. (1958). *Science 128*, 148.

Wald, G. (1960A). The distribution and evolution of visual systems, In: *Comparative Biochemistry*, M. Florkin and H. S. Mason, eds., Academic Press, New York. Volume 1, p. 311.

Wald, G. (1960B). *Circulation 21*, 916.

Wald, G. (1964). *Science 145*, 1007.

Wald, G. (1966). *Proc. Nat. Acad. Sci. U.S. 55*, 1347.

Wald, G. (1967A). *Nature 215*, 1131.

Wald, G. (1967B). *J. Opt. Soc. Amer. 57*, 1289.

Wald, G. (1967–68). *J. Gen. Physiol. 51*, 125.

Wald, G., and P. K. Brown (1950). *Proc. Nat. Acad. Sci. U.S. 36*, 84.

Wald, G. and P. K. Brown (1951–52). *J. Gen. Physiol. 35*, 797.

Wald, G. and P. K. Brown (1953–54). *J. Gen. Physiol. 37*, 189.

Wald, G. and P. K. Brown (1958). *Science 127*, 222.

Wald, G. and P. K. Brown (1965). *Cold Spring Harbor Symp. Quant. Biol. 30*, 345.

Wald, G. and R. Hubbard (1948–49). *J. Gen. Physiol. 32*, 367.

Wald, G. and R. Hubbard (1950). *Proc. Nat. Acad. Sci. U.S. 36*, 92.

Wald, G., P. K. Brown, and P. S. Brown (1957). *Nature 180*, 969.

Wald, G., P. K. Brown, and I. R. Gibbons (1963). *J. Opt. Soc. Amer. 53*, 20.

Wald, G., P. K. Brown, and P. H. Smith (1953). *Science 118*, 505.

Wald, G., P. K. Brown, and P. H. Smith (1954–55). *J. Gen. Physiol. 38*, 623.

Warburg, O. and W. Christian (1932). *Biochem. Z. 254*, 438.

Warburg, O. and W. Christian (1933). *Biochem. Z. 266*, 377.

Willmer, E. N. (1944). *Nature 153*, 774.

Willmer, E. N. and W. D. Wright (1945). *Nature 156*, 119.

Wright, W. D. (1953). *J. Opt. Soc. Amer. 42*, 509.

Yoshizawa, T. (1967). Personal Communication.

Yoshizawa, T. and G. Wald (1963). *Nature 197*, 1279.

Yoshizawa, T. and G. Wald (1964). *Nature 201*, 340.

Yoshizawa, T. and G. Wald (1967). *Nature 214*, 566.

Elwin Marg

The Jigsaw Puzzle
of Visual Neurophysiology*

Visual neurophysiology may seem narrowly specialized compared with visual science in general, but it has a simple general goal: to learn how we see by direct investigation of the very nerve cells involved in vision.

Historically, visual neurophysiology might be said to have started when du Bois Reymond (1849) first measured electrical potentials from the eye and optic nerve of the tench, a European freshwater fish. In a more meaningful and modern sense, the development of our current and rapidly growing understanding of vertebrate visuo-sensory processes can be traced to Hartline (1938). By teasing apart the optic nerve fibers of a frog eye he was able to record from single functional fibers, before the era of microelectrodes. Hartline found that spots of light presented in a certain area of the visual field influenced the firing of the fiber, and this area he termed a "receptive field." Granit and Svaetichen (1939) confirmed Hartline's results with a microelectrode (which was large by today's standards) placed directly on the retina of the open eye.

Thirteen years later Kuffler (1953), working with the newly developed micropipette electrode on the closed eyes of cats, was able to plot the form of the receptive fields of retinal ganglion cells. He found the receptive field topography consisted of a center (disk) and a surround (annulus), one causing the cell to fire more (on-effect which is an excitatory phenomenon) when stimulated by light, and its partner causing it to fire less (off-effect which is an inhibitory phenomenon). In other words, there were on centers with off surrounds or the reverse: off centers with on surrounds.

* I thank Miss Nancy Kuwada for her technical assistance. Supported by a grant from the National Science Foundation and a Research Professorship from the Miller Institute for Basic Research in Science of the University of California, Berkeley.

A similar organization of receptive fields of the frog retina was found by Barlow (1953). However, more complex receptive field properties were found in the frog by Lettvin et al. (1959), properties not found in mammals which appear simpler by comparison.

During the last fifteen years, scores of papers have been published on visual receptive fields from the cells at all levels of the system from the ganglion cell layer of the retina on up into the brain. A list of the important papers would be too lengthy to include here and they can be found in various reference and text books (for example, Granit, 1947, 1955; Graham et al., 1965). The most complete and best known investigations are those of Hubel and Wiesel (1960–1963, 1965; Wiesel and Hubel, 1966), who first worked with cats and later with monkeys. They confirmed and extended Kuffler's retinal data, and then went on to make extensive studies of the lateral geniculate body and the visual cortex.

Hubel and Wiesel's results showed that the receptive fields of the cells of the lateral geniculate body resembled in form those of the retinal ganglion cells. The visual cortex gave a different picture. Here the fields were rectangular or bar-shaped and sensitive to the orientation of a bar-shaped target, for if the stimulus were presented at right angles to the receptive field no response would be elicited. More complex receptive fields were found in which the bar would be effective over a relatively large area or only when one or both ends were limited in length or the target fulfilled other more complex requirements.

The disks, annuli, and bars can be constructed into an organization of how we see, a sort of jigsaw puzzle of visual neurophysiology. The pieces may vary in size but not in shape beyond the limitations already mentioned. Unlike the usual jigsaw puzzle, the pieces in the eye and brain are numbered in the millions and at each level (or order of neuron) the picture is different. Unlike a wooden jigsaw puzzle, the visual one has a high degree of overlapping and superposition of the parts. Colors must be added as well as movement and its direction, and at the cortical level, binocularity in animals with overlapping visual fields. Animal experiments will continue to bring the puzzle into better fit and focus. But can it be finally solved without direct information from the human visual system?

It is necessary to get much of our neurophysiological information from animals (under conditions free of pain) because they do not have to volunteer, their physical integrity can be compromised, their lives can be risked and when necessary they can be sacrificed. If ethics clearly allowed, it would be preferable to obtain data from human beings because most of us are actually interested in human function rather than in, for example, cat function. There are, obviously, fundamental species differences. Furthermore there are important physiological functions (equivalent to psychological functions in the brain) which can be studied only in a drug-free, cooperative, and communicating animal, man.

Neurosurgeons must probe the brains of people for the diagnosis and therapy of certain diseases such as medically intractable epilepsy and Parkinson's disease. The consent of the patient undergoing such procedures makes the collection of scientific data as a by-product of the treatment both highly ethical and an opportunity not to be lost in the extension of our knowledge of mankind for mankind.

With this in mind, Dr. John E. Adams, professor of neurological surgery at the University of California, and I developed rugged microelectrodes which could be used for implantation as part of a program of depth electrography to aid medically intractable epileptics (Marg, 1964; Marg and Adams, 1967). The electrodes, made of straight tungsten wire of 50 microns diameter, are sharpened to a one-micron tip with a short taper and insulated with 15 separately baked coats of an insulating varnish, Isonel 31. A bundle of these electrodes may be used to increase the probability of picking up a unit response, which is further increased by employing an implantable micromanipulator (Fig. 1) controllable from a flexible shaft through the head bandages. The electrodes were connected to the usual follower circuit, amplifiers, loudspeaker, and cathode ray oscilloscope with camera. For use just during surgery deep in the brain at the thalamic level, microelectrodes have been used having a shank diameter of a quarter or a half millimeter (Marg and Dierssen, 1965, 1966).

The patient viewed a large piece of cardboard one meter away with an ink dot or grain-of-wheat lamp for fixation (Fig. 2) (Marg, Adams, and Rutkin, 1968). Hand-held wands with disks or bars of various sizes provided the means of plotting the receptive field as the firing of the cell was heard over the loudspeaker and seen on the cathode ray oscilloscope. The receptive field was outlined in light pencil marks on the cardboard.

Cells were found firing spontaneously in an irregular or "bursty" manner (Fig. 3). Many of them appeared to be uninfluenced by any kind of visual stimulation we could provide such as random disks and bars, gratings, fingers, etc., or the patient just gazing around the room. Some units did exhibit a visual response to certain targets, and showed inhibition of the spontaneous activity when the eyelids were closed. In some of these we could find the position in the visual field of the receptive field and plot it. The electrode would usually maintain the response for the recording session of several hours and sometimes the same unit would be found unchanged the next day.

First we noted whether the receptive field came from one eye or both and then we looked for any dominance of one over the other in the re-

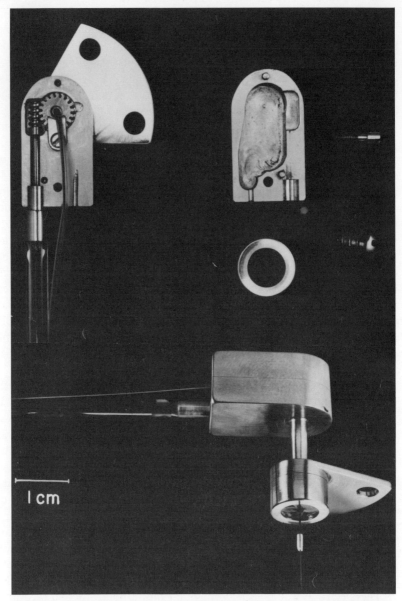

Figure 1. Micromanipulator for implantation in a burr hole. Below: the bundle of microelectrodes project down from a teflon diaphragm on the cylinder which fits in the burr hole. The flange is screwed to the bone and the gear box rests above the scalp. The electrode wires are led off in the small plastic tube through the head bandages. The large

plastic tube contains a flexible shaft which may be turned from outside the bandages, lowering the microelectrode bundle 20μ per turn for an excursion of up to 3 mm. Above: the worm-gear box, lid, diaphragm ring and screws. All metal parts are made of type 316 low-carbon stainless steel.

Figure 2. The supine patient (whose head is just visible at the bottom foreground) is fixating a black dot on the white cardboard sheet 1 m from his eyes. A bar-shaped target on a wand is being manipulated to find the receptive field.

50μV L

5msec

Figure 3. The response of this unit gave the receptive field labeled E in Figure 4.

sponsiveness of the cell. We plotted the receptive field, tried various colors and other tests such as stimulation of other sensory modalities, auditory feedback, or voluntary mental image effects. It was soon evident that the receptive field was always the same whether the target was black on white or white on black, or red, yellow, green, or blue on any contrasting background. Moreover, all attempts by the patient mentally to influence the firing of a visual cortical cell were fruitless. Hue did not seem to be represented in the cells we recorded, nor did mental imagery or other (non-visual) sensory modalities.

All the receptive fields were entirely excitatory as far as we could tell, that is, an increase of firing superimposed on the irregular spontaneous activity was evident whenever a target was brought within the receptive field. When the stimulus exactly matched the field, the response was strong. No inhibitory area surrounding the excitatory receptive field was observed but if it were subtle in effect it could have been missed by the simple method of plotting.

Some of the receptive fields could be plotted repeatedly without any apparent decrement of the response. The responses of other fields faded away until only the irregular spontaneous activity remained. Later they might partially recover, only to fade more quickly the second time. This phenomenon has been called progressive attenuation (Hubel and Wiesel, 1965) or habituation (Horn and Hill, 1966). Different visual stimuli or stimuli of other modalities (such as handclapping) did not cause the response to the original stimulus to return, demonstrating a lack of dishabituation (Horn and Hill, 1966; Horn, 1967).

Only a few of the units observed were plotted, primarily because of the limitations of time, and these are shown in Figure 4. One disk-shaped field (E) was binocular, indicating that the unit being recorded was cortical and not directly from an incoming lateral geniculate fiber, if the human lateral geniculate nucleus is similar in organization to that of the cat and monkey. In one unit (D), a horizontal bar target was effective over a 20 cm vertical range (the limits shown by the dashed lines), similar to the "complex" units found in the cat cortex. In these complex units a distinction must be made between the receptive field (the large area between the dashed lines) and the shape of the target which elicits the response (the bar). All of the fields were from units in the contralateral visual hemisphere except a near-vertical thin bar-shaped field whose unit was in the ipsilateral visual cortex. The horizontal bar-shaped fields seemed to give a greater response to movement relative to steady presentation of the targets than did the others. There was some indication of a shift of size and position in some of the units, which could be characterized as plasticity. This was not caused by poor fixation. Generally fixation was steady during the plotting of the receptive fields and in most fields the edges were sharp and did not wander.

By the time the receptive field of one binocular unit (F) was plotted with one eye, it had habituated and no response could be obtained through

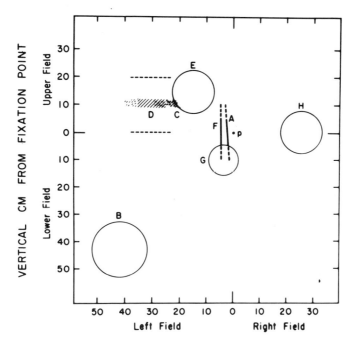

Figure 4. Receptive fields from units in the human visual cortex. "p" is the fixation point at 1 m from the eyes.

a. Left eye dominant, ipsilateral
b. Right eye
c. Right eye
d. Right eye, marked habituation, vertical extent of "complex"
e. Binocularly equal

f. Binocularly equal, marked habituation
g. Binocularly equal, marked habituation
h. Binocularly equal

the other eye. At the next recording session the same unit was still available and the same receptive field could now be obtained from the second eye. This finding would place the habituation phenomenon at the cortical level.

As mentioned earlier, it was possible to inhibit the spontaneous activity of visually responding cells by closing the lids. No other means of inhibiting the cells was found. Occluding the open eyes with a card or reducing the room illumination did not inhibit the cell.

By merely changing the distance of the cardboard with its fixation point, one can plot the receptive field at various fixation distances, if time allows. It is possible to show if the receptive field or "visual grain" changes its angular size with different fixation distance. This presumably would be directly concerned with the basic mechanism of size constancy or scaling (Richards, 1967, 1968).

These preliminary results demonstrate two points: first, that there are species differences in the visual receptive fields between man and experimental animals that can be discovered only by obtaining data directly from man, under, of course, completely ethical conditions; and second, that there are aspects of the function of the visual system that can be elucidated only by this kind of investigation in conscious volunteer human patients. In fact these points can be extended to the functions of the brain generally, sensory, motor, and integrative.

In the span of thirty years, visual neurophysiology has progressed from the first glimpse of visual receptive fields in frogs to the first glimpse of them in man. They are the jigsaw puzzle of vision. It seems reasonable and safe to predict that, as we put together the pieces of the visual receptive fields puzzle in man, it will form a picture which will allow us to understand how we see.

References

Barlow, H. B. (1953). Summation and inhibition in the frog's retina. *J. Physiol.* *119,* 69-88.

du Bois, Reymond E. (1849). *Untersuchungen über thierische Elektricität.* G. Reimer, Berlin.

Graham, C. H., N. R. Bartlett, J. L. Brown, Y. Hsia, C. G. Mueller, and L. A. Riggs (1965). *Vision and Visual Perception.* John Wiley and Son, New York.

Granit, R. (1947). *Sensory Mechanisms of the Retina.* Oxford University Press, London.

Granit, R. (1955). *Receptors and Sensory Perception.* Yale University Press, New Haven, Connecticut.

Granit, R., and G. Svaetichin (1939). Principles and techniques of the electrophysiological analysis of color reception with the aid of microelectrodes. *Upsala Läkaref. förh. 65,* 161-177.

Hartline, H. K. (1938). The response of single optic nerve fibers of the vertebrate eye to illumination of the retina. *Am. J. Physiol. 121,* 400-415.

Horn, G. (1967). Neuronal mechanisms of habituation. *Nature 215,* 707-711.

Horn, G., and R. M. Hill (1966). Responsiveness to sensory stimulation of units in the superior colliculus and subjacent tectotegmental regions of the rabbit. *Exper. Neurol. 14,* 199-223.

Hubel, D. H., and T. N. Wiesel (1960). Receptive fields of optic nerve fibers in the spider monkey. *J. Physiol. 154,* 572-580.

Hubel, D. H., and T. Wiesel (1961). Integrative action in the cat's lateral geniculate body. *J. Physiol. 155,* 385-398.

Hubel, D. H., and T. Wiesel (1962). Receptive fields, binocular interaction, and functional architecture in the cat's visual cortex. *J. Physiol. 160,* 106-154.

Hubel, D. H., and T. Wiesel (1963). Shape and arrangement of columns in the cat's striate cortex. *J. Physiol. 165,* 559-568.

Hubel, D. H., and T. Wiesel (1965). Receptive fields and functional architecture in two non-striate visual areas (18 and 19) of the cat. *J. Neurophysiol. 28,* 229-289.

Kuffler, S. W. (1953). Discharge patterns and functional organization of mammalian retina. *J. Neurophysiol. 16,* 37-68.

Lettvin, J. Y., H. R. Maturana, W. S. McCullock, and W. H. Pitts (1959). What the frog's eye tells the frog's brain. *Proc. Inst. Radio Engr. 47,* 1940-1951.

Marg, E. (1964). A rugged, reliable, and sterilizable microelectrode for recording single units from the brain. *Nature 202,* 601-603.

Marg, E., and J. E. Adams (1967). Indwelling multiple microelectrodes in the brain. *Electroencephal. and Clin. Neurophysiol. 23,* 277-280.

Marg, E., J. E. Adams, and B. Rutkin (1968). Receptive fields of cells in the human visual cortex. *Experientia 24,* 348-350.

Marg, E., and G. Dierssen (1965). Reported visual percepts from stimulation of the human brain with microelectrodes during therapeutic surgery. *Confin. Neurol. 26* (2), 57-75.

Marg, E., and G. Dierssen (1966). Somatosensory reports from electrical stimulation of the brain during therapeutic surgery. *Nature 212,* 188-189.

Richards, W. (1967). Apparent modifiability of receptive fields during accommodation and convergence and a model for size constancy. *Neuropsychol. 5,* 63-72.

Richards, W. (1968). Spatial remapping in the primate visual system. *Kybernetic 4,* 146-156.

Wiesel, T. N., and D. H. Hubel (1966). Spatial and chromatic interactions in the lateral geniculate body of the rhesus monkey. *J. Neurophysiol. 29,* 1115-1156.

Jay M. Enoch

Retinal Directional Resolution

Abstract

A number of techniques have been devised to study several aspects of Campbell's retinal directional acuity effect. The optical properties of (a) a contact lens (serving as an artificial cornea), (b) a human eye bank eye, (c) a glass fiber optics bundle, and (d) rat and squirrel monkey retinas (studied as fiber optics bundles), have been evaluated from this point of view. In general these data support the notion that the effect described by Campbell was largely due to ocular aberrations. I measure a small but statistically significant decrement in retinal resolution when light is incident obliquely at the retina. The resolution decrement is present for all orientations of the resolution target. There may also be a small meridional component. In addition, I have demonstrated that disturbed retinal orientation can result in reduced retinal resolution and light transmission capability. The latter problem is complex. Interrelationships are considered between Campbell's retinal directional acuity effect, the directional sensitivity of the retina, disturbed orientation of central photoreceptors, and measures of foveal resolution.

As a separate consideration, resolution losses recorded in Experiment I are of interest to the contact lens fitter. Contact lens decentration vis-à-vis the eye pupil are common. Experiments such as these can aid in understanding "off center" resolution effects.

Introduction

In 1958 Campbell described a retinal directional acuity effect (Campbell, 1958). He placed a small aperture just in front of the eye, and found that visual resolution was best when the aperture was centered in the pupil. When he displaced the aperture laterally, resolution of grid lines oriented perpendicular to the direction of translation fell sharply, while resolution for bars oriented parallel to the direction of translation was not decreased significantly. When he used a one mm aperture, the loss in resolution of perpendicularly oriented grid targets was of the order of 8X at the edge of the dilated pupil. If one plotted translation data (in mm) against resolution loss, a semilogarithmic function provided a best fit for his data (Fig. 1). In a letter to me he stated that a slightly greater effect

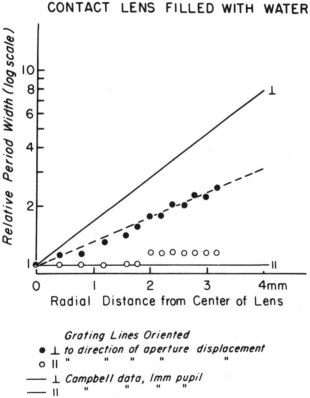

Figure I. These resolution data were obtained by translating an aperture a measured distance (mm) from the center of the contact lens filled with saline. Perpendicular means that the lines of the NBS grating in the target plane were oriented perpendicular to the direction of aperture translation, and parallel indicates that the grid lines were oriented in the same direction as the translation. Campbell's original data on the human eye are superimposed. In both sets of data, an aperture of I mm was employed. Resolution data are presented on a logarithmic scale. The greater the period (2X bar width), the poorer the resolution.

was obtained using a 2 mm aperture. He found the magnitude of the effect fell off with smaller and larger apertures.

From the first I have been interested in this work, since I had just published the first of a number of articles attempting to relate some cases of limited amblyopia (reduced visual acuity) to disturbances in receptor orientation as revealed by studying the directional sensitivity of the retina (Enoch, 1957, 1959A, 1959B; Stiles and Crawford, 1933). At the time, it seemed that we were considering different aspects of the same problem. That relationship is considered in the discussion of this paper.

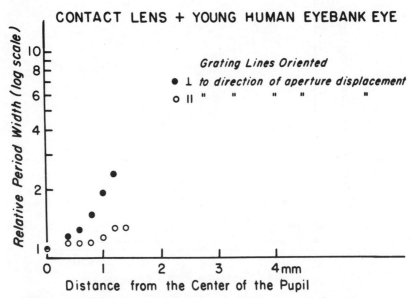

Figure 2. These data are plotted as in Figure 1, and the same contact lens was employed. An eye bank eye with the rear portion removed was placed in the lens. Differences between these data and Figure 1 provide an estimate of the contribution of the eye lens.

Campbell and Gregory (1960) limited the role of ocular aberrations in these measurements by determining retinal resolution interferometrically. They found the phenomenon originally described by Campbell was due largely to these aberrations. In addition, they described a small retinal meridional resolution decrement which measured 1.5X, 1.8X, and 2.9X in three subjects. The same year, I used a glass fiber optics bundle as a model for retinal receptors, and evaluated the effect of varying the angle of incidence of a beam of light on the resolution properties of the bundle (a set of data from that paper appears in Fig. 3) (Enoch, 1960). I found

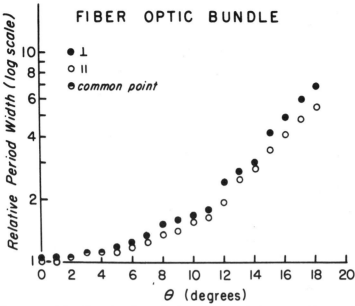

Figure 3.   The directional resolution characteristics of a fiber optics bundle tested under conditions similar to those used on Figures 1 and 2 are presented. These data differ in nature from the functions originally derived by Campbell (Figure 1). In this experiment the fiber bundle rotated about the center of the image formed by a converging lens. However, perpendicular and parallel have the same connotation, and are defined as in Figure 5. The abscissa is expressed in terms of angle of rotation of the incident bundle about the image plane (θ), instead of aperture displacement.

that the function generated with that model was not of the same form as that described by Campbell.

With the passage of time our understanding of these matters has continued to evolve. The most advanced treatment of directional acuity was presented recently by Green (Campbell and Green, 1965; Green, 1967). He studied the effect of aperture decentration using (a) sinusoidal gratings, and (b) a coherent light source to produce retinal interference fringes. He found considerably less directional acuity effect using sinusoidal gratings than Campbell did for bar gratings, and no retinal factor.

Over a period of years, I have employed different techniques in an effort to understand this problem. In general, my studies support the findings of Campbell and his co-workers. That is, I found that ocular aberrations caused most of the measured directional acuity effect. However, my techniques suggest the presence of a small retinal component which is generally non-meridional in nature. This is supplemented possibly by a small meridional factor.

Certain of these techniques can be used to evaluate resolution capability of receptor groups which are disturbed in orientation. I have found that noticeably disturbed receptor orientation results in meaningful decrements in measured retinal resolution capability.

In this report the following five experiments are described:

1. An aperture was displaced in front of a simulated cornea (contact lens filled with saline) and the resolution of the lens unit was evaluated using Campbell's technique.

2. An evaluation of the contribution of the eye lens was made by lowering an eye bank specimen into the contact lens-saline unit and repeating measurements.

3. The directional resolution properties of a fiber optics bundle were determined.

4. The directional resolution properties of dissected flat preparations of retinas were measured.

5. The effect of disturbed retinal orientation on retinal resolution was considered in similar flat preparations.

Experimental

I. DISPLACEMENT OF AN APERTURE IN FRONT OF A SIMULATED CORNEA

A contact lens filled with water simulates the living cornea, but exhibits fewer aberrations than corneal tissue because comparable off-center complex variations in curvature do not take place. Variations in corneal curvature have been reported in many studies over the years; recently contact lens fitters have been active in this area.

In this experiment a standard NBS (National Bureau of Standards) resolution target was placed on a table, a scleral contact lens filled with normal saline was located at a reasonable distance directly above the target, and the image of the target formed by the lens was studied using a simple microscope. The constants of the experiment are listed in Table I.

Table I

Scleral Contact Lens Experiment

1. Radius, front surface, C.L.	8.10 mm
2. Radius, read surface, C.L.	7.92 mm
3. Thickness, center, C.L.	0.75 mm

4. Dioptric Power (air), C.L.	+0.50 D.S.
5. Index of refraction, C.L.	1.49
6. Distance, C.L. apex to resolution target	210.0 mm
7. Apex of C.L. to surface of ground glass of simple microscope	23.8 mm
8. Depth of normal saline in C.L.	7.0 mm
9. Aperture diameter	1.0 mm
10. Magnification of simple microscope	12.5 X
11. Maximum resolution, ½ cycle (C.L. + water)	7.7μ
12. Change in angle of incidence at image plane caused by a 1 mm aperture displacement	2 18'

Note: C.L. = Contact lens

A 1.0 mm aperture, which could be translated laterally, was placed just in front of the contact lens. The plane of focus of the simple microscope was not altered when the aperture was translated. The grating target was moved on the table in order to find the set of lines providing best resolution at the center of the image distribution and microscope field. Relative resolution data obtained using the scleral contact lens are presented in Figure 1. The original findings of Campbell (1958) are superimposed. Here we find a meridional resolution loss of about 3X using the contact lens assuming the aperture had been extended to a 4 mm displacement. In keeping with Campbell's treatment, resolution has been scaled logarithmically. Note the similarity of the functions.

As noted in the abstract, these data are of particular significance to the contact lens fitter, because decentration of the contact lens relative to the eye pupil is common. That is, the result should be similar if the translated aperture is just in front of, or just behind, the contact lens. These effects should be explored in greater depth from this point of view.

II. AN EVALUATION OF THE CONTRIBUTION OF THE EYE LENS TO DIRECTIONAL RESOLUTION

In order to obtain a first order evaluation of the contribution of the eye lens to these functions, a very fresh eye bank specimen with clear ocular media (obtained from a young individual) was placed in the contact lens used in Experiment I. The rear quarter of the eye was removed, the eye was resealed with a microscope slide under pressure, and the pressure in the eye-slide unit was measured, and was maintained within normal values.

As is well known, a contact lens largely eliminates the refractive power of the corneal front surface. In turn, the remaining corneal front surface power is largely nullified by the refractive power of the corneal rear surface. Hence in this experiment the role of the cornea was small, and, in essence, the resolution capability of the water filled contact lens coupled with th eye lens was tested. Results obtained when the same aperture was translated laterally from the center of the contact lens-eye combination are shown in Figure 2. These data suggest that the eye lens also influences image quality in a manner similar to the cornea (estimated in Experiment I). Maximum resolution capability (½ cycle) of the combination was 12.5μ. Unfortunately, the compatibility of the contact lens, eye, test distance relationship, and the effects of dissection and death upon the eye lens and ocular media were not known. We were not able to dilate the iris of this eye or of the many other eye bank eyes tested. The ocular media of other specimens examined scattered more light than the eye whose data are presented. In general, the other eyes were not obtained as soon after death as this one. The role of the eye lens probably can be estimated better by studying directional acuity in both eyes of a unilateral aphakic patient fitted with contact lenses.

III. DIRECTIONAL RESOLUTION OF A FIBER OPTICS BUNDLE

Since the retina is a fiber optics bundle, we may turn to fiber optics bundles as predictive models when studying many retinal characteristics. As indicated above, shortly after Campbell's first publication on this subject (1958), I studied the directional resolution characteristics of a fiber optics bundle which exhibited some frustrated total reflection (Enoch, 1960; Kapany, 1959). The detailed characteristics of this bundle may be found in the original paper (Enoch, 1960). Relevant data obtained during that study are presented in Figure 3. The data are displayed in the same form as in previous figures. The angle of incidence of energy relative to the optic axis of the glass fibers (θ) is indicated on the abscissa. Displacement of a beam of light in the entrance pupil of the eye varies the angle of incidence of light at the retina. In this experiment we rotated the fiber bundle about the grating image. By this means we obtained oblique incidence but eliminated lens aberrations revealed by Campbell and Gregory (1960) and confirmed in Experiments I and II.

Oblique incidence of energy resulted in loss of resolution capability by the fiber bundle in all meridians, but a slightly greater loss was exhibited for gratings oriented perpendicular to the direction of equivalent beam displacement in the pupil (using the same frame of reference). Magnitude of change has little meaning. However, differences in the nature of the functions are of considerable interest. Compare Figure 3 with Figures 1 and 2. One should bear in mind that the dimensions, separations, and indexes of the components of the glass fiber bundle tested, and comparable retinal characteristics, are somewhat different. Are differences in such functions due to dissimilar physical characteristics, or does the retina, acting as a fiber optics bundle, perform in a similar manner?

IV. MEASUREMENT OF RETINAL DIRECTIONAL RESOLUTION CAPABILITIES

In order to evaluate the directional resolution characteristics of the retina *per se,* and to answer the question raised in the previous paragraph, comparable measurements were made on blocks of freshly dissected retinas studied as a fiber optics bundle. At the same time, I also analyzed the resolution characteristics of photoreceptors having disturbed orientation (see Experiment V).

The apparatus and techniques employed were largely the same as described by Enoch and Glismann (1966). In the latter paper some aspects of the role of the retina as a fiber optics bundle were considered. In addition, physical and optical changes occurring in excised retinal tissue were studied as a prelude to the final design of a microspectrophotometer.

In brief, a transparent NBS grating (emulsion to emulsion transfer on high contrast film) was imaged on one end of the retinal fiber optics bundle. The optical system could be rotated about that image plane in order to evaluate the directional resolution capabilities of the sample. The grating could be translated in order to bring different grating images to the retinal area being evaluated. Test $\lambda = 498$ nm. The aperture of the optical system of the instrument limited the incident light rays to a 7.0° cone converging at the retina. This would be equivalent to a 2.8 mm entrance pupil in the human eye (a 1 mm pupil aperture forms an incident bundle of light rays subtending 2.5° at the retina).

The light passed through the excised retina, as it might have *in vivo.* Our optical system served as a schematic eye. The focusing lens and iris aperture played a role similar to the anterior elements of the eye. By rotating the optics rather than translating the aperture, the lens-induced resolution losses were eliminated.

Each retina was placed in a special chamber, and floated in Tissue Culture Medium No. 199 without phenol red (Difco Laboratories, Detroit, Michigan). When albino rat retinas (virtually all rod) were studied, chambers 390μ deep were employed. When squirrel monkey foveas were investigated, 430μ chambers were used. These chambers were slightly deeper than the retinas were thick and were topped by a thin cover slip. The retina in the chamber was placed on a translatable X—Y stage, and resolution at the terminations of the receptor fiber optics bundle was evaluated using a microscope. For resolution studies 200X magnification was used (10X objective $\times$ 1.25X tube factor $\times$ 16 X eyepiece).

Tissue Preparation

The rats were anesthetized with ether, and the monkeys with Nembutal. The retina was dissected (in white light) from its bed as rapidly as possible, zero time being defined as that instant when the optic nerve was clamped with a hemostat. Following clamping, about one minute was required to dissect the rat retina away from the eye cup. Dissection of the squirrel monkey retina usually took somewhat more time, because the retina often adhered quite tightly to the pigment epithelium.

Test Method

In order to treat the retina as a fiber optics bundle (Enoch and Glismann, 1966; Enoch, 1967A) we first focused the microscope on the terminations of well-oriented receptors (Fig. 4). Then the image of the

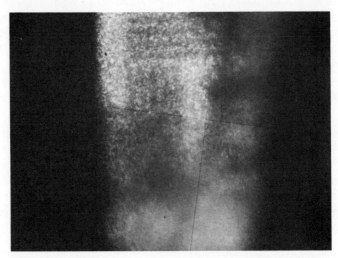

Figure 4. This low-power photograph was taken through our apparatus and shows an NBS grating target imaged on the retina. The microscope was focused upon the rat rod receptor terminations. Groups of receptors which were well and poorly oriented are clearly visible. Note changes which *occur* in transmissivity and resolution when receptor orientation is disturbed. This would be an example of static scanning.

high contrast grating resolution target was focused with the camera lens (schematic eye) in order to obtain sharpest image focus superimposed upon our view of the receptor terminations. The fact that these two planes of focus (that of the microscope, and that of the schematic eye) differed, indicated a fiber optics transfer.* Further, since retinal receptors, disturbed in orientation, transmitted energy in a less efficient manner (Fig. 4) (Enoch and Glismann, 1966; Enoch, 1967B), it could again be inferred that a fiber optics transfer was involved.

It was necessary to distinguish between static and dynamic scanning of the image (Enoch and Glismann, 1966; Kapany, 1957). In static scanning, all the elements in the optical system are stationary. However as the limit of resolution is approached, one's judgment is disturbed by the "noise" of the fiber bundle itself. If the image and detector are held stable and the fiber optics bundle put into motion, resolution is enhanced and more closely approaches the theoretical limit predicted for the interfiber dimensions of the bundle. To achieve dynamic scanning, the retina was translated back and forth on the microscope stage. While we did not reproduce Kapany's ideal random translation (1957) when using this method, gains in resolution achieved were marked.

The dynamic scanning technique was evaluated in Enoch and Glismann (1966). An increase in resolution (dynamic resolution measurements: static resolution measurements) of 1.80 (average deviation, ±0.22 based on 71 measurements) was recorded using rat retinas. Since this value approached Kapany's predicted gain of 2.12 in resolution capability, it was felt that this represented added proof of retinal photoreceptor fiber optics transfer.

The dynamic scanning method was used in this experiment. It was impossible to evaluate the resolution properties of rod-cone mixtures by this method. Hence measurements were limited to all rod rat distributions, and to all cone squirrel monkey foveas.

Figure 5 shows how the rotation of the optical system about the image was defined. This scheme is consistent with the format used in Figures 1-3. A grid oriented perpendicular to the XZ plane would also be perpendicular to the direction of displacement of an aperture in the lens plane.

Results, Normal Incidence

In order to evaluate the effect of oblique incidence of light on retinal resolution, the resolution properties of the retina for normal incidence of light had first to be measured. We first had to determine the best resolution achievable, and second we sought to control artifacts present in the preparation.

---

* That is, they differed when comparing these image planes with those obtained when light passed through a transparent medium of equivalent (approximately) index and thickness.

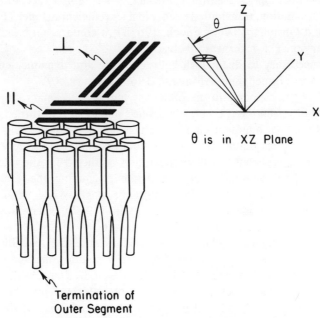

θ is in XZ Plane

Termination of
Outer Segment

Figure 5.   This schematic drawing serves to indicate the orientation
of the grating lines in relation to the axis of rotation of the central
ray of the converging bundle of rays at the retina. The angle θ may be
related to an equivalent displacement of an aperture in the entrance
pupil of the eye. "Parallel" grid bars are aligned in the X direction,
and "perpendicular" grid bars are oriented in the Y direction. This
alignment scheme is common to all experiments described in this
paper.

Figure 6 is taken from Enoch and Glismann (1966). These are typical
rat resolution data. When the preparation was fresh, maximum resolution
was 5.0μ per cycle. Intercell separation in measured specimens ranged
from 1.5μ to 2.2μ. In single preparations, the smallest recorded ratio
(cycle length in microns: twice the mean intercell separation) was 1:18.
A more usual value, taking both measurements into consideration, was
1.45. *Considerable practice* was required in order to discern the finest
resolution targets in these preparations. These data indicate that rods as
a fiber optics bundle have fine resolution properties when well oriented!

Resolution properties of the retina failed rapidly as a function of time
in these dying or dead retinas. As indicated above, time zero was the
moment the optic nerve was clamped during dissection. Various aspects
of this complex problem were considered by Enoch and Glismann (1966).

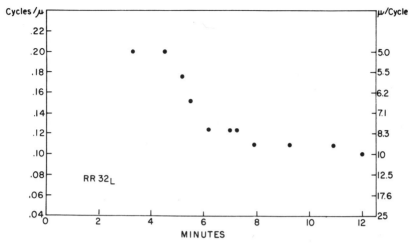

Figure 6. These are resolution data obtained using a flat preparation of freshly excised rat retina. RR32$_L$ indicates rat resolution data, animal 32 in this series, left eye. Two different forms for expressing resolution data are indicated on the ordinate, and the abscissa indicates time in minutes after clamping the optic nerve with a hemostat during dissection.

Since decrements in resolution capability due to post mortem changes often occurred during measurements of resolution for oblique incidence, it was necessary to compensate for these effects in processing our data. Figure 7 shows the rather simple technique developed. Measures of resolution perpendicular to and parallel to the equivalent direction of aperture displacement were taken first for normal incidence by both my assistant and myself. We did not communicate our results until the end of the experiment; as a rule, our data agreed well.

The angle of incidence of the axial ray of the converging bundle of light was then changed by 8° and readings were taken. Then the optical system was set for a second time for normal incidence, and the initial readings repeated.

It was assumed that the resolution function between t_α and t_β was linear (Fig. 5). t_δ was computed and a ratio was taken between measured resolution (8° oblique incidence) γ : computed resolution (normal incidence) δ, for both equivalent perpendicular and parallel orientations of the grating chart as defined in Figure 5.*

* Note in Table II this ratio has been inverted to conform with the format of data presentation used in Figures 1-3.

DATA REDUCTION

A.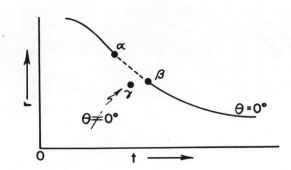

B. Assume α to β is a straight line
C. Determine δ

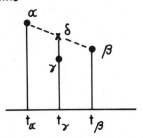

D. Record γ/δ

Figure 7.   This schematic drawing shows the data reduction technique used in Experiment IV. This manipulation of the data was necessary in order to minimize the effects of post mortem changes occurring concurrently. Since measurements were completed rapidly, and t_α-t_β was limited to only a few minutes, the straight line assumption was probably valid. t=time.

TABLE II

Relative Retinal Directional Resolution Capability

Species and Data Set	$\dfrac{\text{Mean Resolution for Normal Incidence}}{\text{Mean Resolution at 8° Obliquity}} = \dfrac{\delta}{\gamma}$		Number of Eyes
	Perpendicular	Parallel	
Rat rods			
1. Best data (JE and assistant)	1.18	1.16	22
2. Best data (JE only)*†	1.25	1.18	5

3. Data, measurement completed in six minutes*‡	1.19	1.16	13
Squirrel monkey fovea			
4. Observer JE only	1.23	1.20	6

* Data in sets 2 and 3 are included in set 1.

† Best data of observer JE are separated from set 1, because only JE made observations on squirrel monkey cones. Thus, his data on the two different preparations may be compared.

‡ The shorter the measurement period, the less the results were influenced by post mortem effects.

An 8° rotation of the 7° converging bundle about its image provided mutually exclusive data for the resolution of the grid lines. In the human, an 8° shift of angle of incidence would correspond to a 3.2 mm displacement from the center of the pupil of a 2.8 mm aperture in the entrance pupil. We compensated for changes in transmission of light for oblique incidence by varying source intensity.

Results, Oblique Incidence

Oblique incidence of light at the retina (8°) resulted in significant decrements in resolution capability (5% level of confidence, t-test) for both orientations of the test gratings (Table II). The difference between the data for perpendicular and parallel grid orientations was not significant at the 5% level of confidence. That the two sets of means determined at 8° were not significantly different from each other is not surprising, since major physical changes were occurring at the time of data accumulation, orientation of receptors was not always perfect in the test area, the observer was pressed to make rapid judgments, and the measured difference was small. However, I believe the small meridional differences measured between the two sets of means are real. Proof must await the development of a more stable preparation. The characteristics measured in this experiment seemed to be closer to the fiber optics bundle function (Fig. 3) than the function originally presented by Campbell (Fig. 1). We collected some data for angles of obliquity of incidence of 4° and 12° which tended to support this contention. Since our findings approximated the format defined in Figure 3, I regard this as one more bit of evidence that the retina serves as a fiber optics bundle.

Ideally, sets of data should have been taken for several angles of oblique incidence on the same specimen. We chose to concentrate at 8° in order to minimize the t_α-t_β error, and to obtain data at a point still within the bounds of the equivalent human pupillary aperture.

Acceptable data were obtained from the foveas of only six squirrel monkey retinas. *Many* more eyes were studied, but it was difficult to find a well oriented area in most foveas. If a block of several million rods was dissected, the chance of finding a well oriented sub-area was relatively good. If, on the other hand, I sought a well oriented fovea, I was seeking good orientation among a few thousand cells. The receptors of the rat and the squirrel monkey are shorter and have somewhat different dimensions from those found in the human retina. Thus, these values cannot be compared directly. In addition, in normal vision, energy may be absorbed over the length of the outer segment. Here, resolution at the receptor terminations was evaluated. Hence, these data provide only an estimate of retinal resolution capability.

V. DISTURBED RETINAL ORIENTATION AND RETINAL RESOLUTION

A glance at any one of the several hundred retinal specimens studied using our apparatus would serve to prove to the reader that disturbed receptor orientation reduces the efficiency of light transmission (rod or cone) and reduces resolution capability of the retina as a fiber optics bundle. Figure 4 shows major variations in a rat retina. Disturbed orientation was the rule, not the exception in these preparations.

The problem was to specify the magnitude of the disturbance in receptor orientation; there was no difficulty in measuring *marked resolution decrements* (Fig. 4). Receptors do not simply bend at the external limiting membrane at an angle of x°. While such disturbance in the orientation may be seen occasionally, far more complex bends are common. Further, receptor orientation varies in groups of cells, and some variance also exists within each group (Enoch, 1967A, 1967B; Safir and Hyams, 1969). Hence, to date I have failed to adequately correlate resolution decrements with disturbance in orientation. When a small or a moderate disturbance in orientation was observed, I noted an overall loss in resolution capability with perhaps a small meridional component.

Since we cannot adequately quantify the effects of oblique or disturbed orientation, we cannot correlate these effects with the directional resolution (retinal component) data presented above. However, it is important for the readers to realize that added factors are involved when receptor orientation is disturbed. Light transmission of directionally sensitive rods (Fig. 4) and cones is altered, the separation of cells becomes disturbed and there is probably greater frustrated total reflection (optical interaction, light leakage between cells, cross-talk) (Enoch, 1960; Enoch and Glismann, 1966; Kapany, 1957) with metabolism, and so forth.

In future studies, the technique discussed relative to Figure 10 in the Enoch and Glismann article (1966; see also Enoch, 1967A and 1967B) may offer a means of specifying the representative area-wide disturbance in orientation, as opposed to trying to indicate the magnitude of individual cell (or group of cells) misalignment. In addition, efforts should be made to extend these studies in order to determine a retinal transfer function.

Discussion

These experiments generally compliment the work of Campbell and his co-workers. There seems little doubt that the effect he first described was largely due to ocular aberrations. The data obtained in this study suggest the presence of a rather small retinal directional resolution decrement with oblique incidence of light at the retina. This component would show a small loss in resolution in all meridians, and perhaps a very small added decrement for grid orientations perpendicular to the direction of aperture displacement. I assume that the human retina acts in a manner similar to our samples. Admittedly the methods employed were difficult, and subject to problems.

Visual researchers must face up to the fact that the retina is a fiber optics bundle, and consider the role played by these *complex optical* properties in visual response. Glass fiber bundles offer accessibility and freedom from dynamic changes occurring during normal response and pathology, and rapid and marked post mortem changes after dissection. The problem is to find fiber bundles having properties approaching those of the retina. These can serve as useful predictive models. This principle was attempted in this study. We expect to receive a fiber bundle shortly which more closely approximates retinal dimensions.

Lastly, it is important to consider the relationship between these findings and amblyopia due to disturbed receptor orientation (receptor amblyopia) (Enoch, 1957, 1959A, 1959B, 1960, 1967A, 1967B). From Figure 4 it is obvious that a disturbance in orientation *can result* in a readily measurable decrement in retina resolution capability. The grating which was resolved by the well oriented group of cells was considerably above the resolution threshold for that portion of the retinal fiber optics bundle (even for static scanning). Some improvement in resolution probably would have been obtained in the other areas by varying the microscope fine focusing, and by increasing the illumination of the source. *However, a marked decrement would remain.* This finding has been repeated endlessly in our laboratory; this is simply an available dramatic example of the effect.

Working with Franz Fankhauser and the late Paul Cibis, I demonstrated changes in the Stiles-Crawford patterns, and visual acuity in the presence of retinal pathology (Fankhauser, Enoch and Cibis, 1961; Fankhauser and Enoch, 1962). These findings indicate that changes in orientation of receptors can occur during life on a substantial scale. It is known that small changes continually occur in directional sensitivity in normal observers (Stiles, 1939; Enoch, 1958, 1963). I have demonstrated the presence of atypical Stiles-Crawford functions in individuals with low grade amblyopias (Enoch, 1957, 1959A, 1959B).

The main question which must be answered is how much disturbance in orientation must there be in the bundle of receptors responsible for finest acuity in order to produce a clinically significant decrement. This question has not been answered, since the retinal areas tested for directionality in earlier studies were not adequately limited to the bundle (or bundles) of receptors responsible for best acuity. In addition, retinal resolution capability has not been evaluated interferometrically at the same time at the same retinal locus. Our understanding of that which is measured in evaluations of the Stiles-Crawford effect (of the first kind) is undergoing change (Enoch and Glismann, 1966; Enoch, 1967A, 1967B). It is not immediately clear how a disturbance in orientation of the central foveal bundle manifests itself in the Stiles-Crawford function (Enoch, 1967A, 1967B) (which is generally tested over a larger retinal area), nor how much disturbance in that function is indicative of significant alteration in receptor orientation. It is necessary to test as small a retinal area as possible for directionality and limiting resolution capability (interferometrically) using techniques which overcome ocular aberrations and translations of the retinal image.

In conclusion, when we speak of disturbed receptor orientation acuity effects (assuming compensation for the Stiles-Crawford effect) we are either dealing with greater obliquities than treated in this paper, or other factors (some listed above) are responsible for the decrements observed.

References

Campbell, F. W. (1958). A retinal acuity direction effect. *J. Physiol.* (London) *143*, 25P.

Campbell, F. W., and A. H. Gregory (1960). The spatial resolving power of the human retina with oblique incidence. *J. Opt. Soc. Amer. 50*, 831.

Campbell, F. W., and D. G. Green (1965). Optical and retinal factors affecting visual resolution. *J. Physiol.* (London) *181*, 576.

Enoch, J. M. (1957). Amblyopia and the Stiles-Crawford effect. *Amer. J. Optom. 34*, 298.

Enoch, J. M. (1958). Summated response of the retina to light entering different parts of the pupil. *J. Opt. Soc. Am. 48,* 392.

Enoch, J. M. (1959A). Receptor amblyopia. *Amer. J. Ophthal. 48* (3), 262.

Enoch, J. M. (1959B). Further studies on the relationship between amblyopia and the Stiles-Crawford effect. *Amer. J. Optom. 36,* 111.

Enoch, J. M. (1960). Optical interaction effects in models of parts of the visual receptors. *A.M.A. Arch. Ophthal. 63,* 548.

Enoch, J. M. (1963). Optical properties of retinal receptors. *J. Opt. Soc. Amer. 53,* 71.

Enoch, J. M. (1967A). The retina as a fiber optics bundle, in N. Kapany. *Fiber Optics, Principles and Applications.* Academic Press, New York (Appendix B).

Enoch, J. M. (1967B). The current status of receptor amblyopia. *Docum. Ophthal. 23,* 130.

Enoch, J. M., and L. Glismann (1966). Physical and optical changes in excised retinal tissue; resolution of retinal receptors as a fiber optics bundle. *Invest. Ophthal. 5,* 208.

Fankhauser, F., and J. M. Enoch (1962). The effects of blue upon perimetric thresholds. *A.M.A. Arch. Ophthal. 86,* 230.

Fankhauser, F., J. M. Enoch, and P. Cibis (1961). Receptor orientation in retinal pathology. *Amer. J. Ophthal. 53* (5), 767.

Green, D. G. (1967). Visual resolution when light enters the eye through different parts of the pupil. *J. Physiol.* (London) *190,* 583.

Kapany, N. (1957). Fiber optics II: image transfer on static and dynamic scanning with fiber bundles. *J. Opt. Soc. Am. 47,* 545.

Kapany, N. (1959). Fiber optics V: light leakage due to frustrated total reflection. *J. Opt. Soc. Am. 49,* 770.

Safir, A., and L. Hyams (1969). Distribution of cone orientations as an explanation of the Stiles-Crawford effect. *J. Opt. Soc. Am. 59,* 757-65.

Stiles, W. S. (1939). The directional sensitivity of the retina and the spectral sensitivities of the rods and cones. *Proc. Roy. Soc.* (London) *127,* 64 (Series B).

Stiles, W. S., and B. H. Crawford (1933). The luminous efficiency of rays entering the eye pupil at different points. *Proc. Roy. Soc.* (London) *112,* 428 (Series B).

Donald G. Pitts

Responses of the On-Off LGN Cell as a Function of Intense Light Flashes[*]

The problem of the effects of intense light on the visual system has been studied by many researchers (Metcalf and Horn, 1958; Jacobson, Cooper, and Najac, 1962; Severin et al., 1963; Fry and Miller, 1964; Miller, 1965; Hamilton, 1965). These efforts have been directed toward determining the threshold energy necessary to produce a retinal burn and the visual recovery from exposure to intense light stimuli. This paper shall be directed toward the analysis of electrophysiological data from ON-OFF single cells in the cat's lateral geniculate nucleus (LGN).

Both the applied and the theoretical approaches have been used in the study of visual recovery from intense light flashes. The theoretical investigator was interested, primarily, in the visual pigments and their relationships to the reciprocity law. The applied laboratory attempted to establish visual recovery times and provide protection for the visual system. Both approaches have provided additional, valuable information in attempting to gain a solution to the problem.

* The research reported in this paper was conducted by personnel of the Ophthalmology Branch, USAF School of Aerospace Medicine, Aerospace Medical Division (AFSC), Brooks Air Force Base, Texas. Further reproduction is authorized to satisfy the needs of the U.S. Government.

This research was performed under Task 630103, Work Unit 6301 03 025 and Program Element 6.16.46.OID, Project 5710, Subtask 03.003, and was partially funded by the Defense Atomic Support Agency.

The author wishes to express gratitude to Sgt. George D. Jones and Sgt. James E. Strong for assistance in conducting these experiments.

Certificate. The animals involved in this study were maintained in accordance with the "Guide for Laboratory Animal Facilities and Care," as published by the National Academy of Science, National Research Council.

The electrophysiologist has been rather slow in utilizing his techniques in studying the effects of high intensity light flashes on the visual system. His reticence is understandable since data need to be established at normal light intensities before higher intensities are approached. The electrophysiologist expresses his data in terms of spike frequency, spike patterns and excitation-inhibition systems which provide information to the organism. Thus it may appear that he is studying something other than flashblindness while he is attempting to unlock the visual system's information code.

Einthoven and Jolly (1908) were probably the first to use intense stimuli in the study of the ERG in the frog's eye. With a 120×10^6 m-cd stimulus the a wave was very large and the b wave reduced as the eye became light-adapted. Cobb and Morton (1952) found in humans a 4-7 msec. ERG latency and a notching of the first three phases of the wave form for 56×10^4 ft.-L. flashes. More recently, animal ERG's were used to study dark adaptation in animals after short flashes of high intensity light (U.S. Army, 1964). The dark adaptation recovery curves were divided into three phases rather than the usual two-phase curve. The initial phase was attributed to neural adaptation. The second and third phases were attributed resynthesis of the photopigments. For the animals studied there was a general rise in the threshold of the ERG with the same flux density as the pulse length increased.

Jacobs (1965) adapted the squirrel monkey to different levels of luminance (21.1 cd/m maximum). Several interesting features were found from LGN broad-band excitatory and inhibitory cells. The nature of the response depended on the luminance of adaptation stimulus, i.e., a cell may fire when adapted to one level but show inhibition on adaptation to another level. The spontaneous discharge rate was related to the adaptation luminance. Excitatory cells increased their spontaneous discharge rate as the adaptation luminance increased, but the reverse was found for inhibitory cells. The range of brightness discrimination was over only a ± 1 log unit range.

Visual cortex responses to a 1000 joule flash bulb 15 cm from the cat's eye have been described by Robertson and Evans (1965A; 1965B). The firing rate of cortical single units increased to a high peak after 10 minutes and declined to the control level after 30 minutes. Repeated flashes produced steplike increases in the firing rate which persist for up to 2 hours. Further, this phenomenon was not due to maintained retinal activity but was cortical. Discussion was given to show that the increased frequency of firing of the cell was within bursts and not to the increased length or frequency of the bursts.

Apparatus and Procedure

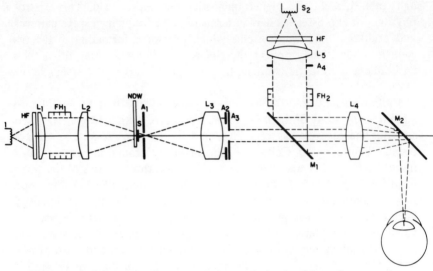

Figure 1. Schematic diagram of the optical stimulator. See text for explanation of the optical components.

APPARATUS

An optical stimulator (Fig. 1) designed for retinal ganglion cell research was used to deliver square wave light stimuli to the eye. The stimulator consisted of two integral optical systems. Source S_2 could serve as a background illuminator but was not used during these experiments. The stimulus light came from source S_1, was collimated by lens L_1, passed through filter holder FH_1, and was brought to focus at aperture A_1 by lens L_2. The stimulus source intensity could be varied by neutral density (ND) filters placed in FH_1 or by the neutral density wedge (NDW). From A_1 the beam was collimated by lens L_3 and passed through apertures A_2 and A_3. Lens L_4 focused the light beam in the entrance pupil of the eye. Apertures A_2 and A_3 deserve special description. Each consisted of 16 separate apertures in two wheel-like discs. In disc A_2 the largest aperture was 14 mm in diameter. The other apertures contain circular opaque centers of 11, 10, 9, 8, 7, 6, 5, 4, 3, 2, 1, 0.75, 0.50, 0.34, and 0.25 mm in diameter. The apertures in disc A_3 are 14, 12, 11, 10, 9, 8, 7, 6, 5, 4, 3, 2, 1, 0.75, 0.50, and 0.34 mm in diameter. The discs were constructed so that their apertures would lock in position and automatically self-center in the optical beam. Thus, multiple spot and annular stimuli could be readily obtained. In these experiments both A_2 and A_3 were set at maximum (14 mm).

Shutter S was constructed of a thin aluminum strip epoxied to a speaker coil. It provided a silent light chopper at the 1×4 mm aperture A_1. The shutter was controlled by Tektronix 160 series wave form and pulse generators. It furnished a square wave-light pulse with a 2.5 msec rise time and a 3.0 msec fall time. A light-dark ratio of 50% and a 2-second pulse length were used throughout the experiment, i.e., one second of light followed by one second of dark. A slight modification to the optical stimulator was made for these experiments. Mirror M_2 was removed and a lens focused the beam into the animal's eye in Maxwellian view.

Source S_1 was a 6.0 volt, 2.5 ampere, tungsten bulb powered by a 6 volt DC battery. The Photometric intensity of the optical stimulator source was calibrated by the method described by Westheimer (1966). The maximum luminance of the source at the plane of the pupil was 2.8 mL. This provided a retinal illuminance of 1.12 lumens/steradian. The irradiance of the source incident on the cornea integrated over 400 to 700 nm was calculated to be approximately 2.76×10^{-6} cal/cm^2-sec.

A Strobonar 65-C xenon flash lamp was used as the flashblindness source. Its calibration procedure was as follows: an SD-100 photodiode was calibrated against an NBS standardized Eppley thermopile and the following relationship obtained:

$$H \left(\frac{\mu W}{cm^2} \right) = 0.385 \text{ Vpd} \left(\frac{\mu W}{mv\text{-}cm^2} \right)$$

where H was the irradiance and Vpd the voltage output of the photodiode, and its output was attenuated by two stainless steel and two copper wire neutral mesh screens to prevent saturation. The transmission of each of the stainless steel and the copper filters was 31.0% and 29.0%, respectively. Therefore, the total transmission through the four filters was 0.81%.

The output of the photodiode was displayed on a Tektronix 502 oscilloscope and photographed (Fig. 2). The average pulse width was 1.75 msec. The effective area of the flash source was 18.1 cm^2. The average voltage from the photodiode was 6.916×10^5 mv. Therefore, the radiant emittance W was calculated to be 86.44 cal/cm^2-sec at the source.

From the expression $H = \dfrac{U}{A\tau}$

H = irradiance
U = energy in calories
τ = pulse duration

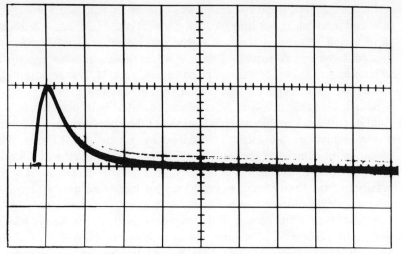

Figure 2.   Wave form of the flashlamp used in calibration. Values are
10 v/cm, 0.5 msec/cm, peak voltage 19.0 v, average voltage 5.69 v,
and pulse-width approximately 1.75 m/sec.

one may calculate the irradiance at any distance since μ is independent
of distance. For the irradiance at the cornea

$$W = \frac{U_s}{A_s{}^\tau s} \text{ and } H_c = \frac{U_c}{A_c{}^\tau c}$$

the following relationship holds:
$$U_s = WA_s\tau_s \text{ and } U_c = H_cA_c\tau_c$$

but $U_s = U_c$ and $\tau_s = \tau_c$

therefore, $H_c = \dfrac{WA_s}{A_c}$

H_c = irradiance at cornea
W = energy at source
A_s = area of source
A_c = irradiated area at the cornea.

The irradiance on the cornea was 3.2 cal/cm²-sec. For those who desire to calculate the energy on the retina, the average dilated pupillary diameter for the cat is 14.6 mm. Vakkur and Bishop (1963) and Vakkur et al. (1963) give additional optical constants for the cat's eye.

PROCEDURE

The cat, *Felis domestica,* was used as the experimental animal. The animal was anesthesized with 25 mg/kg body weight Nembutal, placed in the stereotaxic, and surgery performed. The animal was then Flaxedilized and placed on the respirator with a volume/stroke of 15 ml/kg and a respiratory rate of 20/min. The animal was maintained on constant anesthesia and paralysis by continuous intravenous infusion of 4.2 mg/hr Nembutal and 20 mg/hr Flaxedil in Ringer's solution delivered at the rate of 4.5 ml/hr. The eyes were fully dilated with 1% atropine and 10% phenylephrine HCl eye drops.

Plastic contact lens ERG electrodes were fitted to both eyes. The contact lens electrodes were gold to minimize polarization. The ERG electrodes were filled with a 0.9% NaCl, 1% methylcellulose solution. The animal was placed in a Faraday box and the stimulus light aligned in Maxwellian view in the animal's eye.

Tungsten microelectrodes (Hubel, 1957) with tip diameters of about 0.5μ were used throughout the experiments. The electrode was lowered through the brain to the lateral geniculate nucleus (LGN) until a single cell response was obtained. The response was identified as LGN extracellular type b or c according to the criteria of Bishop et al. (1962). Two different methods were used in locating the single cell spike potentials. One method was to localize the cell as it spontaneously fired. The second involved locating the cell while the visual stimulus was given. Either method satisfactorily isolated a cell for further study. After the preparation had stabilized, the cell was classified an ON, OFF, or ON-OFF according to its response to the maximum intensity of the visual stimulus. It was not unusual to maintain a cell for 6 to 8 hours.

The single cell spike responses were fed through a cathode follower to a low level DC preamp and simultaneously displayed with the light stimulus on a Tektronix 502 oscilloscope. A loud speaker system provided auditory monitoring of the cell. A tape recorder was used to record all data for future analysis. The responses were photographed from the face of the CRO by a Grass Kymograph camera for analysis.

The experimental protocol used for a given cell may be described as follows: after a single unit in the LGN was isolated, its spontaneous light response was recorded; that is, its response to the steady state stimulus light through an ND0.0 filter. The animal was dark adapted for 20 minutes and spontaneous dark responses were obtained. The firing response of the cell was then determined to progressive attenuation of the light stimulus in 0.5 log unit steps until the spontaneous dark firing rate was reached. The cell was flashblinded with the strobe light, and the cell's response was recorded at each different visual stimulus attentuation step. This required a flash for each visual stimulus intensity. The routine for each ND filter was stopped when the auditory firing rate returned to the preflash level. Approximately 20 minutes of dark adaptation was allowed before the next flash was delivered to the eye and another visual stimulus intensity response obtained.

Two shortcomings of this protocol became evident for LGN spike responses after a few experimental sessions. The auditory system was a very poor absolute counter if one depended on the memory of the previous firing rate. Thus, the routine was stopped almost invariably before the firing rate of a cell returned to the preflash level. Second, photographic recording of the spikes required a 100 millimeter per second film rate. This amount of film took up to 4 months for analysis.

Results

Photographic spike records and graphic data for ON-OFF cells are presented. To date, about 15 ON-OFF cells have been isolated but mean data will not be shown because of the wide variability in the activity of cells of the same type. For example, an ON-OFF cell may show a predominant ON response or a predominant OFF response, or respond approximately equal to the ON-OFF of the light stimulus. In addition, the firing rate may vary from 1-2 per stimulus up to 60-65 per stimulus. Thus, the question arises as to how the analysis of such a system should be approached. Can a cell's response be expected to show the same changes to an identical set of stimuli just because it is classified as an ON-OFF cell?

The ON-OFF cell's photographic records for spontaneous dark, spontaneous light, and different visual stimulus intensities, from ND0.0 to ND5.0, in 0.5 log unit steps, are shown in Figure 3. Sixty-second recovery times for the cell to visual stimulus intensities of ND0.0, 0.5, 1.0, 1.5, and 2.0 at 10-second intervals are given in Figures 4 through 8. In each of these figures the preflash cell response to the particular stimulus of interest is shown. These records show typical responses for the predominantly off ON-OFF cell. The difference in spike amplitudes between different figures is the result of a change in oscilloscope sensitivity when photographing and not due to a change in the cell's spike amplitude.

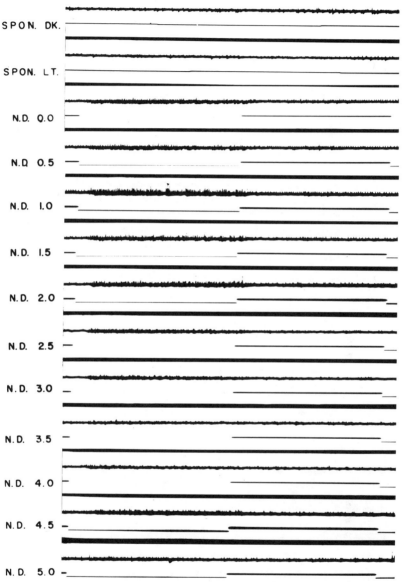

SPON. DK.

SPON. LT.

N.D. 0.0

N.D. 0.5

N.D. 1.0

N.D. 1.5

N.D. 2.0

N.D. 2.5

N.D. 3.0

N.D. 3.5

N.D. 4.0

N.D. 4.5

N.D. 5.0

Figure 3. The response of an ON-OFF cell to different stimulus conditions. Upper trace shows the spike response and the lower trace gives the light stimulus; upward deflection, light on; downward deflection, light off. The light was on for 1 second and off for 1 second. Rate of photography was 1000 mm/sec.

N.D. 0.0

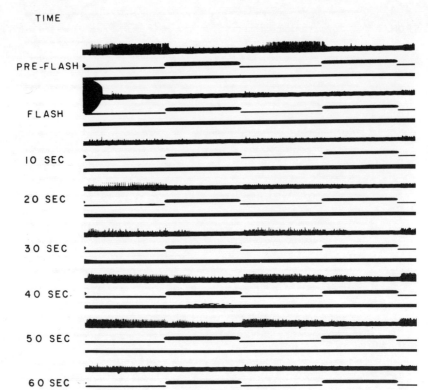

Figure 4. The response of the ON-OFF cell to visual stimulus NDo.o, pre- and postflash. The response for each time period on the ordinate is the first response in each line. The point of the flash is shown in line two as an outline of the oscilloscope face. Stimulus light was on for 1 second, upward deflection of bottom trace, and off for 1 second, downward deflection of the bottom trace.

For graphic data, the number of spike responses were counted for each light and dark portion of the square wave visual stimulus. The ON and OFF spike counts were meaned over 10-second intervals of time after the flash was given. These mean spike counts are plotted in time against the number of spikes per stimulus. Even though mean spike counts were used, the shape of each curve was maintained. In each of these graphs the spike count per stimulus prior to flash is shown at zero time and not joined to

N. D. 0.5

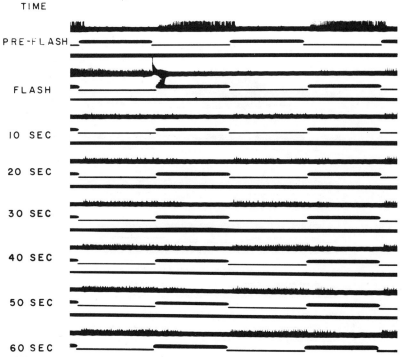

Figure 5.   The response of the ON-OFF cell to visual stimulus NDo.5, pre- and postflash. The point of the flash is shown by the disruption of the spikes in line two. Time and response relationships are the same as in Figures 3 and 4.

N.D. 1.0

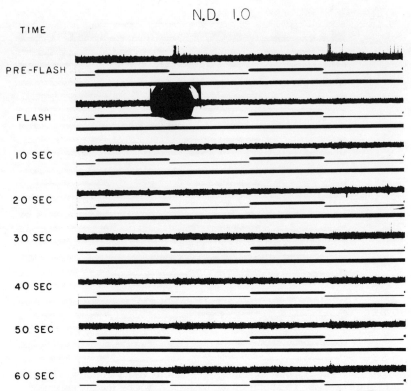

Figure 6. The response of the ON-OFF cell to visual stimulus ND1.0, pre- and postflash. Flash presentation is shown by the circular outline of the oscilloscope face. Time and response relationships are the same as for previous figures.

N.D. 1.5

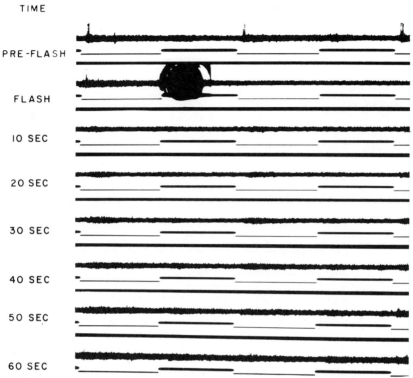

Figure 7.   The response of the ON-OFF cell to visual stimulus ND1.5, pre- and postflash. The point of flash is shown by the oscilloscope face in the second line. Time and response relationships are the same as for previous figures.

N.D. 2.0

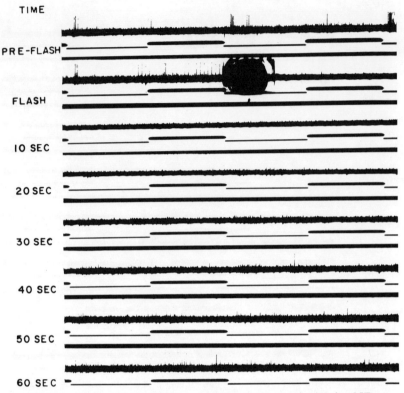

Figure 8.   The response of the ON-OFF cell to visual stimulus ND2.0, pre- and post-flash. The point of flash is shown by the oscilloscope face in line two. Time and response relationships are the same as for the previous figures.

the postflash spike count. This preflash spike count can be used to determine the adaptation rate of the cell.

ON-OFF spike responses for different visual stimulus intensities are plotted in Figure 9. The recovery of the ON-OFF cell to flash #1 is given in Figure 10 for relative stimulus intensity of log 0.0, lower curve, and

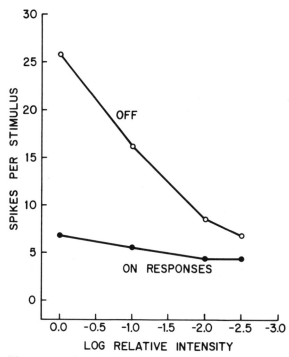

Figure 9. ON-OFF cell responses for different levels of visual stimulus intensity. The intensity at log 0.0 was 2.8 mL incident on the cornea (Cell B-28).

log −1.0, upper curve. The same recovery data for relative visual stimulus intensity log −1.5, lower curve, and log −2.0, upper curve, are illustrated in Figure 11. The ON-OFF spike responses for flash #2 are shown in Figure 12. In this graph the log relative visual stimulus intensities are log 0.0, log −1.0, log −1.5, and log −2.0, from bottom to top, respectively.

Subsequent to the analysis of the previously given data, a CAT 1000 biological computer system has been installed and data analyzed on-line. In an attempt to better understand the effects of intense flashes, frequency histograms have been computed. The frequency of firing prior to a flash

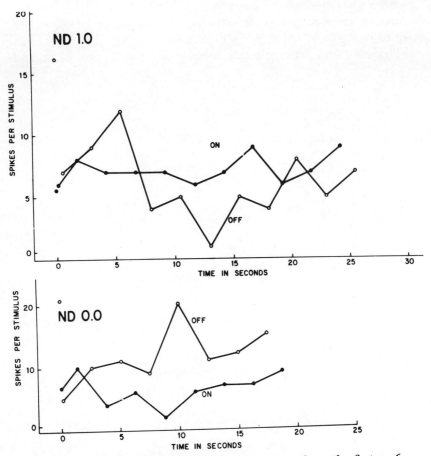

Figure 10. ON-OFF cell response during recovery from the first 3.46 cal/cm²-sec flash incident on the cornea; lower curve, log 0.0; upper curve, log −1.0 relative visual stimulus intensity (Cell B-28).

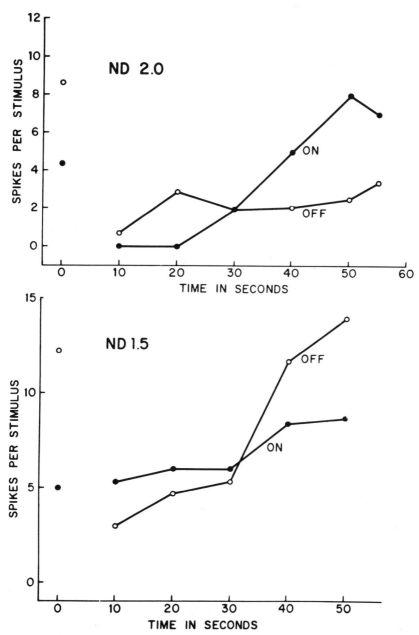

Figure 11.   ON-OFF cell responses during recovery from the first 3.46 cal/cm²-sec flash; lower curve, log 1.5; upper curve, log 2.0, relative visual stimulus intensity (Cell B-28).

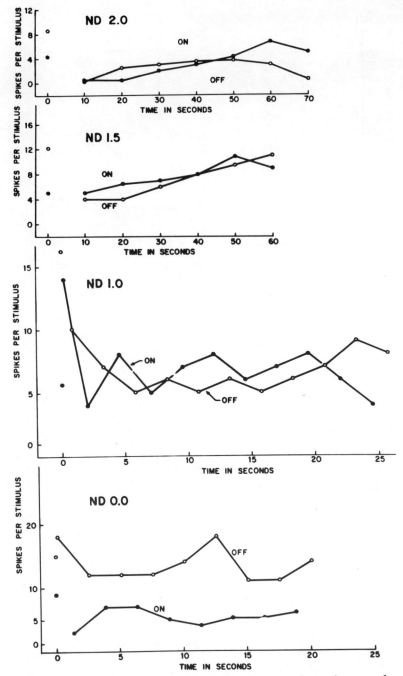

Figure 12. ON-OFF cell responses during recovery from the second 3.46 cal/cm²-sec flash. The lower curve is log 0.0; second curve is log −1.0; third curve is log −1.5 and upper curve is log −2.0 relative visual stimulus intensity (Cell B-28).

for an ND0.0 filter is shown in Figure 13, and the same information post-flash is given in Figure 14. These data are derived from one experiment and one stimulus intensity but reflect an overall impression gained from all experiments. Definitive data must await further analysis.

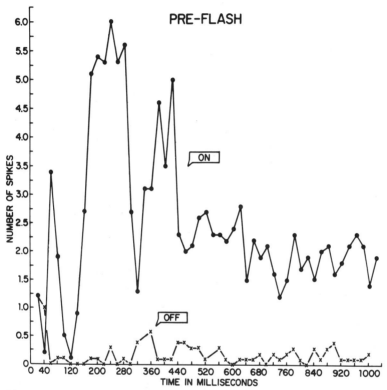

Figure 13. Preflash frequency histogram of the spikes during the 1 second on and 1 second off of the visual stimulus. The ordinate expresses the number of spikes and the abscissa time in milliseconds.

Discussion

Spontaneous activity was not stressed in the results but constituted an interesting part of the research. Many cells failed to show spontaneous activity. Lower spontaneous activity was found when using the visual stimulus to isolate a cell than when isolating a cell without the stimulus. The most common spontaneous activity consisted of groups of 3 to 5 short high frequency bursts, but occasionally groups of bursts fired at regular intervals. Another type consisted of short rapid bursts of spikes

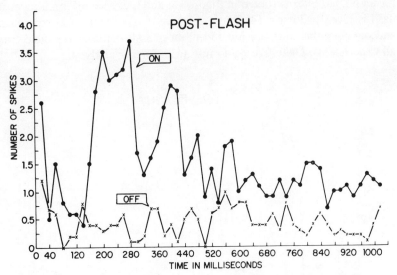

Figure 14. Postflash frequency histogram of the spikes during 1 second on and 1 second off of the visual stimulus. The ordinate gives the number of spikes and the abscissa shows the time in milliseconds.

followed by a long period of silence; then a sudden burst of activity would reappear. This type of activity was so startling that timing of the sequence was not done. The last type of spontaneous activity was single spikes occurring at regular intervals as high as 25 spikes per second. The spontaneous activity appeared to either disappear or synchronize with the light stimulus when it was on. One is referred to Bishop et al. (1962) and Levick and Williams (1964) for references and a discussion on the significance of the characteristics of the LGN spontaneous activity. Bishop et al. (1964) found only a 37% spontaneous activity of 137 cells studied.

The most startling finding was that the classical ON, OFF, and ON-OFF cell classification did not hold at different levels of adaptation. The ON cells gave OFF responses; the OFF cells changed to give ON responses; and ON-OFF cells changed the rate of their ON and OFF responses. The number of the spike responses changed with the intensity of the visual stimulus. This was first thought to be a new finding, but Donner and Willmer (1950) reported similar changes to higher intensities of the stimulus for the ganglion cell. These findings seriously question the validity of classifying cells without specifying the adaptation level and stimulus intensity.

In attempting to analyze neurophysiological data, certain questions arise as to how the data should be handled. For example, the off response in Figure 3 is so dominant that it appears to run into the on portion of the light stimulus; however, close inspection shows a slight delay in the cell's off activity just prior to or just after the on light stimulus began. But where does the true "off" response end and the "on" response begin? If one uses total spike count in analysis, this question becomes critical in data interpretation. It should be added that only a few of the ON-OFF cells studied exhibit this type of behavior while most of them show a distinct cessation of activity before either the on or off light stimulus.

The "normal" responses shown in Figure 3 show a decreased OFF activity and an increased ON activity as the intensity of the visual stimulus is decreased until, at ND5.0, the spontaneous dark rate is reached. None of Figures 4 through 8 shows a return to normal cellular activity. If one studies closely, one can determine points of increase and decrease in both the OFF and ON responses reminiscent of the waxing and waning of an after-image. Certain cells have been followed for as long as 320 seconds without returning to normal and one pure OFF cell remained silent for 17 minutes after the flash.

The graphed data (Figs. 10, 11, 12) show the results of two flashes. The only apparent change brought about by the second flash is that at the highest stimulus intensity (ND0.0, Fig.12) the number of spikes per stimulus corresponds quite well to the original firing level. Other stimulus levels for flash #2 show markedly reduced firing rates and undulating crisscross of the ON and OFF response curves. The data for ND0.0 with the second flash may indicate a reorganization of the neural system with added stress so that the organism may obtain maximum information from the visual stimulus.

The central question is just how the LGN cell receives and conveys meaningful information to the organism. We have attempted three analyses pointed toward answering this question. We noted in the data that for normal stimulus intensity versus number of ON or OFF spikes, the ON-OFF cell firing ratio (ON/OFF) changed but remained fairly consistent. Conversely, after a flash, the ON/OFF ratio is completely erratic and maintains no pattern. This approach has not resulted in a fruitful solution to the problem.

A second method used in analysis is the frequency of firing histogram derived from the CAT 1000 computer. If the pattern of firing changed after flash, the frequency histogram should show a shift in the distribution of firing with time. It is evident when comparing Figures 13 and 14 that there is little or no shift in frequency. About all these graphs show is that the preflash firing rate was above the postflash rate for all frequencies. This is the first attempt in such analysis and further research is required.

We have also used the cell's firing rate per stimulus—the most popular method used—for analysis. It appears that the firing rate varies greatly with both the level of adaptation and the stimulus intensity. At the same time, the LGN system demonstrates a certain adaptation to its environment which appears to provide some meaningful information despite the stress. At the present time, we are inclined to interpret these firing rate changes as a reflection of the change in receptive field responses to different adaptation levels.

Barlow et al. (1957A; 1957B) described changes in receptive fields that occur during dark adaptation. The surrounding zone of the dark adapted receptive field disappears and only the enlarged central ON or OFF responses occur. In light adaptation, the surround always gave an opposite response to that of the center of the receptive field.

Barlow also found that some parts of a receptive field reversed their contribution to the ganglion cell. An explanation of our reversed (ON cell firing to off stimulus, OFF cell firing to on stimulus and reversal of ON-OFF firing) is that when flashed with an intense light, the ganglion receptive field gave a predominant reversal of response to the stimulus. The significance of or the intensity level required to produce this reversal is not known. Ganglion cell experiments are planned to delineate these criteria.

In summary, the responses of ON-OFF single LGN cells to intense light flashes have been given. A full explanation of the significance of these changes is not known, but three methods used in analysis are outlined. Comparison of "normal" and high intensity data shows considerable disruption from "normal." It was suggested that receptive field adaptation effects might explain the reported phenomenon.

References

Barlow, H. B., R. Fitzhugh, and S. W. Kuffler (1957A). Dark adaptation, absolute threshold and Purkinje shift in single units of the cat's retina. *J. Physiol.* (London) *137*, 327-337.

Barlow, H. B., R. Fitzhugh, and S. W. Kuffler (1957B). Change in organization in the receptive fields of the cat's retina during dark adaptation. *J. Physiol.* (London) *137*, 338-354.

Bishop, P. O., W. Burke, and R. Davis (1962). The identification of single units in central visual pathways. *J. Physiol.* (London) *162*, 409-431.

Bishop, P. O., W. R. Levick, and W. O. Williams (1964). Statistical analysis of the dark discharge of lateral geniculate neurones. *J. Physiol.* (London) *170*, 582-597.

Cobb, W., and H. B. Morton (1952). The human retinogram in response to high intensity flashes. *EEG. Clin. Neurophysiol. 4*, 547-556.

Donner, K. O., and E. N. Willmer (1950). An analysis of the response single visual-purple-dependent elements in the retina of the cat. *J. Physiol.* (London) *111*, 160-173.

Einthoven, W., and W. A. Jolly (1908). The form and magnitude of the electrical response of the eye to stimulation by light at various intensities. *Quart. J. Exper. Physiol. 1*, 373-416.

Evans, C. R., and A. D. J. Robertson (1965B). Prolonged excitation in the visual cortex of the cat. *Science 150*, 913-915.

Fray, G. A., and N. D. Miller (1964). Visual recovery from brief exposures to very high luminance levels. SAM-EDR-64-36, U.S.A.F. School of Aerospace Medicine, Brooks Air Force Base, Texas.

Hamilton, J. E. (1965). F-101/F-106 flight simulator flashblindness experiment. SAM-TR-65-82, U.S.A.F. School of Aerospace Medicine, Brooks Air Force Base, Texas.

Hubel, D. H. (1957). Tungsten microelectrode for recording from single units. *Science 125*, 549-550.

Jacobs, G. H. (1965). Effects of adaptation on the lateral geniculate response to light increment and decrement. *J. Optical Soc. Amer. 55*, 1535-1540.

Jacobson, J. H., B. Cooper, and H. W. Najac (1962). Effects of thermal energy on retinal function. AMRL-TDR-62-96. Wright-Patterson Air Force Base, Ohio.

Levick, W. R., and W. O. Williams (1964). Maintained activity of lateral geniculate neurones in darkness. *J. Physiol.* (London) *170*, 582-597.

Metcalf, R. D., and R. E. Horn (1958). Visual recovery times from high intensity flashes of light. WADC Technical Report 58-232. Wright Air Development Center, Wright-Patterson Air Force Base, Ohio.

Miller, N. D. (1965). Visual recovery, Part I and Part II. SAM-TR-65-12. U.S.A.F. School of Aerospace Medicine, Brooks Air Force Base, Texas.

Robertson, A. D. J., and C. R. Evans (1965A). Single-unit activity in the cat's visual cortex: Modification after an intense light flash. *Science 147*, 303-304.

Severin, S. L., et al. (1963). Photostress and flashblindness in aerospace operations. SAM-TDR-63-67. U.S.A.F. School of Aerospace Medicine, Brooks Air Force Base, Texas.

U.S. Army (1964). Use of electroretinography in the study of flash blindness in animals. Contract Report, Contract No. DA-19-129-AMC-97 (N). Pioneering Research Division, U.S. Army Natick Laboratories, Natick, Massachusetts.

Vakkur, G. J., and P. O. Bishop (1963). The schematic eye in the cat. *Vision Research 3*, 357-381.

Vakkur, G. J., P. O. Bishop, and W. Kozak (1963). Visual optics in the cat, including posterior nodal distance and retinal landmarks. *Vision Research 3*, 289-314.

Westheimer, G. (1966). The Maxwellian view. *Vision Research 6*, 669-682.

Antoinette Pirie

The Biochemistry of the Tapetum
Lucidum and the Crystalline Lens

The tapetum lucidum of the eye is a specialized layer lying behind, but adjacent to, the light-sensitive cells of the retina (Fig. 1). It forms a

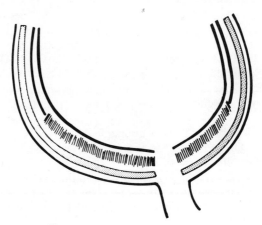

Figure I.   Tapetum diagram

reflecting surface at which any light not absorbed during its first passage through the retina is reflected back again and has a second opportunity to be absorbed. It is the structure responsible for eye shine in animals. We have all caught the eyes of a cat or dog in the headlights of our car.

The eyes glow because the light from our headlights is reflected from the mirror surface of the tapetum lucidum.

There is general agreement that the function of the tapetum lucidum is to increase the sensitivity of the retina by acting as a backing mirror to the light-sensitive cells. Such mirror structures occur, as would be expected, mainly in nocturnal animals and in fishes.

The anatomy of the tapetum in different animals has been thoroughly examined, but Walls (1942), in his book *The Vertebrate Eye,* was the first to tabulate what was known of their chemistry (Table 1). Although the anatomy of the vertebrate eye is basically the same in all the phyla and although the visual pigments are, throughout the whole animal kingdom, based on a single substance, vitamin A aldehyde, yet the chemistry of the tapetum lucidum is quite diverse. Separate chemical methods have been used to evolve a structure which seems to have the same purpose wherever it occurs.

Table 1. Animals in which a tapetum lucidum has been found

Classification	Type of tapetum	Chemistry of tapetum	Example
Arthropods	crystalline	uric acid	Astacus trowbridgii[1]
	crystalline	guanine	Limulus[2]
Fishes			
Elasmobranchs	crystalline, choroidal	guanine	Squalus acanthias
Chondrosteans	crystalline	guanine	Acipenser
Holosteans, Dipnoans and Cladisteans	?	?	Polypterus
Crossopterygians	?	?	Latimeria chalumnae[3]
Teleosts	crystalline, retinal	guanine	Abramis brama
Amphibians			
Toads, Hylas	?		
Reptiles			
Crocodiles	crystalline, retinal	guanine	Alligator mississipiensis
Marsupials	crystalline, retinal	cholesterol	Didelphis virginiana[4]
	fibrous, choroidal	?	Dasyrus thylacinus

Placentals

Bats,			
Megachiroptera	birefringent	?phospholipid	genus Eidelon[5]
Rodents	fibrous, choroidal	?	Cuniculus paca
	?	?	Hystricidae
Carnivores	crystalline, choroidal	Zinc cysteine	dog, fox[6]
Pinnepedes	crystalline, choroidal	Zinc cysteine	Seals[7]
Whales	fibrous, choroidal	?	
Ungulates	fibrous, choroidal	collagen	Cow, horse, sheep
Lemurs	crystalline, choroidal	riboflavin	Galago crassicaudatus[8]
Simians	fibrous, choroidal	?	Aotus

Note: This classification is taken from Walls, *The Vertebrate Eye,* 1942, p. 240, unless another reference is given.

1. Kleinholz (1955)
2. Kleinholz and Henwood (1953)
3. Millot, J., and Carasso, N. (1955)
4. Pirie, A. (1961)
5. Pirie, A. (unpublished)
6. Weitzel et al. (1955)
7. Weitzel et al. (1956)
8. Pirie, A. (1959)

A tapetum lucidum is present in crayfish and lobsters (Kleinholz and Henwood, 1953; Kleinholz, 1955), the Coelocanth, Latimeria Chalumnae, that ancient Crossopterygian type of fish found swimming around about thirty years ago off South Africa (Millot and Carrasso, 1955), as well as in modern elasmobranchs and teleosts; in crocodiles and alligators, seals and whales, in ruminants and carnivores, in marsupials, fruit-bats and lemurs, but not in the pig, not in the higher apes, and not in man (Walls, 1942) (Table 1).

So far as we know there are two large classes of tapeta, those made of fibers and those made of crystals or birefringent fatty material. Herbivores such as the cow, the sheep, the horse, have tapeta that are made of fibers. There is no tapetal pigment, the colors that are reflected from the eyes of a cow are interference color caused by diffraction of light by the fiber grating (Fig. 2). The tapetum in the eye of a horse or cow is a beautiful sight but if the colored part is scraped off the underlying choroid and is then teased out and examined, it yields no pigment, but only fine uniform fibers of collagen (Fig. 3). It is these fibers that reflect and diffract the light. The gizzard of a chicken, another collagenous structure, glistens in

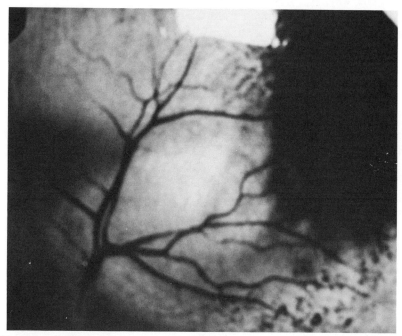

Figure 2. Cow tapetum

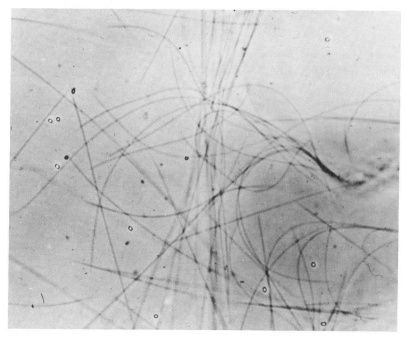

Figure 3. Fibres from tapetum of cow

the same way. All ungulates having tapeta appear to have this type of tapetum fibrosum elaborated from choroidal tissue at the back of the retina.

The next sort of tapetum I wish to illustrate is that of the dogfish, *Squalus acanthias,* Figure 4. This is a crystalline and occlusible tapetum;

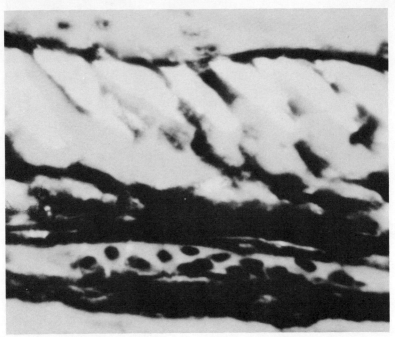

Figure 4.   Dogfish crystals

in bright light the black pigment in the adjacent cells slides over the crystals and so covers the mirror. In dim light the pigment retreats and the crystals can reflect any light that passes through the retina on to them. If the crystals were arranged at random they might reflect the light at such an angle that it fell on a new set of receptor cells. This, although giving increased sensitivity, could lead to a blurring of the image. But Denton and Nicol (1964), in a detailed and fascinating study of the tapetum in cartilaginous fishes, have shown that the crystal plates are arranged at different angles in different parts of the eye and in such a way that the reflected light is not diffused but is either reflected back or directly through the pupil or onto the black back surface of the iris. Figure 5 is a diagram from Denton and Nicol (1964) showing the reflection of the rays of light from the tapetum of *Squalus acanthias* and the different arrangement

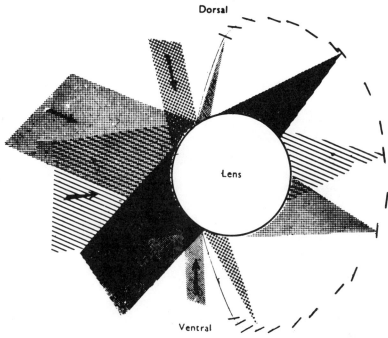

Figure 5.   Orientation of crystals in dogfish (Denton)

of crystal plates at different points on the retina. The plates are always approximately perpendicular to the light that reaches them. Maximal reflectivity occurs between 480mμ and 570mμ, the spectral region in which rhodopsin absorbs maximally. Denton and Nicol (1964) also show that the occlusibility or non-occlusibility of the tapetum by pigment sliding is related to life habits. Spur dogs (*Squalus acanthias*) and blue sharks (*Carcharhinus glaucus*) are active swimmers which hunt herring and mackerel at the surface both by night and day. Their eyes must therefore be able to function over a great range of illumination and they have occlusible tapeta. The rough dogs (*Scyliorhinus canalicula*) and nursehounds (*Scyliorhinus stellaris*), which are bottom feeders of inshore waters, are nocturnal. They have a permanently bright tapetum in the upper half of the eye only and an active pupil to cut out excess light. The deep set squaloids and chimeroids have wide static pupils, a tapetum that extends over the whole eye-ground and that is not occlusible. They need all the light they can get.

The crystals of the tapetum in all the fishes so far studied are made of guanine or its calcium salt. This compound, 2-amino-6 hydroxy purine, is oxidized to uric acid in man and excreted, but in fishes it is laid down both in their scales and in their eyes.

Cartilaginous fishes, the dogfishes, sea-hounds, sharks, and rays, have the most elaborate guanine tapeta but many other fish have brilliant reflecting devices, also made of guanine but of smaller crystals. For example, the bream, *Abramis brama*, has a guanine tapetum in which the crystals fill the cells of the retinal pigment epithelium. Alligators also have a

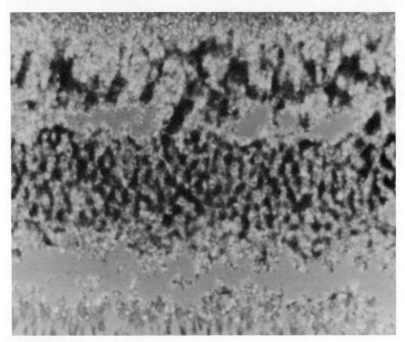

Figure 6.   Bream tapetum

guanine tapetum and Abelsdorf (1897) first saw visual purple fade in light on the white background of the alligator tapetum.

Each large well-arranged crystal of guanine in the tapetum of the shark acts as a separate micro mirror. The much smaller, randomly arranged crystals in the tapetum of the alligator and of the bream may act rather differently. Yoshizawa and Wald (1966) have suggested that they may act by causing light to be reflected back and forth through the outer segments of the rods. The processes of the pigment epithelium in which these crystals lie can surround the outer segments of the rods, and this arrangement, by causing multiple reflections within the rod, could increase the absorption of light by the visual pigments and hence the visual sensitivity. Both alligator and bream live in murky waters so that an increase in sensitivity would be valuable. Welsh (1932) found that the guanine of the tapetum in crustaceans migrated forward to surround the visual cell receptors in dim light and retreated during the day when pigment cell processes replaced the reflecting layer. There is obviously a very close connection between the reflecting tapetum and the visual cells and these small crystals probably do cause back and forth reflections within the individual rod cells.

The tapetum of which we all have experience is the tapetum of cats, dogs, foxes, and other small carnivorous mammals. This tapetum is made up of five to ten layers of cells packed with reflecting material lying behind the retina in the upper part of the eye. Figure 7 is a fundus photograph of a white cat. Figure 8 is a slide of the rods that can be shaken out of the tapetal cells and Figure 9 shows an electron-microscopic picture of

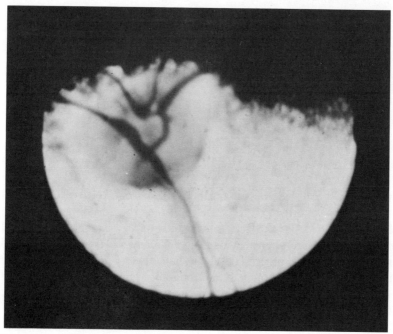

Figure 7.   Cat tapetum in vivo

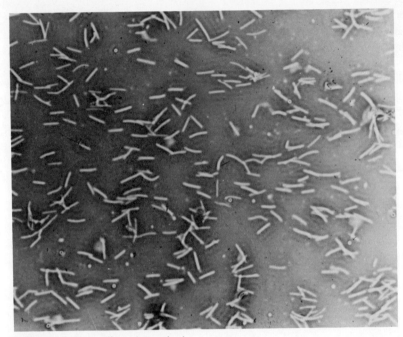

Figure 8.   Cat rodlets (negative)

the rodlets in cells of the cat tapetum taken from Bernstein and Pease (1959), who were the first to take electron-micrograms of tapeta.

Pedler (1963) has made a further study of the cat tapetum and finds that the rodlets are aligned within the cells and in large groups of cells are parallel to each other. The mean diameter is 0.23μ and the average length 4.8μ, but some may be considerably longer and some shorter. From the extracted rodlets the length is calculated as 4 to 8μ. Pedler (1963) considers that the structure of the array of rods is like that of a two-dimensional crystal. It thus should exhibit Bragg reflection from the lattice planes. Calculations of the distances between these planes suggests that the tapetum reflects light predominantly of wavelengths 1206 mμ, 603 mμ and 401 mμ. Weale (1953) found a peak of reflectivity at 600 mμ, 60% of light of 580 mμ to 600 mμ being reflected by the tapetum *in situ*. The absolute threshold is also changed. From behavioral studies Gunther (1951) found the threshold in the cat to be about $\frac{1}{6}$ that in

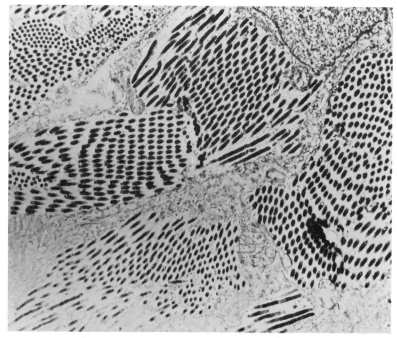

Figure 9.   Electronmicrogram of cat tapetum

man. The cat responded to illumination invisible to Dr. Gunther himself and both Gunther and Weale consider that the very great sensitivity of the cat eye is due to the reflectivity of the tapetum. Dodt and Walther (1958B) compared the size of response to illumination—the electroretinogram—from tapetal and non-tapetal areas of the light-adapted cat retina and found that the intensity required for equal response was 0.45 to 0.75 log units lower for the tapetum.

Unfortunately little is known of the chemistry of the tapetal rods in the cat. Weitzel et al. (1954) found 1 mg Zn/g wet weight in the tapetum and suggest that the material is peptide in nature. From a single acid hydrolysis followed by chromatography I think the rodlets contain protein.

Weitzel et al. (1955, 1956) have made a very thorough study of the chemistry of the tapetum in dogs, foxes, badgers, and seals and have isolated a most interesting compound of zinc and cysteine. This is laid down as crystals but so far has not been prepared in the laboratory in crystalline form and it is difficult to get microscopic pictures.

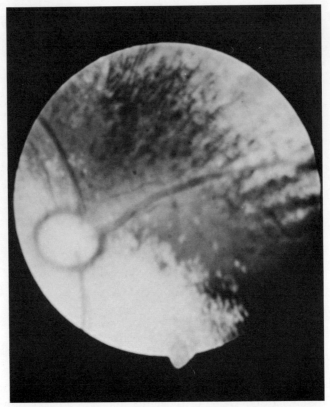

Figure 10. Dog tapetum in vivo

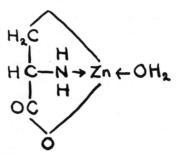

Figure 11. Zinc cysteine

Figure 12. Lemur

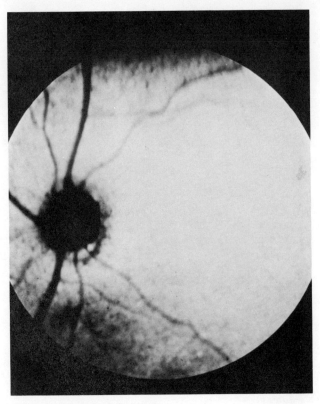

Figure 13.   Lemur tapetum in vivo

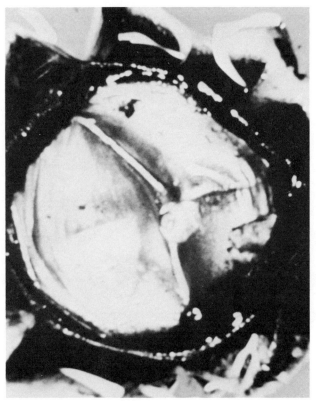

Figure 14. Opened eye of lemur with retina removed

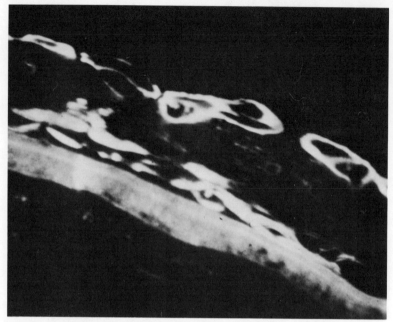

Figure 15.   Section of retina under ultraviolet light

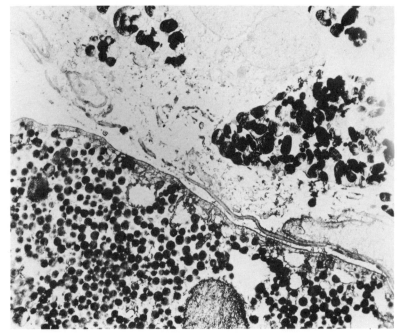

Figure 16. Opossum electronmicrogram of tapetum

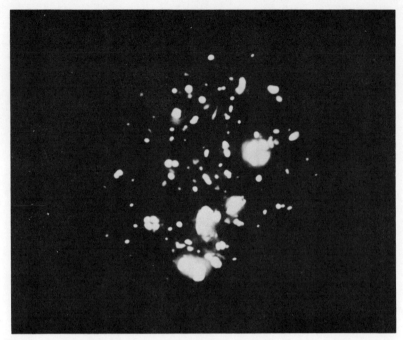

Figure 17.   Crystals from opossum tapetum

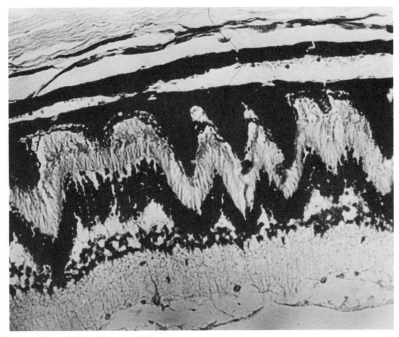

Figure 18. Bat retina section

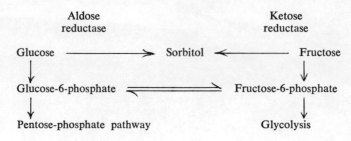

Figure 19. Sorbitol pathway

OXFORD *Sorbitol, glucose, and fructose and inositol in the non-cataractous lens.*

(post-mortem material)

Sex	Age (yr)	Presence or absence of diabetes	Fasting blood sugar (mg/100 ml)	Sorbitol	Glucose (μmol/g lens)	Fructose	Inositol
F	73	—		trace	trace	trace	32
M	55	—		trace	trace	trace	20
M	39	—		0	0	0	32
M	0.5	—		0	0	0	20
M	80	+	215	2.5	2.5	2.5	10
M	70	+	295	2.0	1.5	1.5	35
M	35	+	484	12.5	12.5	5.5	25

Figure 20. Accumulation of sorbitol in human lens

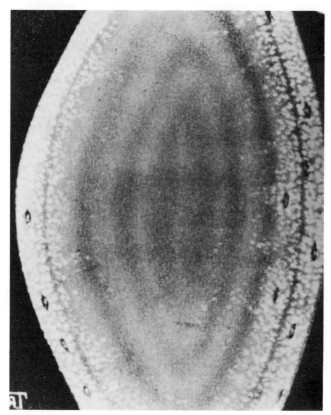

Figure 21. Diabetic lens

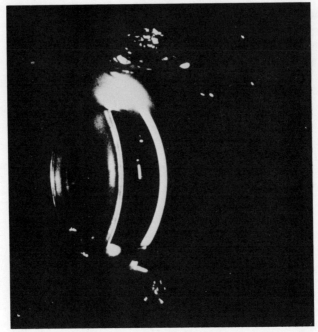

Figure 22. Normal human lens, slit-lamp Zeiss

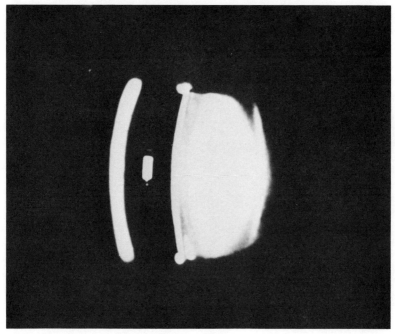

Figure 23.   Senile cataract, slit-lamp Zeiss

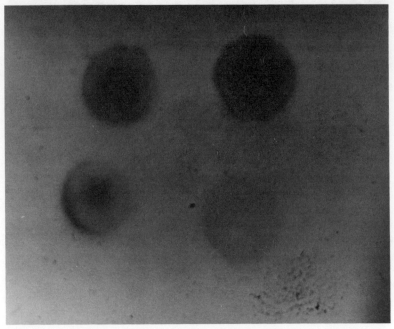

Figure 24. Groups 1-4 (Ruth van Heyningen)

References

Abelsdorf, G. (1897). *Arch. Anat. Physiol. Lpz.* 155.

Bernstein, M. H., and Pease, D. C. (1959). *J. biophys. biochem. Cytol.* 5, 35.

Caird, F. I., Pirie, A., and Ramsell, T. G. (1968). *Diabetes and the Eye*. Blackwell Scientific Publications: Oxford, England.

Crawford, B. H. (1949). *Proc. Phys. Soc. B.* 62, 321.

Dartnall, H. J. A., Arden, G. B., Ikeda, H., Luck, C. P., Rosenberg, M. E., Pedler, C. M. H., and Tansley, K. (1965). *Vis. Res.* 5, 399.

Denton, E. J., and Nicol, J. A. C. (1964). *J. mar. biol. Assn. U.K.* 44, 219.

Dodt, E., and Walther, J. B. (1958A). *Nature, Lond.* 181, 286.

Dodt, E., and Walther, J. B. (1958B). *Pfluger's Arch.* 266, 175.

Duke-Elder, W. S. (1925). *Brit. J. Ophth.* 9, 167.

Elliott, J., and Futterman, S. (1963). *A.M.A. Arch. Ophth.* 70, 531.

Gunther, R. (1951). *J. Physiol.* 114, 8.

Himsworth, H. P. (1932). *Brit. med. J. ii*, 1184.

Hosoya, Y. (1929). *Tohoku J. exp. Med.* 12, 119.

Johnson, G. L. (1901) *Phil. Trans. Roy. Soc. Lond.* 194, 35.

Kleinholz, L. H. (1955). *Biol. Bull. Woods Hole*, 109, 362.

Kleinholz, L. H., and Henwood, W. (1953). *Anat. Rec.* 116, 637.

Millot, J., and Carasso, N. (1955). *C. R. Acad. Sci. Paris.* 241, 576.

Pedler, C. (1963). *Exp. Eye Res.* 2, 189.

Pirie, A. (1959). *Nature, Lond.* 183, 985.

Pirie, A. (1961). *Nature, Lond.* 191, 708.

Pirie, A., and Simpson, D. M. (1946). *Biochem. J.* 40, 14.

van Heyningen, R. (1968). In *The Eye*, vol. 1, 2nd ed. Ed. H. Davson. Academic Press: London.

von Euler, H., and Adler, E. (1934). *Z Physiol. Chem.* 228, 1.

Walls, G. L. (1942). *The Vertebrate Eye*. Cranbrook Institute of Science, No. 19.

Weale, R. A. (1963). *The Aging Eye*. H. K. Lewis Ltd.: London.

Weale, R. (1953). *J. Physiol.* 119, 30.

Weitzel, G. (1956). *Angew. Chem.* 68, 566.

Weitzel, G., Buddecke, E., Fretzdorff, A. M., Strecker, F. J., and Roester, U. (1955). *Z. Physiol. Chem.* 299, 193.

Weitzel, G., Buddecke, E., Fretzdorff, A. M., Strecker, F. J., and Roester, U. (1956). *Z. Physiol. Chem.* 300, 1.

Weitzel, G., Strecker, F. J., Roester, U., Buddecke, E., and Fretzdorff, A. M. (1954). *Z. Physiol. Chem.* 296, 19.

Welsh, J. H. (1932). *J. exp. Zool.* 62, 173.

Yoshizawa, T., & Wald, G. (1966). *Nature, Lond.* 212, 483.

Part II

Neuropsychological and Behavioral Aspects
of Color Vision

Russell L. De Valois

Studies of the Physiology of Primate Vision

In our work on the primate visual system, we have attempted to find relationships between the responses of various types of single cells in the visual pathway and overall behavior of the animal in certain visual situations. Obviously, this requires an analysis on two different levels: information about the visual behavior of the animal, and a study of the responses of the individual cells.

I shall not describe in detail all our behavioral tests of color vision in various primate species; however, I would like to discuss briefly the results of one of our tests with monkeys. Figure 1 shows the results from macaque monkeys, squirrel monkeys, and normal and protonomalous humans in a hue discrimination experiment. The subject is presented with four windows, three of which are illuminated by a light of a single wavelength. The fourth is illuminated by light from a monochromator, which can be varied in wavelength. The subject must, in each trial, select the variable wavelength, which is brought closer and closer to the three standards. This is continued until the subject can no longer discriminate the deviant wavelength. The minimal discriminable separation between the standard wavelength and the variable wavelength gives a measure of hue discrimination at that particular point in the spectrum.

We have obtained these measures at fourteen different points in the spectrum (De Valois, Mead, and Morgan, 19   ). The curve for the normal human observers has the conventional shape which has been found by Wright (1947) and many others; that is, hue discrimination is optimal in the region of about 590 nm, poorer in the middle region of the spectrum, good again around 490 nm, and poor at each end of the spectrum. Compare this curve to the results from the macaque monkey. Although the

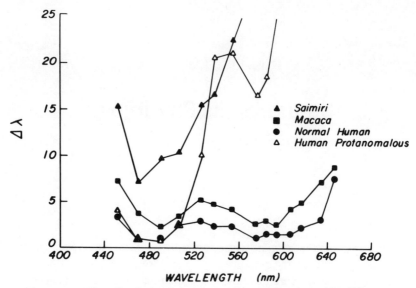

Figure 1. Hue discrimination. The minimum discriminable difference in wavelength ($\Delta\lambda$) at each of 14 wavelengths was determined. Note that macaques and normal human trichromats have two regions of good hue discrimination, whereas squirrel monkeys and protanomalous humans have good hue discrimination only in the short-wavelength region.

monkey's absolute wavelength discrimination is poorer than that of the human at every point in the spectrum, the shapes of the two curves are identical.

In addition to this hue discrimination experiment, we have tested macaques in a number of other visual situations: saturation discrimination tests, photopic and scotopic luminosity tests, anomaloscope measures, and several other tests of brightness and color discrimination (De Valois, Polson, and Morgan, 19 ; De Valois, Hull and Morgan, 19). Macaque monkey has shown functions which are identical in shape to those of normal human observers. On this basis, we concluded that macaques and normal human observers have essentially identical vision. One is therefore justified in drawing conclusions about the physiological basis for human vision on the basis of recordings made from macaque monkeys.

Figure 1, incidentally, also shows data from another species of primate which has quite different hue discrimination functions. The squirrel monkey (*Saimiri sciurus*) has only the one prominent minimum in the short wavelength end of the spectrum. Jacobs (1963) has done extensive behavioral studies of the squirrel monkey and showed it to be severely protonomalous. His electrophysiological recordings from the squirrel monkey, compared with ours from the macaque, have provided considerable insight into the physiological basis for color blindness (De Valois and Jacobs, 1968).

In our electrophysiological studies, we have recorded from single cells with microelectrodes of about 1 micron tip diameter. The electrodes are lowered through a small opening in the skull and into the intact neural tissue down to the lateral geniculate nucleus, showing an electrode track. This nucleus consists of between one and two million cells in the macaque. The cells receive their input from the optic nerve fibers and send their axons to the visual cortex. What we are examining, then, is a sample of the input to the visual cortex in this animal.

We have recorded for some thousands of cells in different areas of the macaque lateral geniculate, attempting to answer a number of questions. Here I would like to concentrate on two of them, one being the question of how many different varieties of cells there are with regard to their responses to certain simple visual parameters, and the other, the examination of how these cells respond in various visual situations that are very similar to some of the behavioral tests described earlier. We shall examine at the unit level the ability of these cells to discriminate between lights of different wavelengths, of different saturation, of different luminance. We hope to identify the number of functionally different pathways there are from the retina to the cortex, and to examine how the cells in each of these pathways respond in situations in which we know the overall response of the animal. Thus we may, hopefully, begin to understand the visual capability of the animal on the basis of some combination of the responses of the individual units.

If one records from a random sample of geniculate cells, one finds quite different responses to flashes of light of different wavelengths put into the animal's eye.

One variety of cell responds in a very conventional manner and is similar to one of the two most common types of cells found in the retina or the lateral geniculate of the cat. These cells respond to light with an increment in firing, regardless of the wavelength of the light. Figure 2 shows the average number of spikes per second given by a large group of such cells in response to flashes of light of different wavelengths. In the absence of illumination or in the presence of any illumination to which the animal has adapted, these cells show a spontaneous firing rate. In response to additional light of any wavelength, the cells show an increment in firing, and they give the largest increment to lights in the middle part of the spectrum. Our examinations of the cat's lateral geniculate and

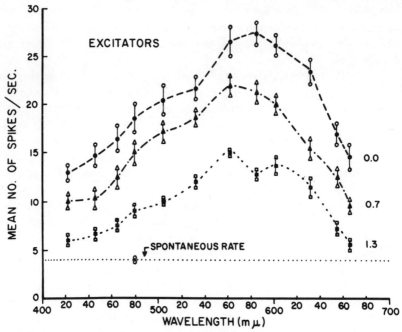

Figure 2. Average firing rates to flashes of monochromatic light of a large sample of non-opponent excitatory cells. Top line: highest intensity. (From De Valois, Abramov, and Jacobs, 1966)

similar studies by Hubel and Wiesel (1961) and others show that similar cells constitute almost half of the cat lateral geniculate population. Studies of the receptive field properties of such cells show them to have an excitatory center and an inhibitory surround. In the case of the macaque monkey, however, such cells constitute far less than 50 percent of the geniculate cell population; they comprise, at the most, about 15 to 20 percent of the total.

Figure 3 shows another variety of cells which are found about equally frequently. These cells also respond in a non-opponent way to lights in different parts of the spectrum, but they respond with inhibition at all wavelengths, giving maximum inhibition in the middle part of the spectrum. Such cells have the same spectral sensitivity as those which respond with excitation to all wavelengths: the amount of light required for a certain decrement in firing in these cells is the same as the relative amounts of light required from different parts of the spectrum to produce a certain increment in firing in the excitatory cells. The spectrally non-opponent inhibitory cells have an inhibitory center and an excitatory surround. In

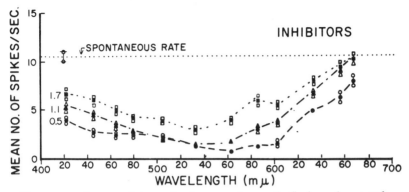

Figure 3. Average firing rates to monochromatic flashes of a sample of non-opponent inhibitory cells. Bottom line: highest intensity. (From De Valois et al., 1966)

the cat, such cells constitute the other half of the cell population; in the macaque, however, the excitatory and the inhibitory non-spectrally opponent cells together constitute about a third of the cell population of the lateral geniculate nucleus (De Valois, Abramov, and Jacobs, 1966).

The other two-thirds of the cells that one sees in the macaque at this level in the visual pathway are quite different in that they give excitatory responses to some wavelengths and inhibitory responses to others. Such cells are found very infrequently in many species of animals; specifically in the cat they are almost nonexistent. In the squirrel monkey they are also far less prevalent than they are in the macaque.

One example of such a cell is shown in Figure 4. We have here some oscilloscope records of superimposed responses of a cell to light of different wavelengths. The record to the left of the first line is before the light. Between the two lines the light is on, and after the second line, the light is off again. One can see that the cell, which is firing spontaneously before the light, shows an increment in firing rate to a light of 505 nm, for instance. The rate decreases again after the termination of the light. This cell gave a similar response to all of the short wavelengths, although it gives its largest response in the region of 500 and 480 nm. This looks very much like the response one would get from an excitatory cell, such as that shown earlier. However, if one stimulates with a long wavelength light, say 630 nm, the cell which is firing spontaneously is inhibited all the time the light is on. These responses of excitation to the short wavelengths and inhibition to the long wavelengths are produced by the cell independent of the intensity of the light. Such cells we term +G−R (green-excitatory, red-inhibitory) cells.

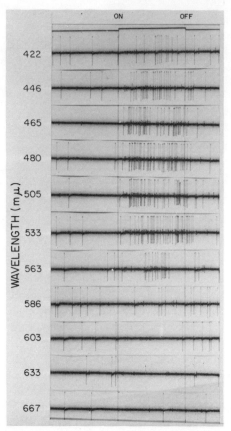

Figure 4. An example of the responses of a spectrally-opponent cell. These are superimposed records of the responses to various mono-chromatic lights. A line has been drawn through the records at left to indicate the time of light onset; the line at right marks light offset (one second duration). It can be seen that this green-excitatory, red-inhibitory cell fires to light of wavelengths below 560 nm, and inhibits to long wavelengths.

One finds considerable variety among such spectrally opponent cells; we have presented evidence (De Valois, Abramov, and Jacobs, 1966) that these cells fall into four different categories in terms of the spectral locations of the excitatory and inhibitory peaks. In Figure 5 is a summary of the average responses for a large number of such cells to three different intensities of light. As can be seen, the response sizes vary with intensity, but the spectral regions at which excitation and inhibition are shown remain roughly constant.

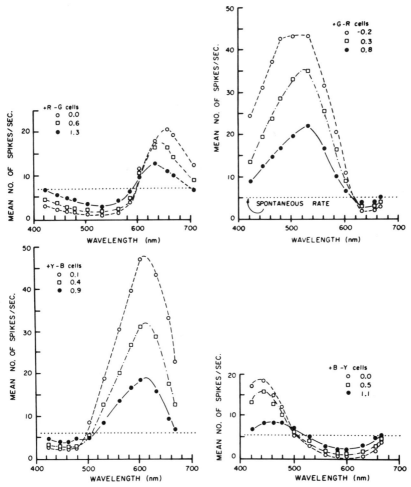

Figure 5. Average firing rates of samples of the four types of spectrally opponent cells. (From De Valois et al., 1966)

Found equally frequently with the green-excitatory red-inhibitory cells are what are in some ways mirror images to them: cells which show excitation to the very long wavelengths and inhibition to the short. With higher intensities of light there is more excitation to long wavelength stimuli and more inhibition to short wavelengths. The cross over from excitation to inhibition in these +R—G cells is about 600 nm, corresponding to the cross point of the +G—R cells. The cells which have been described above cross over from excitation to inhibition at wavelengths beyond 560 nm, with the average at 600. There are other cells, however, which cross over from excitation to inhibition at around 500 nm. These we have termed yellow-excitatory blue-inhibitory, and their mirror images, blue-excitatory yellow-inhibitory cells.

We believe, then, that there are six principal varieties of cells in the lateral geniculate, with regard not to the organization of the receptive field or the spatial analysis done by the cells, but rather with their responses to different wavelengths and intensities of light. There are four varieties of spectrally opponent cells, with maximum excitation at different regions of the spectrum, and two varieties of spectrally non-opponent cells. In Figure 6 we have a representation of what actual information would be

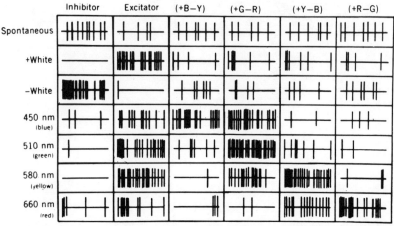

Figure 6. A drawing of the responses of each of the six cell types in the primate visual pathway to a variety of different lights. The number of spikes in each case is based on the average firing rates of cells of that type. (From De Valois and Jacobs, 1968)

coming to the cortex up these six different channels in response to white light and to light from different spectral regions.

The second question I would like to discuss is how these cells respond in various discrimination situations. That is, what is going on in the visual pathway when the animal is in, say, a hue discrimination situation. We know a macaque can distinguish between 590 and 600 nm. What classes of cells are providing the information upon which his discrimination behavior is based? Do all the cells aid in this discrimination or are there certain classes of cells that discriminate in certain parts of the spectrum, and others in other parts of the spectrum? Are there certain classes of cells that are responsible for hue discrimination, and other classes of cells that are responsible for brightness discrimination?

We have approached this in a very direct way, namely by recording from a cell while we go through various discrimination experiments, and determine whether or not the cell shows a change in response when we alternately present two stimuli which we know are discriminably different to the animal.

A first question we have asked in these discrimination experiments is whether or not the cell in question has color vision, that is, whether it can discriminate between white and monochromatic light of equal luminance (De Valois and Marrocco, 19). In the previous experiments we stimulated with flashes of light of different wavelengths. While we were producing a change in the color of the light, we were also producing a luminance change; that is, we were introducing an increment in luminance in each of these stimuli. We now ask ourselves what cells of the various classes would do if the luminance were kept constant and only the color were changed. We shift from a white preadapting light to various monochromatic lights adjusted for equal luminance (on the basis of phychophysical measures of the photopic sensitivity of macaques). In Figure 7 are presented the average firing rates of cells of the 6 classes to such an equal-luminance color shift. It can be seen that the inhibitory cells show no discrimination at all: one can shift from white light to any monochromatic light and the cells show no change in firing rate. The same is roughly true of the excitatory cells, which also show no discrimination between white and monochromatic light. It should be remembered that the excitatory cells are very responsive to a flash of monochromatic light, but a flash of monochromatic light produces both a luminance and a color change. Here we are producing just a color change, keeping the luminance constant, and these cells show no response change at all.

On the other hand, the four classes of opponent cells are very responsive to a pure color change. The $+$R$-$G cells, for instance, fire to long wavelengths and inhibit to short wavelengths when there is just a color change. Incidentally, the firing rates of the various classes of opponent cells are quite different under these circumstances from what they are in response to a flash of light. For instance, the yellow-excitatory, blue-inhibitory cells are extremely responsive to a flash of light, much more so than the $+$B$-$Y cells. But in response to an equal-luminance color change the $+$B$-$Y cells are as responsive as the $+$Y$-$B cells.

It is clear from this experiment that the spectrally opponent cells are providing the organism with its color vision capabilities, whereas the excitatory and inhibitory non-opponent cells do not discriminate between

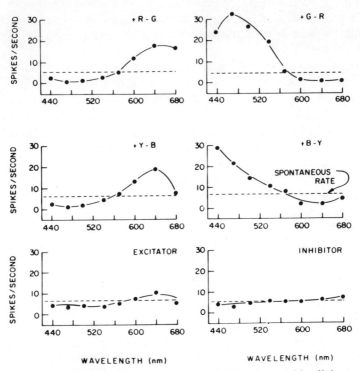

Figure 7. The average firing rates to a shift from a white light to a monochromatic light of the same luminance. The dotted lines mark the maintained firing rate to the white adapting light from which the shift was made. It can be seen that the non-opponent cells cannot distinguish between white and monochromatic lights of the same luminance, whereas the opponent cells all do, except at their neutral point. (From De Valois and Marrocco, in press)

white and monochromatic light, just as the similar cells in the cat, for instance, would not discriminate between white and monochromatic light. Thus we have an achromatic system in the non-opponent and a chromatic system in the opponent cells. This chromatic system not only differentiates between white and monochromatic light, but, in certain cases, differentiates better if the brightness is kept constant.

In another experiment (De Valois, Abramov, and Mead, 1967) we have examined the hue discrimination behavior of the different types of cells. In Figure 8 is shown a record from one cell in such a situation. We

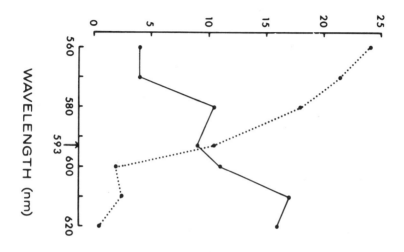

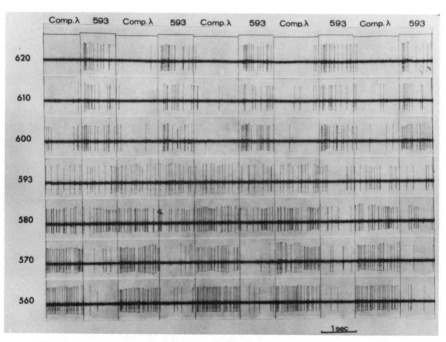

Figure 8. An example of the responses of a +G—R cell to shifts in wavelength, the luminance being held constant. (From De Valois, Abramov, and Mead, 1967)

are shifting between 593 nm and some other wavelength. The eye is adapted to 593 nm, and then we shift from this to some other wavelength (again equated for luminance). Thus we record from the cell while the eye is put in the same situation as that of a normal human observer who was looking back and forth between two fields of different wavelengths and trying to decide whether they are different colors or not. When the shift is from 593 nm to 585 nm the +G−R cell in this example shows an increment in firing. Thus it is discriminating this 8 nm shift in wavelength. When the light is then shifted back to 593 nm, the cell inhibits. In response to a larger wavelength difference, between 593 and 570, the cell shows a large increase in firing when the shift is to the 570, a decrement when the shift is to 593, and an increment again when one shifts back to 570. A shift to wavelengths longer than 593 nm produces inhibition, and again one can see that the cell distinguishes a very small wavelength difference. This same experiment is then carried out at each of a number of other wavelengths with this same cell.

From such data over a large population of cells, one can find out what wavelength difference at each point in the spectrum would be required to produce a certain criterion change in the firing rate for each class of cell. These data, combined over cells and with the mirror-image cells lumped together, is shown in Figure 9. It can be seen that the RG cells (+R−G

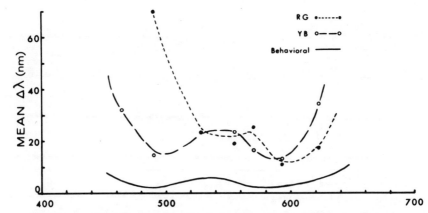

Figure 9. The wavelength discrimination of the different cell types compared with the overall human wavelength discrimination function. (From De Valois et al., 1967)

and +G−R) discriminate wavelength differences very well in the long wavelength end of the spectrum. They do a superb job of discriminating around 570-600, but much poorer in the short wavelength end of the spectrum. They show almost no ability to discriminate between 490 nm, say, and nearby longer or shorter wavelengths. On the other hand, the YB cells (+Y—B and +B—Y) do very well in the short wavelength end of the spectrum. They clearly provide the organism with the ability to discriminate wavelengths in this portion of the spectrum.

One can see, then, that the overall bimodal wavelength discrimination function of the organism is the consequence of macaques' (and normal human observers') having two types of opponent cell classes, with the RG cells providing the organism with the capability of discriminating among long wavelengths, the YB cells giving the organism the capability of discriminating among short wavelengths. So in effect we have a double dichromatic color vision system, a tritanopic one that gives us a wavelength discrimination capability among the long wavelengths, and a deuteranopic one that gives us the capability of discriminating among the short wavelengths.

In another discrimination experiment (De Valois and Marrocco, 19) we have examined the extent to which cells of the various varieties can discriminate luminance differences. While recording from a geniculate cell, we adapted the eye to a white light, and then shifted to a white light of higher or lower luminance. We know that a monkey can discriminate fine luminance differences. Again we ask the question: which cell type(s) give the animal this capability, and what are the natures of their discriminative responses?

Figure 10 shows the results of this experiment. Clearly the excitatory and inhibitory non-opponent cells are very sensitive to small luminance differences and show changes in firing rate which are approximately linear with log luminance. The excitators fire to luminance increments and they inhibit to luminance decrements; the inhibitors do just the opposite. The +R—G and +G—R cells, on the other hand, show little response to luminance changes: they respond far less to a very large luminance increment than to a slight color change, for instance. The +Y—B and +B—Y cells are intermediate between the RG and the non-opponent cells in their sensitivity to this type of stimulus change.

We may conclude from this luminance discrimination experiment that all the cell types contribute to our brightness discrimination; the non-opponent system the most, then the YB system, and finally the RG system the least.

Finally, I would like to discuss briefly a different type of experiment, in which we are looking not at the relationship of the outputs of these various cells to visual behavior, but rather at the relation between these various cell types and cone pigments. A +G—R cell, for instance, clearly has an input from an excitatory mechanism which is more sensitive in the

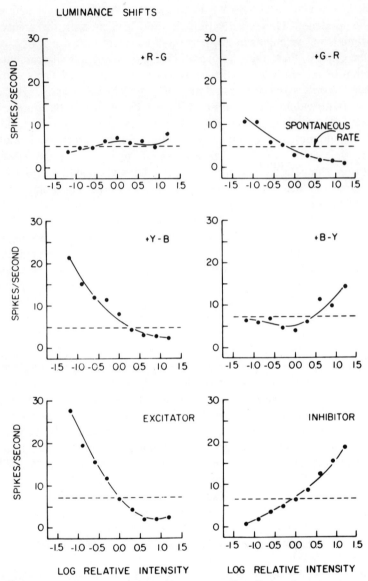

Figure 10. Responses of various types of cells to increments and decrements in the luminance of a white light. All of the cell types (with the possible exception of the +R-G cells) detect luminance changes, but the non-opponents are more responsive than the opponent cells in this situation. (From De Valois and Marrocco, in press)

short wavelength part of the spectrum, and receives inhibition from a mechanism that is more sensitive to the long wavelength part of the spectrum. We have used chromatic adaptation to isolate these inputs and determine their spectral sensitivity. If we turn on a very long wavelength adapting light of, say, 680 nm, this light will be absorbed almost exclusively by the longest wavelength pigment in the retina. This should selectively adapt or desensitize that mechanism, thus revealing the shorter wavelength mechanism in isolation. That, indeed, is what happens. When one turns on a very long wavelength adapting light while recording from a $+$G$-$R cell, one gets much more excitation in the middle portion of the spectrum than one got before, and the maximum excitation is produced not around 500 but around 540 nm; longer wavelengths which previously produced inhibition now produce varying amounts of excitation. What we have done is to eliminate the inhibitory mechanism and thereby reveal an excitatory mechanism sensitive across almost the whole spectrum, with maximum sensitivity at about 540 nm. The reverse situation, for reasons that are to be expected from the nature of the absorption curves of the photopigments, namely, that they all absorb in the short wavelengths, is not equally successful; but nonetheless one can, by the use of short wavelength light, selectively depress the short wavelength mechanism and reveal in somewhat greater isolation the long wavelength mechanism. This mechanism now shifts in sensitivity considerably toward the shorter wavelengths and turns out to have its maximum somewhere below 590 nm. We have good reason to think that, if we could completely eliminate the short wavelength system, the long wavelength system would shift to an even shorter wavelength, perhaps 570 nm.

We have done this adaptation experiment with all of the different classes of opponent cells, in an attempt to determine what the photopigment input is to the various classes of spectrally opponent cells that we see at the level of the lateral geniculate. In Figure 11 are summarized the results of these experiments. At the top are the results from the RG cells. At top left are the data from the short-wavelength systems isolated from the $+$R$-$G and $+$G$-$R systems respectively. The line is not an attempt to fit the data, but is the Dartnal nomogram with a peak at 535 nm, which is approximately where Marks, Dobelle, and MacNichol (1964), and Brown and Wald (1964) find evidence for peak absorption by the middle wavelength absorbing photopigment in the monkey eye. And, one can see, there is very good agreement between the spectral sensitivity of the isolated short wavelength mechanism from the RG system and the absorption curve of the photopigment as measured by spectrophotometry of individual cones. There is good reason, then, to think that the short wavelength mechanism, that is, the excitatory mechanism in the $+$G$-$R cells, and the

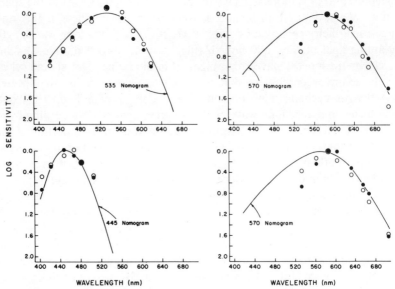

Figure 11. Spectral sensitivity curves of the components isolated with the chromatic adaptation technique shown in Fig. 13. Top left: short-wavelength component from the RG cells. Top right: long-wavelength component from the RG cells. Bottom: short and long wavelength components from the YB cells. In each case, the filled circles are the excitatory and the open circles the inhibitory components from the respective mirror-image cells. (From De Valois and Abramov, in press)

inhibitory mechanism in the $+R-G$ cells, has its origin from the middle wavelength photopigment in the eye. If one does the same experiment to try to isolate the long wavelength mechanism from the RG cells (top right), one cannot completely eliminate the shorter wavelength mechanism, but one can nonetheless see what the slope is at the long wavelength end. This slope, in the Dartnall nomogram, varies as a function of the long wavelength peak of the photopigment. The data from the $+R-G$ and $+G-R$ cells both reveal a good fit at the long wavelength end to a 570-nomogram pigment. The points are off from the nomogram at the short wavelength end because the inhibitory response is not completely eliminated; the cells thus give a smaller response than one might expect. We therefore conclude that the $+R-G$ cells have an excitatory input from a 570 pigment and an inhbitory input from a 535 pigment; and that the $+G-R$ cells have an excitatory input from a 535 and an inhibitory input from a 570 system.

If one does the same experiment with the BY cells, isolation of the short wavelength mechanism reveals a 445 system in the case of both the $+B-Y$ cells and the $+Y-B$ cells. Isolation of the long wavelength system feeding into the YB cells reveals a 570 input to both the $+B-Y$ cells and $+Y-B$ cells.

In summary, our recording evidence indicates the presence of three different cone types, in agreement with both the psychophysical literature and the results of microspectrophotometry. We find that these three cone types are hooked up in multiple ways in the various channels leading from the receptors to the cortex. In one set of channels, the $+G-R$ and the $+R-G$ opponent cells receive an excitatory input from the 535 pigment and an inhibitory input from the 570 pigment, or vice versa. These cells are then subtracting the relative outputs of these two cone types resulting in cells which are relatively unresponsive to white light or to luminance changes, but discriminate very small color differences, particularly among long wavelengths. The YB cells are comparing the relative outputs of the 570 and 445 pigments, to give another system with color vision. The YB color system optimally distinguishes among short wavelength lights, and is somewhat more responsive to white light and luminance differences than is the RG color system. Finally, the excitatory and inhibitory cells are adding together the outputs of these different types of cones, at least that of the two long wavelength cones. This gives a white-black system which is very responsive to luminance changes but cannot distinguish between white and colored lights.

References

Brown, P. K., and G. Wald. Visual pigments in single rods and cones of the human retina. *Science, 144*: 45-52 (1964).

De Valois, R. L., and I. Abramov. Cone systems underlying spectral opponent cell activity. In preparation.

De Valois, R. L., I. Abramov, and G. H. Jacobs. Analysis of response patterns of LGN cells. *J. opt. Soc. Amer., 56*: 966-977 (1966).

De Valois, R. L., I. Abramov, and W. R. Mead. Single cell analysis of wavelength discrimination at the lateral geniculate nucleus in the macaque. *J. Neurophysiol. 30*: 415-433 (1967).

De Valois, R. L., Elaine M. Hull, and H. C. Morgan. Psychophysical studies of monkey vision: III. Macaque saturation discrimination and anomaloscope tests. In press.

De Valois, R. L., and G. H. Jacobs. Primate color vision. *Science: 162*: 533-540 (1968).

De Valois, R. L., and R. Marrocco. Response of LGN cells to color, luminance, and purity shifts. In press.

De Valois, R. L., W. R. Mead, and H. C. Morgan. Psychophysical studies of monkey vision: II. Macaque hue discrimination and neutral point. In press.

De Valois, R. L., Martha C. Polson, and H. C. Morgan. Psychophysical studies of monkey vision: I. Methods and macaque luminosity tests. In Press.

Hubel, D. H., and T. N. Wiesel. Integrative action in the cat's lateral geniculate body. *J. Physiol., 155*: 385-398 (1961).

Jacobs, G. H. Spectral sensitivity and color vision of the squirrel monkey. *J. comp. physiol. Psychol., 56:* 616-621 (1963).

Marks, W. B., W. H. Dobelle, and E. F. MacNichol, Jr. Visual pigments of single primate cones. *Science, 143*: 1181-1183 (1964).

Wright, W. D. *Researches on Normal and Defective Colour Vision.* St. Louis: Mosby, 1947.

Sherman L. Guth

Heterochromatic Luminance Additivity

Photometry and colorimetry have classically been considered as separate areas within the visual sciences. Photometry has been concerned, with the specification of the magnitude or quantity of visual stimuli, whereas colorimetry has dealt with the specification of the chromaticity or quality of lights. Each is also based on different principles, with Grassmann's laws of color mixture forming the heart of colorimetry, and Abney's law of luminance additivity being the basis of modern photometry. In this paper I intend to show that the luminance additivity law is about as incorrect as it can possibly be (at least for predictions of threshold or brightness matching judgments), and that a marriage of colorimetry and photometry seems possible, at least at light levels near foveal threshold.

The Concept of Luminance

As all visual scientists know, the physical radiance of a light tells us little about the effect of that light on the human visual system, for the eye is not equally sensitive to all wavelengths. However, knowledge of the function which relates stimulus wavelength to foveal visual sensitivity allows the application of an appropriate correction factor to the radiance of a monochromatic light in order to adjust for visual sensitivity. That is, if P_λ is the relative radiance of wavelength λ, and if V_λ expresses relative sensitivity to wavelength λ, then its relative luminance, L_λ, can be defined as,

$$L_\lambda = P_\lambda \, V_\lambda. \tag{1}$$

125

There are various ways to measure V_λ, but I shall limit my present discussion to the threshold technique. That is, it is possible to measure foveal visual sensitivity by determining the radiance required to make a monochromatic light, of say 45′ visual angle, barely visible. Then, V_λ is taken to be inversely proportional to the radiance required, with maximum sensitivity being assigned the value 1.0.

Now that monochromatic luminance has been defined it remains to specify the luminance of a polychromatic light. A sensible way of specifying the luminance of a mixture of wavelengths is simply to add the luminances of its components. That is, let the luminance of a mixture, L_m, be,

$$L_m = \sum_{\lambda=0}^{\infty} P_\lambda \, V_\lambda \, \Delta\lambda, \qquad (2)$$

where $\Delta\lambda$ is some arbitrarily narrow wavelength band. Equation 2 is a mathematical expression of Abney's law. Since the very definition of the luminance of a light depends on the validity of the law, many experiments (see Dresler, 1953, for a partial literature review) have been completed to test it, especially at photopic levels.

My own tests of the additivity law were initially performed at threshold levels. The general rationale of the studies can perhaps be best understood by considering the following example: imagine that we determine the relative radiance required to bring a red stimulus, λ_1, and a green λ_2, individually to threshold. Now suppose that we present to a subject an additive mixture which consists of one-half of the red threshold radiance plus one-half of the green threshold radiance. As you might expect, the additivity law demands that the mixture should be at threshold.

Two criticisms often are raised regarding the threshold prediction. The first relates to the idea that if a subthreshold stimulus does not activate a particular chromatic mechanism, then the combination of two such subthreshold stimuli would not be visible if each were specific to a different mechanism. There are several responses to this criticism. First, all of the data to be presented here were collected using a simple adjustment method where subjects defined for themselves what "threshold" means. In our experience, subjects in this situation adopt a strict criterion for threshold and produce stimuli whose radiances can be cut in half and still be seen with rather high probability in, say, a forced-choice situation. Second, there is nothing crucial here about the threshold level. Threshold is merely a convenient criterion response level. We have also used a near-threshold standard against which, say, red and green matches were made. The relevant question here was, for example, whether or not one-half of the red plus

one-half of the green matched the standard. In general the results were
the same as with the threshold criterion. This is not to say that level has
no effect. We will show later that it most certainly does. Rather, the point
is that there is nothing special that we should expect to happen at threshold
as opposed to near-threshold levels.

A second question often arises because of a confusion between luminance and brightness. Abney's law predicts that one luminance unit of a
red plus one luminance unit of a green should equal two luminance units;
the law does not say that the mixture should appear twice as bright as
one of the components. Said in another way, if a red and a green are each
made to match a unit amount of a white standard, then for additivity to
hold, a mixture of the red and the green would match two units of the
standard. It does not at all matter how much brighter the two unit standard
appears to be than the original unit amount.

It is clear then, that the additivity law demands that an additive
mixture of two half-threshold lights should itself be at a just visible level.
Also, the addition of a one-quarter threshold light plus a three-quarter
threshold light should yield a just-visible stimulus. *In general, the additivity
law predicts that the mixture of any two lights whose luminances (in
threshold units) sum to unity should be at threshold.*

Threshold Data

Our basic experiment involving light mixtures at threshold has already
been published (Guth, 1967), but we shall here reproduce a description
of the method and results for the convenience of the reader.

Method. Figure 1 schematizes the apparatus which has been described
in detail elsewhere (Guth, 1965). Its essential features were two Bausch
& Lomb High Intensity monochromators (A_1 and A_2) with quartz-iodine
sources which illuminated the interior of an integrating sphere (C) from
opposite sides. From their position in a darkened booth, subjects viewed
a ¾ ° visual angle circular area of the interior of the sphere through the
sphere exit hole (G) which was positioned about 20 inches from the
viewing aperture (N). Small achromatic fixation points were superimposed on the field to allow fixation on the stimulus which appeared in a
completely dark surround. The light entering the sphere from either monochromator was measured by reference to the output from suitably positioned 1P39 phototubes. The light was exposed for ½ second intervals
by means of the rotating semicircular disc (H).

The subjects used an adjustment method to bring the field to a just-
visible level by turning a wheel which caused a change in the position of
a circular wedge (b) mounted in front of either source. Data were reported

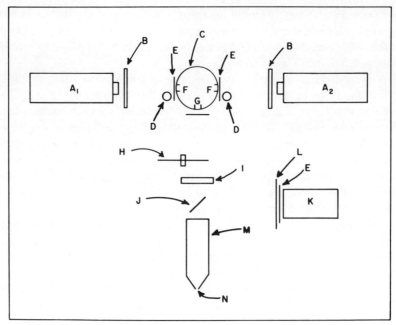

Figure 1. Schematic diagram of apparatus: A, monochromator; B, wedges; C, integrating sphere; D, phototubes; E, opal-glass screens; F, sphere entrance holes; G, sphere exit hole; H, sectored disc; I, shutter; J, partial mirror; K, source for fixation display; L, fixation display mask; M, viewing tube; N, viewing aperture. (From Guth, 1967)

for three pre-trained paid subjects, each of whom passed a near-point visual acuity test and color vision screening with the American Optical Company HRR Isochromatic Plates.

The initial trials for a given subject during a given session involved the use of source A_1 alone to estimate the radiance required to bring a monochromatic light, e.g., red, to threshold when no mixing was involved. That radiance was then reduced by ½, so that when the monochromator light was unrestricted, a red subfield of about 0.50 luminance units was presented to the subject.

To verify experimentally that the subfield was actually ½-threshold the following technique was employed. Source A_1 was occluded and A_2, which had been set to produce the same red as the subfield, was adjusted to again produce a just-visible field. The subfield was then unoccluded and the subject again adjusted the red from source A_2, which was this time being added to a subfield of identical wavelength, until threshold was reached. If the attempt to establish a ½-threshold field had been successful, then the amount of red required from A_2 when the subfield was present should have been ½ of the amount required when the subfield was not present. (Assume for the moment that a ½-threshold field was obtained.)

Source A_2 was then set to emit another wavelength, say a green, and the radiance required to bring it alone to a just-visible level was determined. The final step was to ascertain the amount of green from A_2 which had to be added to the red subfield in order to bring the mixture to threshold. If scalar additivity held, then the amount of green which had to be added to the red should also have been half of the radiance required to bring itself to threshold when no mixing was involved. In terms of threshold units, if 0.50 units of red had to be added to the red subfield to bring the stimulus to threshold, then 0.50 units of green should have been similarly required.

With this procedure, the accuracy with which the subfield was established was not critical, since the condition where a wavelength was added to itself empirically specified the luminance, in threshold units, of the subfield. For example, had the original estimate of the amount of red required for threshold been incorrect and the subfield been in fact only 0.40 units, then it would have been found that 0.60 units of red would have been required to bring the subfield to threshold. In this case, the additivity law would predict that 0.60 units of green would also have been required for a threshold response.

Wavelengths of 420, 435, 475, 500, 525, 550, 575, 600, 635, and 685 nm were used, with each of the ten serving both as a subfield and as a wavelength-to-be-added. That is, each of ten wavelengths was added to each of ten subfields, making a total of 100 major experimental conditions for each subject. At the beginning of each 45 minute session, subjects dark-adapted for at least eight minutes and then made ten practice threshold settings which included five observations upon which the subfield radiance for that day was based. During the balance of the session the subfield remained constant in all respects, and relative radiances of each of seven wavelengths required to bring the field to a just-visible level were determined three times with the subfield present and three times with the subfield absent. The subfield, the wavelengths-to-be-added to that subfield, and the ordering of trials within a session were all chosen randomly with certain minor restrictions which ensured the successive completion of full experimental replications within and across conditions.

The dependent variable for any particular wavelength in a given session was the median of the three radiances of that wavelength which had to be added to the subfield and the median of the three radiances required when no addition was involved. For any particular data point (i.e., a given wavelength added to a given subfield), the total of seven median radiances required under additive conditions were averaged and divided by the average of the corresponding median radiances required with the subfield blocked. A total of 117 sessions per subject were required to complete the experiment.

Results. Figures 2, 3, and 4 show the results of adding each of ten

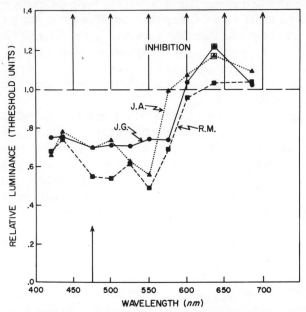

Figure 2.   Additivity functions for each of three subjects showing the mean relative luminance which was added to a 475 nm (indicated on the abscissa by an upright arrow) subthreshold field in order to bring the mixture to a just-visible level. (From Guth, 1967)

wavelengths to subfields of 475, 575, and 635 respectively. The solid lines of Figure 5 are mean curves derived from Figures 2, 3, and 4, as well as similar summary curves from the other conditions of the experiment. To the extent that any particular curve does not approximate a horizontal line through the data point associated with the condition where a wavelength was added to itself, the additivity law fails. For example, in Figure 2, the curve for subject J. G. shows that he required about 0.70 luminance units of 475 nm to bring a 475 nm subfield to threshold. It is clear, however, that in general more than 0.70 units were required. In some cases, such as at 635 nm, the curve goes above 1.0, indicating an inhibitory effect in that in order to bring the 475 nm subfield to

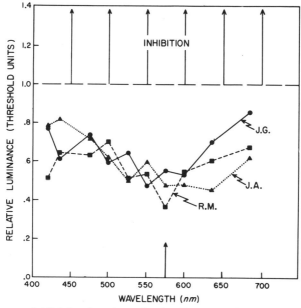

Figure 3.   Additivity functions (as in Figure 2) for a 575 nm sub-threshold field. (From Guth, 1967)

threshold, more radiance of 635 nm had to be added to it than was required to bring 635 nm to threshold when no subfield was present.

A Vector Model for Bichromatic Mixtures at Threshold

As discussed in an earlier paper (Guth, 1967), Cohen and Gibson (1962) and Swets (1960) have suggested vector models as alternatives to Abney's law. Their particular models do not conform to the facts of luminance addition at threshold, but the idea that vector summation might be an alternative to Abney's law remains a possibility. The general notion is that any monochromatic stimulus can be represented by an origin-bound vector whose length and direction indicate its luminance and wavelength, respectively. Addition is performed as in mechanics where the resultant of two forces is a joint function of the vector lengths and the angle between them. Within such a framework it is almost always possible to account for the luminance of a mixture of two wavelengths, with 0° indicating perfect additivity, 180° complete inhibition, and all other outcomes falling somewhere between. For our data, the determination of an angle which characterizes additivity between any two wavelengths, say λ_i and λ_j, is a

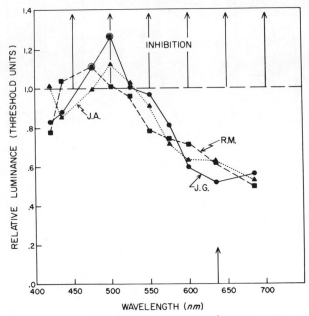

Figure 4. Additivity functions (as in Figure 2) for a 635 nm sub-threshold field. (From Guth, 1967)

simple matter. We first determine the luminance, L_i, (in threshold units) of a subfield of wavelength λ_i by subtracting from unity the luminance of λ_i (the same wavelength) which had to be added to the field to yield a threshold response. We then determine the luminance, L_j (also in threshold units) of λ_j which had to be added to the subfield of luminance L_i in order to produce a just-visible stimulus. Since the resultant is always of unit length (i.e., at threshold), it is then only necessary to use,

$$\text{Cos } \theta \; (_{i, \; j}) = \frac{1 - L_i^2 - L_j^2}{2L_i L_j}$$

to find the cosine of the angle between λ_i and λ_j which would yield unity as the vector sum of L_i and L_j. We have published such cosines (Guth, 1967; Table 1). (Slight adjustments of a few data points were necessary to eliminate cosines greater than 1.0 which occur whenever the estimated subfield value was *greater* than some other estimated value on the same function. The adjusted values are indicated with plus signs in Figure 5.)

Although it is true that an angle can almost always be derived to account for the luminance of the mixture of two wavelengths, it is not neces-

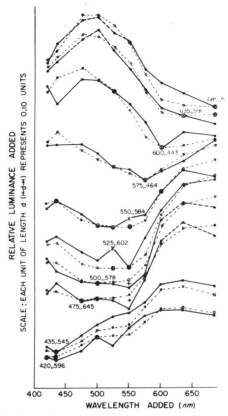

Figure 5. *Closed circles:* Grand mean additivity functions for each of ten subthreshold fields. The coordinates given for the small "x" on each curve indicate the wavelength of the subfield for that curve and the relative luminance which had to be added to that subfield for a threshold response when the subfield and the stimulus being added were identical in wavelength. Broken portions of the curves indicate inhibition as described in the text. *Open circles:* Predicted functions using a vector model as described in the text. Ordinate values for all functions can be reproduced using the scale unit "d" together with the subfield data point as a reference value. (From Guth, 1969)

sarily true that data involving many different two-component mixtures can be accounted for within a simple two-dimensional vector space. For example, consider that there are available estimates of angles, θ, which account for the luminance of each two-component mixture which can be made from wavelengths A, B, and C. First suppose that $\theta_{AB} = 60°$, $\theta_{AC} = 90°$, and $\theta_{BC} = 45°$. It is immediately obvious that such results cannot be handled within a two-dimensional space, but that three dimensions will be sufficient. On the other hand, suppose that $\theta_{AB} = 60°$, $\theta_{AC} = 90°$, and $\theta_{BC} = 10°$. In this case we see that even a three-dimensional vector space cannot adequately describe the data; that is, if θ_{AB}, θ_{AC}, and θ_{BC} are considered to be fixed (for purposes of explanation), then θ_{BC} would have to fall somewhere between 30° and 150° if a three-dimensional vector space is to fit the data.

Fortunately, the nature of the vector space which can accommodate a matrix of cosines such as derived from Figure 5 is a problem ideally suited for solution by means of *factor analysis,* a numerical procedure developed specifically for the purpose of deriving dimensionalities of vector spaces.

A factor analysis typically provides, for each vector (or variable), the loadings on each of n orthogonal dimensions which are required to account for a set of data. Whenever the square root of the sum of the squares of the loadings on each dimension sum to the vector length (unity in our case) for each vector, all of the variance of the data is said to be accounted for. The result of our own factor analysis was that three dimensions were sufficient to account for 93 percent of the data variance. That is, the sum of the sums of squares of the loadings on each of the three dimensions for our ten vectors was 9.3 out of a possible 10.0.

Given the results of the factor analysis (i.e., the loadings of each color vector on each of three dimensions) it was possible to demonstrate how well the model fitted the raw data by simply determining the predicted luminance of any wavelength which would have to be added to a subfield of given wavelength and luminance in order to yield a mixture with unit luminance. These predicted values are shown by the dotted curves in Figure 5.

The vector model appeared so successful in accounting for the data from the three subjects of Figure 5 that we decided to apply it to all of the foveal (45′) additivity data which had ever been collected in our laboratory in order to derive a model for threshold luminance additivity for the "average observer." This was accomplished in the following manner: we first found that it was possible to summarize all of our data using 11 different wavelength ranges of not more than 10 nm wide. For example, additivity data involving mixtures of blue-greens of 470, 475, 476 or 480 nm with reds of 625 or 635 nm were grouped into a single category identified as mixtures of 475 and 630 nm. We then summarized data from each of 14 subjects by first converting the additivity data of each experiment in which they participated to cosines, and then entering these cosines in the appropriate one of 55 possible cells into which two-component mixtures from 11 wavelengths could fall. Cosines within each cell were then averaged for each subject. Finally, averages were taken across subjects for each of the 55 cells. Table 1 shows the resulting mean cosines, together with the number of subjects upon which each cosine is based.

Table 1

Each element in the upper-right triangular submatrix gives the derived cosine between wavelengths λ_i and λ_j whose mixtures were studied in various threshold luminance additivity experiments. Each element in the lower-left submatrix gives the number of subjects upon which the cosine in the corresponding symmetrical cell is based. (Guth, 1969).

λ_i

λ_j	420	435	475	500	525	553	572	600	630	647	685
420	1.000	.858	.683	.456	.682	.496	.370	.147	.137	.176	.132
435	5	1.000	.548	.611	.508	.453	.283	.133	.040	—.020	.140
475	5	5	1.000	1.040	1.195	.912	.541	—.029	—.189	—.241	—.317
500	5	5	5	1.000	1.194	1.117	.603	.185	—.116	—.073	—.224
525	5	5	5	5	1.000	1.325	.942	.261	—.097	—.080	—.055
553	5	5	5	5	5	1.000	.902	.377	.250	.338	.109
572	5	5	5	5	5	7	1.000	.734	.650	.560	.369
600	5	5	5	5	5	5	5	1.000	.827	.800	.752
630	5	5	10	10	5	11	9	8	1.000	.926	.837
647	13	0	14	10	0	6	0	0	3	1.000	.890
685	5	5	5	5	5	5	5	5	5	0	1.000

Note: Values greater than 1.0 were assumed to be due to experimental error and were adjusted to 1.0 for the factor analysis of the matrix. An entry of zero in the lower-left submatrix implies that the associated cosine was estimated by interpolation.

The result of the factor analysis of these data is shown with closed symbols in Figure 6. The sums of the squares of the loadings of the 11 vectors on the first three factors summed to 10.55 out of a possible 11.0, again indicating an excellent fit to the data. The extent to which the sums of squares of the loadings on each dimension for a particular vector did not equal unity was assumed to be error variance, and the remainder was therefore distributed among each of the three factors in proportion to the absolute value of the initial loadings on each factor. The adjusted loadings are shown as open symbols in Figure 6. The smooth curves of Figure 6 were drawn to provide a reasonable fit to the adjusted data points and to have the property that the square root of the sum of the squares of the loadings on each factor sum to unity ±5 percent for each wavelength at 10 nm increments across the spectrum. In addition, the curves were drawn to be consistent with color mixture data as described below. These final loadings are shown in tabular form in Table 2.

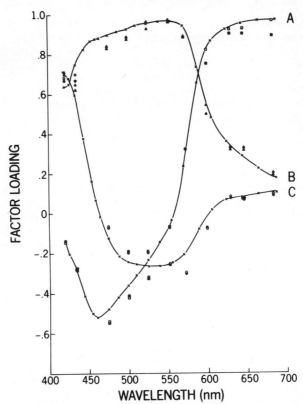

Figure 6.   Results of an eleven-variable factor analysis showing obtained (solid symbols) and adjusted (open symbols) loadings of each
of eleven origin-bound color vectors on each of three (A, B, and C)
orthogonal dimensions. Loadings given by the "X" 's on the smoothed
curves have been further adjusted to have their sum of squares add to
unity and to be consistent with colorimetric data as described in the
text. (From Guth, 1969)

Table 2
Loadings of Unit-Luminance Wavelength Vectors on Each of Three Orthogonal
Axes. Based on Adjusted Results of a Factor Analysis of Threshold Luminance
Additivity Data. (Guth, 1969).

Wavelength (nm)	A	Axis B	C
420	—.140	.640	.710
430	—.240	.680	.650
440	—.360	.800	.460

Table 2 (Continued)

450	—.470	.850	.250
459	—.520	.880	.070
460	—.520	.880	.070
470	—.500	.890	—.080
480	—.460	.910	—.160
490	—.400	.920	—.220
500	—.360	.930	—.240
510	—.310	.950	—.250
520	—.260	.955	—.260
526	—.230	.960	—.260
530	—.200	.965	—.260
540	—.140	.970	—.260
550	—.080	.970	—.250
560	.020	.965	—.235
570	.200	.950	—.200
580	.460	.860	—.150
590	.700	.710	—.070
600	.830	.550	—.010
610	.890	.460	.030
620	.920	.400	.060
630	.940	.340	.075
640	.955	.310	.080
645	.960	.290	.085
650	.965	.280	.090
660	.970	.245	.095
670	.975	.215	.100
680	.980	.190	.110

Figure 7 shows a photograph of a vector model which was constructed using dimensional loadings given in Figure 6 and Table 2.

Since the latter vector model is based on the pooled data from many subjects and experiments, it is not easy to compare the model with raw data as was done for three subjects in Figure 5. Alternatively, we chose to compare the model with the data in the following manner: given the cosines of Table 1, it is possible to ask how much of each of various wavelengths would have to be added to a 0.50 unit subthreshold field of a given wavelength in order to yield a resultant with unit length. (In fact, not all subjects added wavelengths to subfield of 0.50 units, but we used the obtained cosines to predict what the raw data would have looked like if they had all used the 0.50 subfield.) These "artificial" data are shown with solid lines in Figure 8.

It is now possible to compare the predictions made by the model with the artificial raw data. The predictions are shown with dashed lines in Figure 8. (Recall that the model was based on an adjustment and smoothing of the factor-analytically generated data points of Figure 6. Since the latter data points are directly dependent upon the cosines of Table 1, the

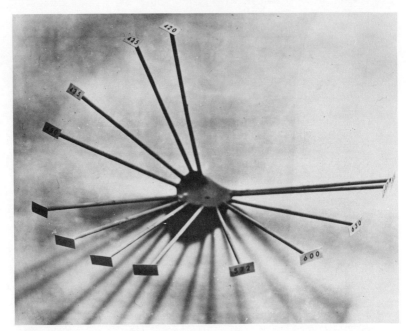

Figure 7. Photograph of a vector model constructed from dimensional loadings given in Figure 6. From upper center counterclockwise the wavelengths are 420, 425, 435, 450, 475, 500, 525, 535, 572, 600, 630, 647 and 685 nm. (From Guth, 1969)

curves of Figure 8 can also be interpreted as showing the effect of the adjustments and smoothing.) The brackets above and below a sample of data points of Figure 8 show the standard deviations of the data points obtained by using the cosine from each subject who contributed to the mean cosine for that particular wavelength combination to generate an artificial raw datum, and then to compute the standard deviation of a set of predicted values. For the most part, the predicted values are well within the range of variability of the artificial raw data.

On the Ability of the Vector Model to Account for Color Mixture Data

If the vector space as defined by the loadings given in Figure 6 is at all general, and if Grassmann's basic law of color-mixture is true, then metamers must occupy the same point in the space. That is, Grassmann's law states that metamers behave identically in mixture. This demands that metamers be additive in their luminances as well as their chromaticities.

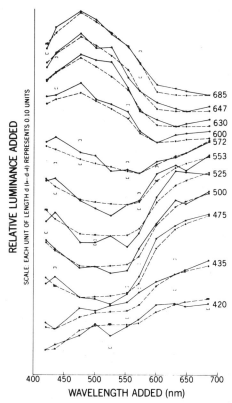

Figure 8. Predicted additivity functions (dashes) based on the vector space defined by Figure 6 versus obtained (solid lines) functions similar to those shown in Figure 5. However, the obtained data here are artificial in the sense that they show what would have happened if all subjects had added each of 11 wavelengths to each of 11 subthreshold fields of 0.50 luminance units, and if the mean cosines shown in Table 1 are valid. (That is, since all subjects did not work with the same subfield values, we could not average raw data. Instead, we averaged cosines based on raw data, and constructed imaginary data points based on those cosines.)

Therefore, the vectors associated with each member of a set of metamers not only must have the same length (i.e., have the same luminance), but also must be colinear, because luminance additivity only holds for vectors which are colinear. Since a color-matching experiment yields a pair of metamers for each spectral standard being matched, its data should allow a test of the generality of our vector model, providing that the matching was accomplished at near-threshold levels. Fortunately, Richards and Luria (1964) have published such data.

Initially it should be recalled that color-matching data can also be summarized in terms of a vector space, with unit radiance of each of three primary wavelengths representing unit amount of a vector which is an axis of the space. The data are then interpreted as specifying the amount of each of these primary vectors which is required to locate in the space

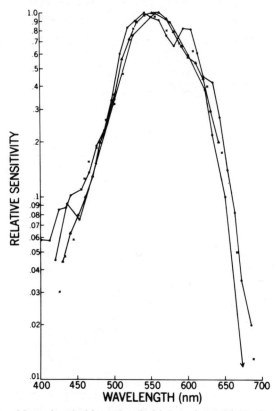

Figure 9. Near-threshold or threshold foveal sensitivity functions obtained by Kinney (triangles) who used direct brightness matching in a 2° field with a 0.01 mL standard, and by Sperling and Lewis (open circles) who used the threshold method with a 45′ field. Also shown with filled circles is a sensitivity function based on a non-selective sample of subjects who participated in various published and unpublished luminance additivity studies completed in our laboratory. Values indicated with unconnected "X"'s are the luminosity coefficients used in transforming the unit luminance vector space to a unit radiance vector space. (From Guth, 1969)

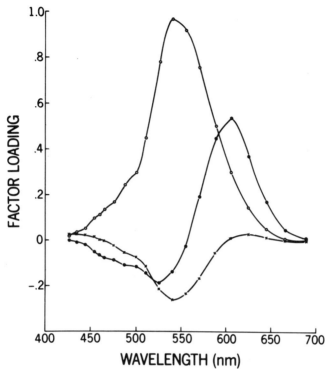

Figure 10. Loadings (tristimulus values) of unit radiance wavelength vectors on each of three orthogonal dimensions. Obtained by transforming the unit luminance loadings of Figure 3 using the luminosity coefficients given in Figure 9. The dimensions are referred to in the text as **R** (closed circles), **G** (open circles), and **B** (crosses), and the associated tristimulus values are called *R, G,* and *B*. (From Guth, 1969)

each standard color vector of unit radiance. The amounts of the primaries are called tristimulus values. To compare the Richards and Luria colorimetric vector space with our own luminance vector space we proceeded as follows: our unit luminance vectors were first transformed into unit radiance vectors using luminosity coefficients shown with "X" 's in Figure 9. The resulting vector space is defined by the loadings given in Figure 10. The new loadings may be thought of as tristimulus values for three imaginary orthogonal primaries which we shall call **R**, **G**, and **B**. It is possible to transform these tristimulus values to a set based on the Richards and Luria primaries of 645, 526, and 459 nm which we shall call **R′**, **G′**, and **B′**, respectively. If the tristimulus values associated with these two sets of primaries are called R, G, and B, or $R′$, $G′$, and $B′$, then we can follow directly the transformation equations given by Wyszecki and Stiles (1967). Our Figure 10 gives the amounts of **R**, **G**, and **B** required to match one unit of either **R′**, **G′**, or **B′**. These values were obtained from the loadings given for 645, 526, and 459 nm in Table 2 and from the luminosity coefficients (0.175, 0.820, and 0.125, respectively) for these wavelengths as shown in Figure 9. The equations are:

$$\mathbf{R'} = 0.1680\,\mathbf{R} + 0.0507\,\mathbf{G} + 0.0148\,\mathbf{B}$$
$$\mathbf{G'} = -0.1886\,\mathbf{R} + 0.7872\,\mathbf{G} - 0.2132\,\mathbf{B}$$
$$\mathbf{B'} = -0.0650\,\mathbf{R} + 0.1100\,\mathbf{G} + 0.0087\,\mathbf{B}$$

The transposition and inversion of the matrix formed by the coefficients of the **R**, **G**, and **B** terms yields the following basic transformation equations which allow the transformation of tristimulus values R, G, and B given for any wavelength in Figure 10 to corresponding tristimulus values $R′$, $G′$, and $B′$ in the Richards and Luria system.

$$R' = 4.7735\,R + 2.4341\,G + 4.8051\,B$$
$$G' = 0.1896\,R + 0.3635\,G - 3.4141\,B$$
$$B' = -3.5406\,R + 5.2319\,G + 22.3977\,B$$

Figure 11 compares the Richards and Luria tristimulus values with those based on our vector space.

The similarity between the two sets of functions shown in Figure 11 is impressive, but conclusions should be drawn cautiously. That is, our choices of factor loadings for the original vector space (Figure 6) were made in order to yield a reasonable fit between the models. The choices were well within the range of variability of the data, but more precise data might well reveal important differences between the luminance and colorimetric spaces.

In regard to the latter issue, it is of interest to observe how variations in the choice of luminosity coefficients and factor loadings can affect the apparent identity of the luminance additivity and color-mixture spaces. In particular, we can first derive luminance additivity vector spaces based upon either raw or adjusted luminosity coefficients and factor loadings, and then compare those spaces with the colorimetric space based on the Richards and Luria data. Figures 12, 13, and 14 provide these comparisons. The obvious conclusion that can be drawn from Figures 12 and 13 is that the transformation to the Richards and Luria system is much more dependent on small variations of the color vector factor loadings in the

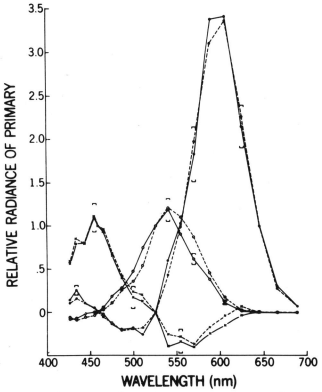

Figure 11. Tristimulus values required to match unit radiance stand-ards at low levels obtained by Richards and Luria (1964) (solid curves) and corresponding values predicted by the vector space derived from luminance additivity data as described in the present paper (dashes). (From Guth, 1969)

luminance additivity space than on small variations of the luminosity coefficients which were used in the luminance-to-radiance transformation.

It also should be recalled that the observational conditions of the ex-periments were not identical. That is, the Richards and Luria test field was 2° whereas ours was 45′. However, in view of the fact that Richards and Luria obtained only small differences between their color-mixture functions obtained with either centrally or parafoveally (3°, 20′) po-sitioned fields, it is not expected that their foveal data would differ much from those which would be obtained using 45′ fields, at least in terms of differences caused by rod intrusions. A more difficult problem caused by the field size differences was the choice of luminosity coefficients which

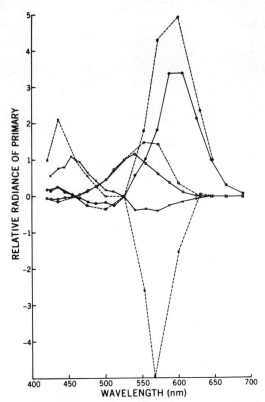

Figure 12. Similar to Figure 11, except that the luminance additivity vector space was based on a transformation which used the obtained luminosity coefficients shown with the filled circles in Figure 9 and the obtained factor loadings shown with closed symbols in Figure 5.

were used in transforming our luminances to radiances. Again, our choices were arbitrarily made to yield an adequate fit between the two vector spaces. The reasonableness of our choices is illustrated in Figure 9 which includes luminosity functions obtained by several investigators with various near-threshold foveal displays. [The luminosity function obtained by Richards and Luria themselves was not used, because much of it is based on flicker-photometric procedures which, as explained elsewhere (Guth, 1969), we consider to be an inappropriate measure of spectral sensitivity for the present purposes.] In addition, it is known that luminance additivity functions are different for 2° and 45′ fields (Guth, 1968), but the observed differences were small, and might not have much of an effect on the correspondence of the models.

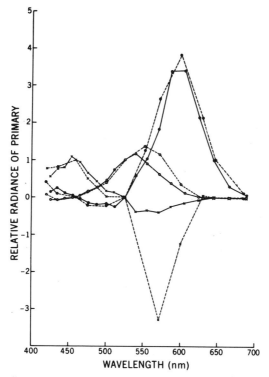

Figure 13. Same as Figure 12 except that the transformation was based on adjusted luminosity coefficients (the "X"'s of Figure 9) and the obtained factor loadings (closed symbols of Figure 6).

Nonadditivity at Suprathreshold Levels

Kohlrausch is usually (e.g., Judd, 1958) credited with the earliest systematic studies of luminance nonadditivity. (See Kohlrausch, 1935, for a partial summary of his early work.) The general finding was that a mixture of a red and a green, each of which had been matched to a standard white, tended to be judged less bright than a white which was equal to twice the original standard. The phenomenon has been called the Helmholtz-Kohlrausch effect (Urbanek and Ferencz, 1942).

We have completed several studies at suprathreshold levels, and have found that nonadditive effects are not the same at high levels as they are at threshold levels. This is nicely demonstrated in one of our experiments in which four subjects were tested with violet-plus-green, red-plus-green, and green-plus-green (as a control) mixtures. The basic approach for this

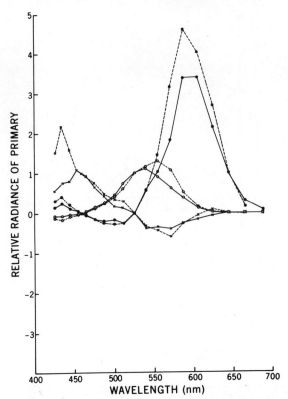

Figure 14. Same as Figure 12 and 13 except that the transformation involved the obtained luminosity coefficients and the adjusted factor loadings.

experiment was to present to subjects an achromatic standard and require an adjustment of the radiance of various comparison stimuli until a match with the standard was obtained. The comparison stimulus consisted of light from either one or both of two monochromatic sources.

Method. The actual procedure was formally identical to that used in the threshold studies described earlier. For example, we defined the relative luminance of a particular standard as one unit and determined that radiance of, say, a 521 nm green from source number one which was required to yield a match with that standard. We then reduced that 521 nm radiance by about half, making available from source number one a green stimulus which had a relative luminance of about 0.50 units. We refer to this latter field as a subfield since it is directly analogous to the subthreshold field which we use in our threshold studies.

The paradigm for an experimental check on the luminance of the subfield was as follows. Source number one was occluded, and source number two was used to provide an identical comparison stimulus of 521 nm. The subject adjusted the radiance of the comparison from source number two until a brightness match with the standard was obtained. Following this, we opened source number one which should have been presenting a 0.50 unit (approximately) subfield, and required that the subject add 521 nm light from the source number two to the 521 nm subfield until a match was obtained. If the subfield were actually one-half unit, the subject should have then required only one-half of the radiance that he required when the subfield was not present. We next changed the wavelength of source number two to, say, a red and determined the radiance required to match in brightness that red with the standard. For the additivity test, source number one was opened, and we determined the amount of red which had to be added to the green 0.50 subfield to bring the mixture to the match-point. If additivity holds, then the radiance of the red which had to be added to the green subfield should have been one-half of that required when the subfield was blocked. Similar tests were made with a violet added to the green subfield. (The various kinds of matches were actually distributed randomly within a session.) In general terms, if additivity holds, then a given subfield should reduce the radiance required to bring any wavelength to a match by exactly the same factor that it reduced the radiance required to bring itself to a match.

Four subjects with normal vision were each tested using achromatic standards at 0.10 ("low") or 6.0 mL ("high"). In addition, the usual threshold criterion was used. The standard appeared in the right half of a vertically partitioned circular bipartite field of either 1.5° or 45' visual angle which was presented in Maxwellian view with a dark surround. Since there were three intensity levels and two field sizes, there were six major experimental conditions. The standard source was a 6.6A/T2-1/2Q/CL—45w Sylvania quartz-iodine lamp operated at 115 volts. Intensity adjustments of the standard were made with Wratten neutral filters, and the various luminance levels of the standard were measured by successively matching the Maxwellian field with an extended flashed opal glass field which was housed in a portable apparatus designed to resemble the actual test display in all respects. The luminance of the extended field was then determined with a MacBeth illuminometer.

The two-channel comparison field on the left could be filled with monochromatic radiation using a Bausch and Lomb High Intensity monochromator and/or Balzer's interference filters, both with about 10 nm half-peak bandwidths. The rectangular images of the various fields were superposed at the center of the natural pupil and were at most 1.8 mm wide by 0.7 mm high. The handwheel which subjects used to adjust the brightness of comparison stimuli was linked to neutral wedges via a Selsyn system and could be electrically switched to control the brightness of either one or both of the comparison light channels. The relative radiance of any comparison stimulus was determined directly by deflecting that light alone to a suitably positioned phototube.

Head positions were maintained using dental impressions mounted on individually adjustable bite-bars. The entire display was seen through a 10.8 mm diameter viewing aperture which was 8.25 cm from the right eye. The observation booth was painted white and was illuminated with a single incandescent lamp which yielded an overall booth luminance of either about 2.0 mL or .2 mL for the high and low conditions, respectively. The booth was dark during threshold observations.

For each of the two standards and the threshold ·conditions, wavelengths of 422, 521, and 647 nm were added to a 521 nm subfield of about 0.60 luminance units. Within each session, the summary measure for each wavelength was the mean radiance required for a match when adding to the subfield divided by the mean radiance required when no adding was involved. The actual dependent variable was the logarithm of that ratio. Within each experimental session, matches to the standard (or threshold settings) were made five times with each of the three wavelengths added to the subfield, and five times with each of the three wavelengths alone. Two of the subjects were tested for four sessions per condition, and the other two for eight sessions. Therefore, each data point for each subject was based on a total of either 40 or 80 observations, with one-half of the observations being required for the amount of that stimulus required when it was added to a subfield, and the other one-half for estimates of the radiance of that stimulus required to match the standard (or for a threshold response) when no adding was involved.

Results. The similarity between the solid and dashed line in Figure 15 reflects the fact that no reliable difference could be found between the two field sizes used. Further generalizations will therefore be made without reference to that variable. It is clear from the Figure that red-plus-green subadditive effects exist at levels ranging from threshold to about 6.0 mL. Statistical analyses also support the observation that the red-plus-green nonadditive effects in this experiment were reliably ($p < .05$) greater at the high level than at threshold. Furthermore, violet-plus-green mixtures are subadditive at threshold, whereas they are slightly (but reliably) *superadditive* at the high level. That is, it takes *less* violet than predicted by the additivity law to bring a green to unit luminance. In other words, the sum of the components is less than the unit length resultant.

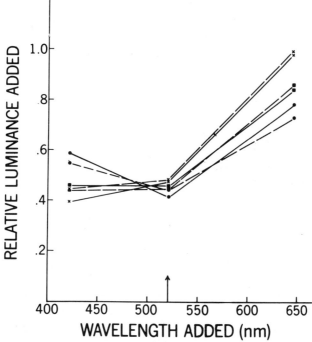

Figure 15. Additivity functions showing mean relative luminances of 422, 521, and 647 nm which were added to 521 nm fields of 1½° (solid lines) or 45′ (dashes) angular subtense in order to bring the field to unit luminance. Unit luminance was either threshold (circles), 0.20 mL (squares) or 6.0 mL (crosses).

Discussion

The congruence of our luminance vector space with the Richards and Luria colorimetric vector space suggests that when near-threshold stimuli are considered, the appropriate photometric and colorimetric models are identical. It is simply that the position of a vector in the color space represents its chromaticity and its length represents its luminance. Starting from a colorimetric standpoint, this should seem elegantly concise. That is, the classical colorimetric view is that the position of a vector is its chromaticity, and its length is *proportional* to its luminance. Why not allow a vector length to represent its actual luminance? Seen in this way, a vector photometry is more parsimonious than one based on Abney's law.

Specific suggestions regarding a system for low-level photometry and colorimetry are probably premature at this time, but there would seem to be an advantage in using orthogonal primaries to facilitate computations of lengths of several vectors. Also, if in an orthogonal space the tristimulus values were first computed for any polychromatic stimulus using the usual colorimetric procedures, then the luminance of that stimulus could be estimated by simply calculating the square root of the sum of the squares of the luminances of those tristimulus values. (Note that direct tests of the validity of the latter procedure have not yet been made. All our experiments to date have involved bichromatic mixtures. Whether or not the luminance of, say, three-component mixtures can be predicted using a vector space derived from data on bichromatic mixtures is a question which remains to be answered.)

As for higher brightness levels, it appears that the photometric and colorimetric models must remain separate. The fact that violet-plus-green enhancement effects occur at higher levels means that a vector model is not appropriate, for it cannot be true in a metric space that the resultant of two vectors is greater than their algebraic sum. An obvious question is why does there exist a correspondence between photometry and color-imetry only at low levels? Is it perhaps true that gain control mechanisms are affecting the total output (brightness) of the various components of the visual system in different ways than their relative (chromaticness) outputs? Also, given that a vector model cannot be completely satisfactory for high level photometry, might it still provide enough of an improvement over an algebraically additive system to warrant its use under some circumstances? We hope that future work will answer these questions.

References

Cohen, J., and W. A. Gibson (1962). Vector model for color sensations. *J. Opt. Soc. Am. 52*, 692-697.

Dresler, A. (1953). The nonadditivity of heterochromatic brightness. *Trans. Illum. Engng. Soc.* (London) *18*, 141-165.

Guth, S. L. (1965). Luminance addition: General considerations and some results at foveal threshold. *J. Opt. Soc. Am. 55*, 718-722.

Guth, S. L. (1967). Nonadditivity and inhibition among chromatic luminances at threshold. *Vision Res. 7*, 319-328.

Guth, S. L., J. V. Alexander, J. I. Chumbly, C. B. Gillman, and M. M. Patterson (1968). Factors affecting luminance additivity at threshold among normal and color-blind subjects and elaborations of a trichromatic-opponent colors theory. *Vision Res. 8*, 913-928.

Guth, S. L., N. J. Donley, and R. T. Marrocco (1969). On luminance additivity and related topics. *Vision Res. 9*, 537-576.

Judd, D. B. (1958). A new look at the measurement of light and color. *Illum. Engng. 53*, 61-71.

Kohlrausch, V. A. (1935). Zur Photometrie farbiger Lichter. *Das Licht 5*, 259-279.

Richards, W., and S. M. Luria (1964). Color mixture functions at low luminance levels. *Vision Res. 4*, 281-313.

Swets, J. A. (1960). Color vision. *Quarterly Progress Report, Research Laboratory of Electronics*. Massachusetts Institute of Technology, Cambridge, Massachusetts.

Urbanek, J., and E. Ferencz (1942). Comité des études sur la photométrie visuelle. Rapport du secrétariat. (Comité hongrois). *10th Session—Commission Internationale de l'Eclairage, Scheveningen (1939)*. Volume 1, 44-91.

Wysezecki, G. and W. S. Stiles (1967). *Color Science. Concepts and Methods. Quantitative Data and Formulas*. Wiley, New York.

W. D. Wright

Small-Field Tritanopia
A Re-Assessment

Introduction

The blue-blindness of the central fovea was first reported by König and Köttgen in 1894. This observation was ignored and then forgotten until it was rediscovered by Willmer in 1944 and investigated quantitatively by Willmer and Wright in 1945. Further observations on color confusions occurring with small fields viewed by retinal areas other than the central fovea were reported by Hartridge in 1945 and additional quantitative studies were published by Thomson and Wright (1947), for small fields fixated both centrally and at retinal positions displaced by 20 minutes and 40 minutes of arc from the foveal center.

While many other experiments have been carried out which have a bearing on the interpretation of the phenomenon of small-field tritanopia, to some of which reference is made below, direct interest in the subject seems to have waned until the Frederic Ives Medal Address by Wald on "Blue-Blindness in the Normal Fovea" delivered to the Optical Society of America in 1966 (Wald, 1967). In this paper Wald assembles some very significant evidence on the distribution of the blue receptors in the retina, especially in the foveal area, and gives some important data on the variation of blue sensitivity with size of field, which he shows declines more rapidly with decrease in size of field than is the case with red or green. His discussion leads him to raise again some of the key questions, for example, "how large an area is blue-blind?," "is the blue-blindness of small central fields associated primarily with their small size or with their position in the fovea?," "why a blue-blind fixation area?," and so on.

Wald's approach is essentially direct and his conclusions are based primarily on evidence from receptor sensitivities determined by threshold

152

measurements recorded under different conditions of adaptation. These are related to the anatomical structure of the retina and the known distribution of rods and cones across the fovea. He clearly distrusts the evidence based on color mixture experiments, makes virtually no use of color discrimination observations, and is not much concerned with the color sensations themselves as generated under small-field conditions. It is entirely reasonable that he should pay special attention to the retinal receptors themselves, but there is, of course, much more to color perception in general and to small-field tritanopia in particular, than this. This present paper, then, is concerned with some of these further questions and since I may seem critical of some of Wald's ideas, I wish to acknowledge without reservation that I found his paper a most valuable stimulus to fresh thinking.

It is in any case very timely to reexamine the significance of small-field tritanopia because of the relevance of new research on vision which has been reported since the 1940's. Here I would include the studies on eye movements and on the stabilized retinal image in relation to color perception (Ditchburn, 1963), studies on peripheral adaptation and Troxler's effect by Clarke (1960), investigations on peripheral color vision by Moreland and Cruz (1959), experiments by McCree on color confusions produced by voluntary fixation (1960A), and the new concepts that are emerging on the coding of neural information in the retina as a result of electron microscope studies of the retinal tissue (Pedler, 1965). It is also clear that the determination of the color sensitivity curves in tritanopic vision is closely related to the wider problem of the relation between the color mixture curves or color matching functions of normal trichromatic vision to the sensitivity curves of the receptor processes as determined by Stiles' two-color threshold technique (Wyszecki and Stiles, 1967).

First, however, we must examine what we mean by tritanopia and the evidence for the occurrence of tritanopia under small-field conditions.

Definition of Tritanopia

Observers may be classified by their color vision under three broad headings of monochromats, dichromats, and trichromats. A subject is placed in one or other of these groups depending on whether he needs to use one, two or three controls in color matching. The dichromats can themselves be divided into three clearly differentiated types known as protanopes, deuteranopes, and tritanopes. The protanopes and deuteranopes confuse colors in the red-yellow-green range of wavelengths in the spectrum, whereas the tritanopes confuse colors in the green-to-blue range. These confusions incidentally determine the stimuli that have to be used in measuring the dichromatic color mixture curves. Thus with the protanopes and deuteranopes, one stimulus must be a blue and the other must be chosen from the longer wavelength, green-to-red, range. With the tritanope, on the other hand, one stimulus must be a red and the other must be chosen from the shorter wavelength, green-to-blue, range. A further distinguishing feature is in the spectral sensitivity curve (the V_λ curve), since the protanope's curve shows a marked loss of sensitivity at the red end of the spectrum compared with the normal, and the deuteranope's curve is more or less normal. The tritanope's curve is possibly a little below normal at the blue end but this effect tends to be obscured due to varying degrees of yellow pigmentation in the eye.

The tritanope's characteristics can therefore be summarized by the data illustrated in Figures 1 through 4 (Wright, 1952).

Figure 1 shows the tritanopic V_λ curve compared to the normal, both

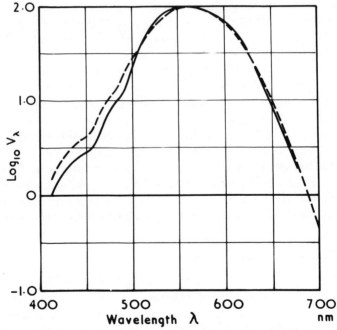

Figure 1. Spectral sensitivity of a tritanope compared with a normal observer. Log V_λ plotted against wavelength λ. Solid line, tritanope; broken line, normal.

curves being plotted on a logarithmic basis; Figure 2 shows the colors which are confused by the tritanope as defined by the tritanopic con-

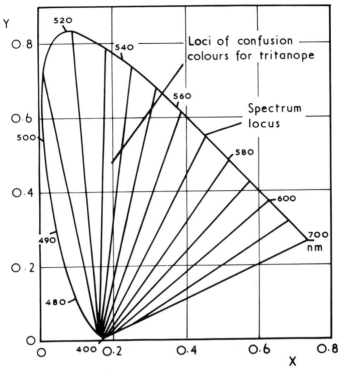

Figure 2. C.I.E. chromaticity chart showing loci of confusion. Colors for tritanope.

fusion loci plotted in the normal chromaticity chart; this includes the information given by the wavelength discrimination curve, which is also shown separately in Figure 3; Figure 4 shows the dichromatic chromaticity coordinates of the spectrum for the tritanope. As an alternative, these coordinates could have been plotted in the normal chromaticity chart to give a spectrum locus lying on a straight line passing through the points representing the red and green matching stimuli. They can also be combined with the V_λ curve and the relative luminosities of the matching stimuli to give curves which would be directly related to the sensitivity curves of the retinal receptors. However, the curves as plotted in Figure 4 are more informative so far as color matching relations are concerned.

On the above basis, identification of tritanopia depends on color matching and color discrimination and not on the absence of blue receptors. There is indirect evidence that the blue receptors are absent, or are at least inactive, in the convergence of the confusion loci to a common point in

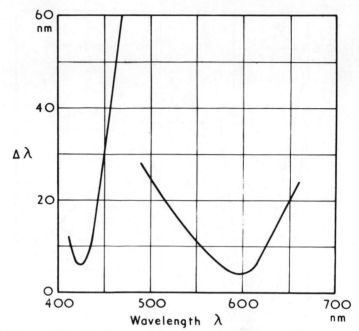

Figure 3. Wavelength discrimination of tritanope. Just noticeable wavelength difference $\Delta\lambda$ plotted against λ.

the blue corner of the chromaticity chart in Figure 1 (Helmholtz, 1867). In considering possible causes of small-field tritanopia we have to bear in mind the possibility that the blue receptors might be present but that their response might in some way be "switched off." We also have to be careful how we use the term "blue-blindness." Does blue-blindness mean the absence of blue receptors or the inability of the retina to give any response to wavelengths at the short-wave end of the spectrum? These are not necessarily the same thing. Or does blue-blindness mean the inability of the visual system to generate the sensation of blueness? This ambiguity suggests that it would be best to discontinue the use of the term.

One use of Figure 4 should be mentioned at this stage, namely as a means of identifying wavelengths that are confused by the tritanope. Thus any pair of wavelengths, such as λ_1 and λ_2, at which a horizontal line intersects the r_λ (or g_λ) curve have the same chromaticity coordinates and therefore appear to have the same color to the tritanope. This suggests that in the far violet the color is approaching very closely to that of the yellow part of the spectrum and therefore to a second neutral point (but not a neutral range as Wald suggests).

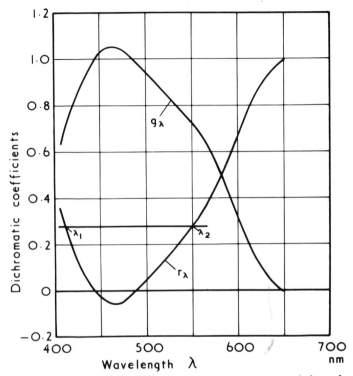

Figure 4. Dichromatic chromaticity coordinates, r_λ, g_λ, of the colors in the spectrum for a typical tritanope. Wavelengths such as λ_1, and λ_2 which have the same value of r_λ (or g_λ) will be confused by the tritanope.

Evidence for Small-Field Tritanopia in the Center of the Fovea

The basic evidence for small-field tritanopia in the foveal center is that with a small field subtending 15 minutes of arc it is possible, as in congenital tritanopia, to match all colors in the spectrum with a mixture of two wavelengths, one from the red end and one from the blue, when the field is viewed in the center of the fovea.* Further, wavelength discrimination curves show a maximum discriminating ability in the yellow wavelengths

---

* See results of L. C. Thomson and W. D. Wright, *J. Physiol, 105,* 316, 1947. These results are being quoted in preference to the earlier paper by Willmer and Wright because for this investigation the experimental conditions had been improved by the provision of a lens to correct for the chromatic aberration of the eye. Also comparable observations were made to one side of the foveal center.

and virtually no discrimination between 520 mm and 480 mm. It is doubtful whether the color matching and discrimination curves are identical to those found in congenital tritanopia, but there is no doubt at all of the general similarity of their main features.

The viewing conditions in the experiments by Thomson and Wright were rather carefully described. Thus in the color matching experiments the instrument controls were adjusted to give an approximate match, the field was viewed momentarily, and then the observer looked away. Adjustments to the controls to improve the match were then made and the field was viewed again. This procedure was repeated until the two halves of the field were seen to match for the particular fixation conditions under test. There was, therefore, no question of prolonged viewing being required to secure a dichromatic match. This can be confirmed by tests with selected pairs of surface colors chosen from a given tritanopic confusion locus and having the same luminance. These are seen to match under immediate foveal fixation when viewed under small-field conditions.

Evidence for Small-Field, Off-Center Tritanopia

The experiments of Thomson and Wright may again be quoted to confirm that with a small field subtending 15 minutes of arc, it is possible to match all colors in the spectrum with a mixture of two wavelengths, 650 nm and 460 nm, when the field is viewed at displacements of 20 minutes and 40 minutes of arc from the foveal center. The color matching curves are again generally similar to those found in congenital tritanopia. The retina however, is significantly more sensitive to the blue wavelengths in the off-center positions. An interesting change in color appearance was also reported, namely that a wavelength of 510 nm looked green when viewed centrally but turned blue in the 40 minute position. This in spite of the fact that in both positions blue-green confusions occurred.

Actually the wavelength discrimination curves at the 20 minute and 40 minute positions gave rather less convincing evidence of tritanopia, especially for observer L.C.T. Part of the uncertainty arose from the rather poor discrimination in the yellow, presumably due to the generally less critical judgments that are an inevitable corollary to the lower visual acuity away from the visual axis. By comparison, therefore, the loss of discrimination in the blue-green was relatively less striking.

Conflicting Factors in Assessing the Evidence for Small-Field Tritanopia

The orthodox interpretation of tritanopia is that only two types of color receptor are functioning in the retina, the blue receptor being missing or inactive. We must now examine possible effects which could lead to the simulation of tritanopia even though the blue receptors might still be functioning.

One factor which has to be taken into account is absorption by the yellow macular pigment. Wald (1967) gives a very good summary of the distribution of the pigment over the central area of the retina and the consequent loss of sensitivity to light at the short-wave end of the spectrum. This to some extent obscures any variations in sensitivity due to differences in population density of the blue receptors, but the yellow pigment cannot in itself induce an apparent tritanopia. Thus, the absorption of the blue light can always be compensated by an increase in energy and so remove any apparent abnormality of color matching or color discrimination.

Another factor which might tend to induce tritanopia is the adaptation of the retina. When a patch of light is fixated on a given area of the retina, the sensitivity of that retinal area falls and the apparent brightness of the patch diminishes. This effect is so marked and so rapid in the extra-foveal and peripheral areas of the retina that the patch may disappear altogether. This is known as Troxler's effect, to which reference has already been made (Clarke, 1960). There is some reason to believe that the blue receptors are especially liable to this loss of sensitivity and, if so, this could lead to a simulation of tritanopia under small-field fixation. This is certainly a factor which cannot be ignored in interpreting the evidence for small-field tritanopia, but it is usually assumed that the foveal area is very resistant to Troxler's effect. It could be, however, that this resistance is more apparent than real, since the high acuity at the center of the fovea may require a steadiness of fixation that is normally impossible to achieve because of involuntary eye movements, if Troxler's effect is to be manifest in the fovea. Certainly with a stabilized retinal image, rapid fading of the visual response does occur at the fovea.

The influence of eye movements, in fact, is the third conflicting factor which has to be taken into account. The studies on eye movements and the stabilized retinal image, for example by Ditchburn (1963) and his colleagues, suggest that with partial stabilization, color discrimination degenerates leading to color confusions which are symptomatic of tritanopia. Ditchburn has suggested that color information is generated at the boundary between one colored area and another by the fine scanning movements of the eye. It might be, then, that small-field tritanopia under conditions of fixation could be attributed to the elimination of those eye movements which appear necessary for the transmission of blue information from retina to brain.

This is certainly the implication of McCree's studies (1960B), since he found that some observers, of whom he was an outstanding example, were able to fixate with sufficient steadiness to produce marked loss of color discrimination, even with a field subtending 80 minutes of arc. Moreover, the type of color confusion that occurred most readily was in the blue-green, corresponding to tritanopia. One important feature of his observations was that an exact brightness match between the two halves of the field being compared was necessary for the confusions to occur. Does this imply that uniformity of luminance across the field in some way inhibits the transmission of the color-difference information to the visual cortex?

What, Then, Is Meant by the Blue-Blindness of the Central Fovea?

This question is a quotation from Wald's paper, and he answered by saying that it means "little more than that in sufficiently small fields within and beyond the photopic area, color-matching experiments reveal tritanopic confusions, though perhaps only after appreciable intervals of fixation."

I would personally want to stress that the tritanopia of the foveal center does represent a unique visual situation. This region of the retina is the one with which we make all our most critical judgments. It defines the peak value of visual acuity and hence the visual axis on which we depend for our sense of direction and binocular fusion. It is the area which we use for recording all the detailed information in the scene before us. It is the area which has the maximum capacity for luminance discrimination and luminance matching, and for red-green color discrimination and matching. In the light-adapted eye, it is the area which has maximum light sensitivity for the detection of point sources of white light, and also of red, yellow, and green lights. Yet in this one respect of low sensitivity to blue light and inability to distinguish blue from green or pink from yellow, to take two examples of tritanopic confusion colors, the foveal center is ineffective.

Wald (1967) himself provides convincing evidence for the absence of blue receptors in a central area of the fovea subtending a visual angle of only 7 or 8 minutes of arc. This in itself is almost sufficient to explain the tritanopic color confusions observed with a centrally fixated field subtending 15 minutes of arc, since the number of blue receptors in a 3 or 4 minute annulus immediately surrounding this central 8 minute area is likely to be quite small. Their contribution to any blue discrimination in a 15 minute field is likely to be minimal and might well be below threshold relative to the response from the much larger number of red and green receptors in the same area. This is especially likely in view of the steeper decline in blue response with blue receptors than the corresponding decline with red and green receptors. If it were possible to measure the blue response in comparison with the red and green responses in this central 15 minute field, it would therefore almost certainly be smaller than the ratio of the number of blue receptors to red and green receptors.

Once we move away from the central area of the fovea, two changes occur. The visual acuity falls off so that the two halves of a 15 minute field are seen less sharply. Color discrimination between the two halves is therefore impoverished. On the other hand, the number of blue receptors is increased, although they are still likely to be significantly fewer than the number of red and green receptors for the first degree or so from the foveal center. Small-field color matching and color discrimination can hardly be studied much beyond 2 or 3 degrees from the fovea because of the drop in visual acuity and need not, therefore, concern us. But within this distance the smaller number of blue receptors, coupled with contributions from Troxler's effect and the confusion effect reported by McCree, could be sufficient to explain the off-axis tritanopic tendency reported by Thomson and Wright. On this basis, small-field tritanopia away from the foveal center is something of an artifact.

If this analysis is correct, we can claim, I believe, that the tritanopia of the foveal center represents a unique visual phenomenon of no little significance. Wald suggests that its purpose may be to reduce the effect of chromatic aberration of the lens system of the eye and hence enhance the visual acuity in the preeminently important area of the central fovea. Almost certainly, though, there is some other mechanism also at work, since under normal conditions of observation the red fringes associated with chromatic aberration are as invisible as the blue fringes.

The Sensitivity to Blue and the Perception of Blueness

However we may interpret the evidence for small-field tritanopia, there is ample evidence to show that the eye responds differently to blue light from the way it responds to the other colors in the spectrum and that this often leads to at least a partial manifestation of tritanopia. Reference has already been made to the difference in area-threshold relationship for blue light which Brindley (1960) associated with extensive convergence of the blue cones on single optic nerve fibers. Wald (1967) however thinks the effect can equally well be attributed to the sparsity of the blue cones, so that a bigger area of the retina has to be stimulated before an effective number of cones can be brought into action.

Then there is the loss of blue-green discrimination and the absence of the sensation of blueness under conditions of steady fixation and image stabilization. This surely implies some special coding of "blueness information" for transmission to the visual cortex.

Quite recently West (1967) has reported some very significant results on luminance discrimination with a stabilized retinal image in which the effects of discrimination of sharp and blurred boundaries separating the fields under test have been studied. From our point of view the item of greatest interest is that the effect of blurring the boundary is very different for blue light compared with red, yellow, and green. West concludes that the effective size of receptor fields in the fovea (not just the foveal center) is about 2 minutes of arc for red, yellow, and green, but for blue is of the order of 7 minutes of arc. He attributes this to greater spatial interaction with blue light.

Another piece of evidence that the blue process is in some way very different from that of the red and green is in congenital tritanopia itself. If tritanopia is due to the absence of the blue receptors and protanopia due to the absence of the red receptors, why is tritanopia so very much rarer (perhaps about one in 13,000) than protanopia (one in 100 males)? And why is the genetic pattern of tritanopia, with its more or less equal incidence of males and females (Wright, 1952) so very different from the pattern for protanopia and deuteranopia (Kalmus, 1965)?

Perhaps the study of acquired color defects may be a promising route to a solution of these problems. There is some evidence, at least, that diseases affecting the outer layers of the retina, such as choroidal lesions and diseases of the visual cells tend to produce defects similar to the congenital tritan group. On the other hand, if the site of the lesion is in the nerve fibers or inner layers of the retina, a red-green color defect tends to develop (Birch and Wright, 1961; Verriest, 1963, 1964).

There is clearly more in this than just the presence or absence of blue receptors. Perhaps we have to await further information on the retinal network which the electron microscope is revealing. And we have to remember that the conditions which govern the perception of blueness also govern the perception of yellowness. If Wald had entitled his paper "Yellow-Blindness in the Normal Fovea," would he have assessed this in terms of the presence or absence of yellow receptors?

References

Birch, J., and W. D. Wright (1961). Color Discrimination. *Phys. in Med. and Bio. 6*, 3.

Brindley, G. S. (1960). *Physiology of the Retina and Visual Pathway.* Edward Arnold, London.

Clarke, F. J. J. (1960). A study of Troxler's effect. *Optica Acta 7*, 219.

Ditchburn, R. W. (1963). Information and control in the visual system. *Nature 198*, 630.

Helmholtz, H. v. (1867). *Handb. der Physiologischen Optik.* Leipzig. English translation by J. P. C. Southall (1924). Physiological optics. *Opt. Soc. Am. 2*, 148.

Kalmus, H. (1956). *Diagnosis and Genetics of Defective Color Vision.* Pergamon, London.

König, A., and E. Köttgen (1894). *Sitz. Akad. Wiss.* Berlin, p. 577.

König, A. (1903). *Ges. Abh. Physiologischen Optik.* Barth, Leipzig, p. 338.

McCree, K. J. (1960A). Color confusion produced by voluntary fixation. *Optica Acta 7*, 281.

McCree, K. J. (1960B). Small field tritanopia and the effects of voluntary fixation. *Optica Acta 7*, 317.

Moreland, J. D., and A. Cruz (1959). Color perception with the peripheral retina. *Optica Acta 6*, 117.

Pedler, C. M. H. (1965). CIBA Foundation Symposium on Color Vision. Churchill, London, p. 52. Also *Spectrum 16, 9*, 1965.

Thomson, L. C., and W. D. Wright (1947). Color sensitivity of retina within central fovea of man. *J. Physiol. 105*, 316.

Verriest, G. (1963; 1964). Les deficiences acquises de la discrimination chromatique. *J. Opt. Soc. Am. 53*, 185; *Mem. Acad. Royale de Medicine de Belgique 4*, 37, 1964.

Wald, G. (1967). Blue-blindness in the normal fovea. *J. Opt. Soc. Am. 57*, 1289.

West, D. C. (1967). Brightness discrimination with a stabilized retinal image. *Vision Res. 7*, 949.

Willmer, E. N. (1944). Color of small objects. *Nature 153*, 774.

Willmer, E. N., and W. D. Wright (1945). Color sensitivity to fovea centralis. *Nature 156*, 119.

Wright, W. D. (1952). The characteristics of tritanopia. *J. Opt. Soc. Am. 42*, 509.

Helen M. Paulson

Color Vision Testing
in the United States Navy*

This paper discusses the color vision tests currently used by the United States Navy and demonstrates how these tests work. It also presents some unpublished data on two color-vision tests which are not used by the Navy but are of interest because one, an Army color vision test, was proposed for Naval use and the other is often used in nonmilitary practice for advising color defective servicemen.

The Navy, like the other military services in this country, routinely tests its men for color vision. It is well known that color is used frequently in the Navy for purposes of signalling, identification, and communication. A few examples are color-coded electrical wires, ammunition, electronic components, gas cylinders, navigational lights, consoles and panels, and even color-coded file copies of Naval letters. Until September, 1951, normal color vision was a requirement for enlistment in the Navy and for appointment to its commissioned branches. Today all color defectives are accepted into the Navy. However, officer training programs and over half of the seventy-six enlisted men's training schools *do* have a color vision requirement for acceptance. The requirement is imposed because the men will be called upon to make critical judgments in the duties they will perform following the particular training.

In testing for admittance to these training programs and schools, the Navy uses two color-vision tests: various sets of pseudoisochromatic plates and the Farnsworth Lantern (hereafter referred to as FaLant). The FaLant was developed by the late Commander Dean Farnsworth,

* The opinions or assertions contained herein are the private ones of the author and are not to be construed as official or reflecting the view of the Navy Department or the Naval Service at large.

when he was attached to the Medical Research Laboratory at the Submarine Base in New London, to minimize the number of color defectives disqualified. Whereas the plates were designed to pass normals and fail *all* color defectives, the FaLant was designed to pass normals *and* the mild color defectives and to fail the moderate, severe, and dichromatic color defectives. With the introduction of the FaLant as the final test for admittance to these training programs and schools, 30 percent of the color defective population is being salvaged for training and duty (that is, the FaLant passes three of the ten men in a hundred who are color defective) by rating them as safe for making accurate color judgments.

Several hundred FaLants are now in use at Naval hospitals, training schools, air stations, and at many Armed Forces Entrance and Examining Stations. If an applicant passes the Plate Test, he need not be tested on the FaLant. However, if he fails the Plate Test, administration of the FaLant Test is required. In other words, failure on the Plate Test is not a cause for disqualification. Color vision disqualification for this training may be determined only by a FaLant test.

These two tests, the plates and the FaLant, are used at the Submarine Medical Research Laboratory (SMRL) in New London as part of its Color Vision Test Battery. They will be discussed again later in this paper when we describe this battery.

First, I want to explain what we mean by color defective vision. We mean that defect which is inherited, occurs in 10% of men and less than 1% of women, does not change in type or degree with age, has no relation to other visual defects, and for which there is no known cure or remedy. Many unsuccessful attempts to develop a cure have been studied by SMRL in the past—vitamin injections, training on plates, special filters, hypnotism, etc. Currently, there is some pressure for us to investigate the newest "cure"—an electronic device.

Again, what do we mean by color defectives? We do not refer to monochromats or achromats. Color defectives are *not* color blind, as they are so *very* often mislabelled. They *do* see color. Their problem lies in the fact that they cannot distinguish as many colors as a normal can; certain colors which appear different to normals look alike to them.

Figure 1 is Farnsworth's iso-color diagrams. The diagrams were drawn to illustrate typical color confusions and color differentiations made by the protan ("red" defective) and by the deutan ("green" defective). For the sake of brevity, the tritan ("blue" defective) will not be considered in this presentation. At New London we do have tests for tritans and we have examined eight of these rarely occurring individuals. As Dr. Wald said in his paper, the inheritance pattern is different for tritans from that for protans and deutans; we have a father-son and a father-daughter

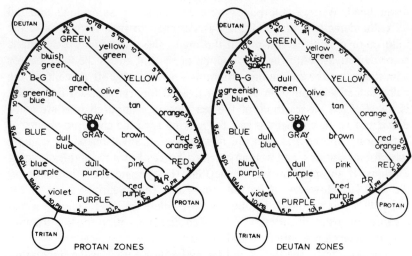

Figure 1. Farnsworth's Iso-Color Diagrams, showing the color confusions and the color differentiations which characterize two types of color defectiveness

occurrence, with the parent in each case being defective to a greater degree than the offspring.

Returning to Figure 1, two dimensions of color are represented: hue, running circularly on the diagrams (bluish green to green to yellow-green, etc.), and saturation, moving out from gray (dull green to green, dull purple to purple, etc.). Unrepresented on these two-dimensional diagrams is the third dimension of color, brightness. The lines in the two diagrams represent the family of lines which in theory converge on a single point. The point for protans lies off the diagram to the lower right, that for deutans lies off to the upper left. To protans and deutans, colors along any line passing through their respective points are maximally confusable. Confusions are not restricted to colors lying on the lines. To a color defective, any color may look like others lying in its neighborhood. The areas between the lines drawn in Figure 1 represent typical ranges of confusion for color defectives, and these areas are called zones.

The mild color defective confuses a few colors located close to each other in a zone or on a line; the severe color defective confuses many more colors, colors located close to each other and also colors located far apart in a zone or on a line. For example, the mild deutan confuses olive and tan, whereas the severe deutan confuses not only olive and tan but also green and red. The moderate deutan confuses gray and red-purple and gray and bluish-green, whereas the severe deutan confuses those, plus bluish-green and red-purple.

To facilitate the discussion of how the color vision tests work, I have labelled in Figure 1 the first two zones "#1" and "#2." Stimuli for screening tests employ colors in Zones 1 and 2 because these confusion zones are common to both the protan and the deutan. The remaining zones are not common to the two types and are the zones used in constructing tests which will differentiate color defectives as to type.

Now we move to the SMRL Test Battery for Determining Degree of Color Vision Defect. Table 1 presents this battery. The tests used are pseudo-isochromatic plates, the FaLant, the D-15, and the H-16. Normal trichromats pass all tests. The anomalous trichromats are categorized in our battery as mild, moderate, and severe. The mild color defective fails the plates, but passes the remaining tests. The moderate fails the plates and the FaLant. The severe fails the plates, the FaLant, and the D-15. The dichromat fails all four tests. To determine type of defect (protan or deutan), an anomaloscope is used for the mild and the moderate and the D-15 is used for the severe and dichromatic.

Table 1

SMRL Test Battery for Determining Degree of Color Vision Defect

	Pseudo-isochromatic		D-15 Test	H-16 Test
	Plates	FaLant		
Class I				
Normal Trichromat	Pass	Pass	Pass	Pass
Class II				
Mild Anomalous Trichromat	Fail	Pass	Pass	Pass
Class III				
Moderate Anomalous Trichromat	Fail	Fail	Pass	Pass
Class IV				
Severe Anomalous Trichromat	Fail	Fail	Fail	Pass
Class V				
Dichromat	Fail	Fail	Fail	Fail

In a moment, these four tests will be shown and explained but, before we leave this table, let me mention some incidence data based on the screening of high school boys in the New London area and the referral of some three hundred color defectives for test on this battery. Thirty percent of the color defectives were mild, 20 percent were moderate, and 50 percent were severe or dichromatic. The ratio of deutans to protans was 2.6 to 1.

Figure 2* is a pseudo-isochromatic plate in which a normal sees a number "5" and a color defective sees a number "2." The background dots are orange and the number "5" is composed of yellow-green dots. Orange and yellow-green lie in confusion Zone 1 and hence the "5" is not distinguishable to protans nor to deutans. Also in the plate is a number "2" composed of light pink dots. As can be seen by referring to Figure 1, pink and orange are readily distinguishable to protans and deutans, and hence they see the "2." Another plate is shown in Figure 3. To this plate, normals respond with a "3" and color defectives with a "5." Again, the background dots are orange. The lower half of the "3" is composed of blue-green dots and is distinguished from the orange dots by protans and deutans. The upper half of the "3" is composed of yellow-green dots and is not distinguished from the background dots by the color defectives. They instead follow the pink dots in the upper half of the figure and respond with a "5." For both of these plates, the normal responds to the larger color difference, orange and green, rather than the less striking difference, orange and pink.

Only the so-called reversible number plates have these "hidden" numbers. Many plates consist of numbers and backgrounds whose colors lie in the same confusion zone. In such plates, the color defective sees no number and responds with "nothing." The advantage of there being a few reversible-number plates in a set is that it aids detection of malingerers; a person *must* respond with either a "5" or "2" to the plate depicted in Figure 2.

Before we move to the next test in our battery, let me mention a few important facts about the plate test: (1) not all color defectives fail all the plates, and hence the entire selection must be administered; many of our Navy color defectives report that they passed the plates when only one or two plates were used; (2) the plates must be administered under Illuminant "C" (simulated daylight). One half of the test is the book of plates and the other half is the illuminant. Recently we found half of the testing facilities surveyed to be lacking the daylight lamp. As has been well-documented in the literature, deutans frequently pass the test when it is administered under incandescent light or sunlight. (3) A small percentage of normals will fail a set of plates, but such failures can be readily differentiated from color defective failures. They are caused by a tendency to complete numbers ("3" to "8," "5" to "6," etc.), by overall poor discrimination (as indicated by the Farnsworth-Munsell 100-Hue Test), by visual defects other than color, etc.

---

* Editor's note: Figures 2 through 11 have been omitted since to reproduce them in black and white would have been meaningless.

Figure 4 is a picture of the FaLant. It presents specially selected pairs of red, green, and white lights. Since two reports from our laboratory have recently been published on the FaLant, I shall explain how it works very briefly.

The red, green, and white lights are confusable to a severe color defective or dichromat because the stimuli were selected to lie in a confusion zone common to protans and deutans—Zone 2 in Figure 1. The severe or dichromat, therefore, tends to guess and make nonreproducible responses to these stimuli.

The moderate color defective does not confuse the strong red and green stimuli in the FaLant because they lie too far apart in Zone 2. He fails the test because he makes two types of reproducible errors: (1) he tends to call the white light the complement of the color with which it is presented (i.e., red-white is called red-green, green-white is called green-red, etc.); and (2) when lights of the same color, but differing in brightness by 50% (as in the FaLant), are presented to the moderate color defective, he tends to call them different colors. For example, a pair of green lights is called green-white, with the brighter green being called white. This occurs because his hue perception is so reduced that he resorts to brightness cues to help him judge colors.

The mild color defective's hue perception is reduced but slightly, and he does not resort to brightness cues and does not call the white light the complement of the color with which it is presented. The pairs of lights in the FaLant are correctly identified by him.

Our recent report presents data which show that the FaLant Test is a valid and reliable test, and so these topics will not be covered here. Before we move to the next test in our battery, however, I would like to show you a table from one of these reports which shows the FaLant error score as it is related to type and degree of color vision defect. In the FaLant Test, a pass score is an error score of 0–1 and a fail score is 1½–9. Table 2, based on the error scores obtained by 391 color defectives who failed the FaLant test, shows that the FaLant has excellent internal validity in that (1) the median error score systematically increases as the degree of color vision defect increases and (2) the median error score for each degree of defect for protans (who are additionally handicapped by their reduced luminosity curve) is always a little higher than that for deutans of the same degree of defect.

Table 2

Frequency Distribution of Scores of 391 Color Defectives Failing the FaLant
as Related to Type and Degree of Color Vision Defect

Color Vision Defect	N	1½	2	2½	3	3½	4	4½	5	5½	6	6½	7	7½	8	8½	Median
Deutan Type																	
Class III	96	11	12	11	14	13	12	7	6	8	2	0	0	0	0	0	3
Class IV	47	3	0	5	4	6	10	6	8	2	2	1	0	0	0	0	4
Class IV-V	41	2	0	0	4	4	5	7	4	3	5	3	1	2	1	0	4½
Class V	76	1	1	0	1	6	2	3	17	16	7	10	5	3	3	1	5½
Protan Type																	
Class III	45	5	8	6	3	8	4	3	4	2	1	0	1	0	0	0	3½
Class IV	35	0	2	1	2	4	5	4	7	0	4	4	1	0	1	0	4½
Class IV-V	30	0	1	0	6	0	3	0	3	4	2	4	3	4	0	0	5½
Class V	21	0	0	0	0	1	1	0	2	0	3	5	2	4	2	1	6½

$\Sigma N = 391$

Notes: 1. No subject obtained a FaLant score of 9.
2. All medians were computed to the nearest real score.
3. Class IV-V consists of men who were not administered test to determine whether they were Class IV or Class V.

Our next test in our battery is the Dichotomous-15 Test. Figure 5 shows how a severe or dichromatic protan would arrange the colored buttons by "hue." Normals, milds, and moderates arrange the buttons in color order from the starting button blue, to greenish-blue to bluish-green to green to yellow-green, etc., ending with purple. The severe or dichromatic protan puts the last button (dull purple) next to the starting button (blue) because they are in one of his confusion zones and look alike to him, whereas greenish-blue and blue are not and look different to him, and hence will not be placed contiguously. By referring to Figure 1, the remainder of the button arrangement by the severe or dichromatic protan, as depicted in Figure 5, can be readily understood. The severe or dichromatic deutan would arrange the D-15 buttons in another fashion, as can be deduced from the deutan zones diagram in Figure 1. To him, the greenish-blue and purple look alike and would be placed contiguously, etc.

Figure 6 shows the H-16 Test as the normals, mild, moderate, and severe color defectives arrange the buttons. The chromaticities of these pigments lie farther out on the color diagram than those of the D-15 test, and hence the test is easier for color defectives to pass. Figure 7 shows how a deuteranope (that is, a dichromatic deutan) would arrange the buttons. Again, comparing the arrangement depicted in Figure 7 with the deutan zones diagram in Figure 1, it is readily understood why he arranges them in such a fashion. Again, the arrangement that the protanope would make can be deduced from the protan zones diagram in Figure 1.

We have found good agreement between failure on this H-16 Test, which is rapid and easy to administer, and two anomaloscope tests for dichromacy. In classical usage of the anomaloscope, a color defective is defined as a dichromat if he matches the entire range of various mixtures of red and green filtered light appearing in one half of the test field to the yellow light which appears in the other half of the test field. The subject is able to adjust the brightness of the latter field. The other anomaloscope test is an SMRL innovation in which (1) the yellow filter is removed and replaced with the green filter and (2) a blue filter is inserted so that varying amounts of it may be mixed with the red filter for the purpose of desaturating the red. Dichromats find a match on this latter anomaloscope by adjustment of the brightness of the green field and of the saturation of the red field and will call this green-to-red match a yellow-to-yellow match.

In justification of our acceptance of milds as "safe for making accurate color judgments" and our rejection of moderates (and severes and dichromats) as "unsafe," another investigation was recently undertaken by SMRL. Figure 8 shows, on the C.I.E. (The International Commission on Illumination) Chromaticity Diagram, the 136 different chromaticities (each chromaticity shown at three brightness levels: 100%, 50%, and 25%) which were judged by normals and by color defectives. In this experiment, we used 40 normals, 10 mild protans, 10 mild deutans, 10 moderate protans, 10 moderate deutans, and 8 severe and dichromatic protans and deutans. This study was run in conjunction with the Naval Ship Research and Development Center, Annapolis, whose interest in the data is for a different purpose. Although detailed statistical analyses of the data from this experiment have not been completed, it appears they will support our acceptance and rejection standards.

Now let us return to the two tests mentioned in the introductory paragraph. Figure 9 shows the Armed Forces Vision Tester (an Ortho-Rater) with the Bausch and Lomb Slide Number 71-21-21 in position. This slide is used at the Armed Forces Entrance and Examining Stations (AFEES's) by the Army to test the color vision of those men who fail the plates. In the spring of 1966, the Bureau of Medicine and Surgery, Department of the Navy, informed us that, commencing 1 July, 1966, Naval enlistees would be examined by Army medical personnel due to the critical shortage of Naval medical officers. The Bureau desired our opinion of this B & L Color Vision slide. We purchased one and tested 44 color defectives on it in the few weeks we had before the deadline for decision. Though the N was small, the results on the 44 color defectives were so overwhelmingly indicative of a poor color vision test that we felt free to state that it was unsatisfactory for use on Naval enlistees and urged the continuation of placement of FaLants in the AFEES's.

The original N of 44 has now grown to 200 and exactly the same types of data emerge as emerged with the first 44. Table 3 shows the data on the 200 color defectives. It fails only 11% of the color defectives (or 1.1% of the male population). Note that it passes 87% of the severe deutans and 87% of the severe protans and it passes 57½% of the dichromatic deutans and passes all of the dichromatic protans.

Table 3

Results of 200 Color Defectives Tested at SMRL on the
Bausch and Lomb Ortho-Rater Slide Number 71-21-21

	Type of Defect				
	Deutan (126)		Protan (74)		
Degree of Defect*	Pass	Fail	Pass	Fail	N
Mild	50	0	21	0	71
Moderate	21	0	28	1	50
Severe	13	2	13	2	30
Dichromatic	23	17	9	0	49

* Based on SMRL Color Vision Test Battery

The comments we made in 1966 still apply. The Bausch and Lomb Ortho-Rater Slide Number 71-21-21 is an unacceptable test for the Navy's use for the following reasons:

1. Although it passes the color defectives who pass the Farnsworth Lantern, it also passes three-fourths of the severe anomalous trichromats and dichromats.
2. Of the 22 who failed the test, 18 failed because they called the red an orange or red-orange (all in the "red" family) and such failures would be vulnerable to the attack of those who claim the subject failed the test because of slight errors in color naming instead of defective color perception.
3. Although Army Regulation 601-270 states that the examinees are not to be advised in advance as to the colors used in the test, the knowledge that the three colors in the test are red, green, and yellow is bound to be easily spread, and this knowledge will enable most all color defectives to pass the slide, because they then promptly call "red" those three colors that look orange to them and look different to them from the three greens and the three yellows.
4. The Bausch and Lomb Slide is easily memorizable because the colors

cannot be exposed and judged individually; the colors are numbered and are permanently arranged in one set way.

5. It appears to be especially incongruous for protans since a protanope (dichromat) is more likely to pass it than a somewhat less severe protan, probably because to a protanope (with his characteristic reduced luminosity function) the red appears too dark to be called "orange" (the typical error of all who fail the test).

For presentation here I have also selected data on the "A.O. H-R-R Pseudoisochromatic Plates for Detecting, Classifying, and Estimating the Degree of Defective Color Vision," more familiarly known as the Hardy-Rand-Rittler (H-R-R). The published reports on this test involve a relatively small number of color defectives. This fact, coupled with the widespread use of the test in civilian practice, resulted in my decision to describe our unpublished data obtained from approximately seven hundred color defectives. We in New London aided Dr. Hardy, Dr. Rand, and Miss Rittler in the early work on these plates by testing several hundred color defectives on prototype models and by making resultant recommendations for improvement. The data shown today, however, are based on the production set only.

Before we look at the data, let me demonstrate, with the aid of Farnsworth's diagrams (Fig. 1), how this unique test works. Figure 10 is an H-R-R plate consisting of a background of gray dots in which there is imbedded a circle composed of red-purple dots and a triangle composed of pink dots. To the protan pink and gray look alike, but red-purple is differentiated from gray. He therefore responds with "circle." Figure 11 is another plate from the same test with a bluish-green circle and a blue-green triangle on a gray background. The deutan responds with "triangle" and the protan with "circle," as would be predicted from Figure 1.

The H-R-R test has been designed to estimate the degree of defective color vision by the amount of saturation required by the examinee to differentiate a color from gray. Failure on the less-saturated plates only identifies a mild; failure, as well, on the medium-saturated plates is assumed to be characteristic of a moderate color defective; failure on the highly-saturated plates, as well as the others, is indicative of a severe color defect.

In the discussion that follows, it is assumed that the results of a valid battery of tests, such as SMRL's, can be used to evaluate a single brief test, such as the H-R-R. In presenting our data, we are equating our SMRL mild category with the H-R-R mild category, our moderate category with the H-R-R medium category, and our severe and dichromatic categories with the H-R-R strong category. (The latter is permissible because it is not claimed that the H-R-R's strong category separates severe anomalous trichromats from dichromats.) It is recognized that the H-R-R's and SMRL's categorizations of color defectives into three degrees of defect would never be expected to agree precisely, but the classifications should agree to a large extent because the authors of the H-R-R made use of our battery in validating early versions of their test.

Table 4 shows how the H-R-R results relate to those from the SMRL battery classification in terms of degree and type of defect. Note that we have not as yet tested any normals on the H-R-R, and so we cannot make any statement as to how it works for them. As can be calculated from Table 4(A), 70% of SMRL milds were diagnosed as mild by the H-R-R, 62% of SMRL moderates were diagnosed as medium, and 51% of SMRL's severes and dichromats were diagnosed as strong. The obvious reason the H-R-R does not work better than it does in estimating the degree of defect is the small number of plates—five in the mild category, three in the medium category, and two in the strong category. There are other, but less important, reasons for the discrepancies between the H-R-R and SMRL classification with respect to degree of defect.

Table 4

A. O. Hardy-Rand-Rittler Test and the SMRL Battery (685 Color Defectives)

A. DEGREE OF DEFECT RESULTS

SMRL Classification	Hardy-Rand-Rittler Plates				
	Normal	Mild	Medium	Strong	N
Mild	25	176	49	2	252
Moderate	1	38	123	36	198
Severe and Dichromatic	0	10	105	120	235
Total N's	26	224	277	158	685

B. TYPE OF DEFECT RESULTS

SMRL Classification	Hardy-Rand-Rittler Plates				
	Normal	Deutan	Protan	Unclassified	N
Deutan	21	364	10	75	470
Protan	5	3	179	28	215

The data in Table 4(B) for H-R-R diagnosis of type of defect show that there is better agreement with the SMRL battery here than for degree of defect. As can be calculated from the data, 77% of our deutans were classified as deutan by the H-R-R and 83% of our protans were classified as protan by H-R-R.

Table 4 shows how the H-R-R results relate to those from the SMRL battery in terms of degree and type of defect separately. How did the H-R-R compare with our battery classification in terms of degree and type together? Of the 685 color defectives tested on the H-R-R, one half received the same classification on the H-R-R as they did on our SMRL battery of tests.

Incidentally, since the H-R-R production set was issued in 1955, we have tested three tritanomalous persons and one tritanope on the B-Y Defect Plates. The results were that all four tritans passed the two screening plates and the four diagnostic plates.

Test-retest reliability of the H-R-R Plate Test is presented in Table 5. Our N is not very large; 61 color defectives were retested on the H-R-R Test. Again, the data are presented separately: reliability with respect to degree and with respect to type. Here we see that the probability of like results on two administrations of the H-R-R is better with respect to degree than to type of defect (72% versus 59%, as can be calculated from the data in Table 5). For type and degree of defect combined, we find that, of the 61 retested, 26 received the same classification on retest as they had on test.

Table 5

Reliability of the A. O. Hardy-Rand-Rittler Test (61 Color Defectives)

A. DEGREE OF DEFECT RESULTS

	Retest			
	Mild	Medium	Strong	N
Test				
Mild	21	4	0	25
Medium	10	19	1	30
Strong	0	2	4	6

B. TYPE OF DEFECT RESULTS

	Retest			
	Deutan	Protan	Unclassified	N
Test				
Deutan	12	0	5	17
Protan	6	19	3	28
Unclassified	6	5	5	16

Using our battery classification for comparison, of the 61 color defectives retested on the H-R-R, 13 agreed with SMRL classification as to type and degree on both test and retest, 11 agreed on test but failed to agree on retest, 11 did not agree on test and did agree on retest, and 26 did not agree with the SMRL classification on either H-R-R test or retest.

In summary, we have found that the results from the H-R-R plates are not in good agreement with those produced by our battery. This is not surprising, in view of the fact that a battery of tests is developed and used, because to date no one single color vision test has appeared which will handle the complex task of classifying color defectives by type and degree of defect. We have reached the conclusion that the H-R-R certainly does not qualify as a substitution for our battery, nor does it make a unique contribution to our battery.

The other purpose of the H-R-R Test is to detect color defectives. Although, as stated previously, we have not examined any normals on the H-R-R, when we examine our data to see how well the H-R-R identifies color defectives as color defective, we find that 96% of the subjects classified as protan or deutan by our battery are classified as defective by the H-R-R plates. Thus, as a test for detecting color defective vision, the H-R-R has considerable merit.

Part III

Behavioral Aspects of Vision

Clarence H. Graham

Movement and Space Perception*

The topic of movement and space perception is, I believe, a most representative one for a program that deals to a considerable extent, as does the present one, with psychological subject matter. From the days of the classical psychophysicists to those of Wertheimer (1912) and the Gestalt theorists, movement has greatly preoccupied the thoughts and doctrines of psychologists, and the discussion of relations between space and movement, as taken up, for example, by Bourdon (1902), has set a very high standard of psychological analysis.

I have dealt with the topic of movement and space perception earlier in discussions (Graham, 1963, 1965) that differ from the present one in both data and concepts, and a most recent one (Graham, 1968) covers essentially comparable material. All of the accounts involve stimulus situations that range from the very simple to the quite complex. My present discussion presents new data obtained under simple conditions of movement in experiments directed to such questions as the following: what are some variables that govern movement under threshold conditions? Another discussion will consider movement in the frontal place and some of its consequences for space perception. In these latter matters I shall deal variously with some new, recent, and fairly old investigations in which I have participated.

Differential Movement and Real Movement

Despite the complexities which may enter into the various aspects of movement, it is nevertheless true that elementary components of this topic

* This paper was prepared under Contract Nonr 266(46), between Columbia University and the Office of Naval Research. The work is part of the program of Project No. 142-404. Reproduction in whole or in part is permitted for any purpose of the United States government. Support has also been provided by the Institutional Scientific Research Pool of Columbia University.

179

are analyzable ensembles of stimulus conditions. Differential movement is, in fact, the prevalent characteristic movement with which we are concerned. In the absence of differential movement between a test and reference object, one sees in darkness a phenomenon that occurs with a single, small, illuminated, stationary object: the autokinetic effect. This effect is probably a source of factors that raise movement threshold by an order of magnitude or more when no reference stimuli are present in the field (Pearce and Matin, 1966). The presence of a reference line near the moving stimulus is an important aid to the discrimination of real movement of, for example, a not too bright line. The reference line may move or it may remain stationary. As to the object itself: if the eye follows it as it moves, then the retinal conditions of stimulation (except for irregularities in following) are comparable to the situation that exists with the object stationary. In any case, further analysis of simple conditions beyond the realm of a totally dark field takes us to topics that underlie basic and elementary principles of real movement.

The specific experiments here discussed are concerned with the real movement provided by single visual stimuli under conditions of constant rate. Certain influencing parameters of both displacement and velocity threshold have been examined. For help in these experiments I am especially indebted to my colleagues Barbara Mates and Yun Hsia.

Velocity Threshold

In a first sequence of experiments, determinations were made on the velocity threshold. The subjects in the experiments were presented with a vertically descending moving stimulus whose rate of descent could be varied by means of a hydraulic regulating system. A circular background field, 13° in diameter, was arranged (in the main experiment) so as to be in the same frontal plane as the moving stimulus; the latter appeared on the background as a brighter small rectangle on the large, dim, background field. In the main experiments the luminance of the moving stimulus pattern was sixteen times the background luminance. Each trial was for a period of six seconds. The subject fixated the object as it moved well into the visual field; he regarded the leading edge, responding on the presence or absence of movement any time during the remainder of the six second stimulus duration.

Following a series of preliminary experiments, we performed more elaborate ones on the following topics.

Number of reference lines. The first experiment concerned the way in which the number of stationary reference lines influenced the threshold. The reference marks in the field shown in Figure 1 consisted of illuminated

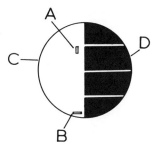

Figure 1. Conditions for determining how the number of stationary reference lines influences the threshold. (A represents the moving stimulus; B, the fiducial mark not used in these experiments; C, outline of visual field; D, design of reference slits and their background.)

slits, each of width five minutes, on a dark straight-edge separated by intervals of 0.83, 1.66, and 3.32 degrees. The slits were cut perpendicular to the straight edge and were extended to intersect the circumference. In addition, the straight-edge without slits was used in one experimental condition. The moving stimulus was a vertical rectangle (6.9 × 17.2 min) whose horizontal separation from the straight-edge side to each vertical pattern was about ½ degree. The results indicate that the velocity threshold decreases as the number of fiducial lines, in fact objects, in the field increases. This result, which might be taken to be in line with some results, in a somewhat different context by Leibowitz (1955B), is interpreted to mean that the probability of appropriate comparisons of moving stimuli with reference marks increases as the number of reference marks increases. It also seems reasonable to assume that the improvement is a function of interaction effects.

Length of stimulus line. The second experiment involved determinations of the velocity threshold as influenced by the length of stimulus line. In this experiment it was found that varying the length of moving line, from about 17 to 52 minutes, gave a change in threshold by a factor of about 2, depending on the subject. Figure 2 gives average data for five subjects. Threshold rate increases as stimulus length increases. This result is roughly comparable to some findings on the effects of area in a different reference by Brown (1931A).

Luminance. The third experiment showed, in line with recent experiments by Leibowitz, that velocity threshold decreases as stimulus intensity increases (Fig. 3).

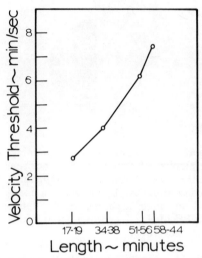

Figure 2. Velocity threshold as a function of length of stimulus line. (Threshold rate increases as stimulus length increases.)

Displacement Thresholds

The experiments on displacement threshold were conducted by means of a cathode ray equipment. Arrangements were such that a background field of white light (color temperature 5500°K) at about 32 ft.-lamberts and with a diameter of 14.3 degrees, had superimposed upon it, four bluish dots approaching the color of the oscilloscope trace. They formed the tips of an imaginary stationary square (3.7 degrees on the diagonal) that served for fixation. The subject regarded the center of the fixating area and responded on the basis of just-discriminable movements of the small, circular cathode ray trace (usually of 1.42 min dia.).

Effect of rate of displacement and luminance. The experiment involved the determination of a threshold displacement of the cathode ray trace from a stationary condition. The results show that (1) as rate increases through a range from 8.2 min per sec to 69.4 degrees per sec, threshold displacement decreases to a minimum value, about 1 min at a rate of 2.0 degrees per sec, beyond which value displacement remains unchanged (see Fig. 4). The displacement versus rate curves are displaced on the threshold axis so that the higher intensity condition of the moving stimulus gives a curve that is lowest on the threshold axis; lower intensities give successively higher thresholds.

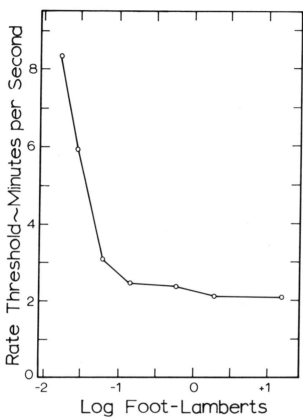

Figure 3. Velocity threshold as a function of stimulus intensity.

Effect of length of stimulus line. Experiments were also done on length of stimulus line as it influences displacement (Fig. 5). Conditions were the same as for previous background situations. The stimulus length was provided by varying the length of vertical sweep (sweep frequency 1325 cycles per sec) and displacements were provided by voltage changes on the vertical plates. Stimulus lengths varied between about 2.8 and 45 min of visual angle. At a prevailing rate of displacement of either 8.2 min or 7.1 degrees per sec little variation of displacement threshold due to length of line could be shown (see lowest pair of lines in Fig. 5). However, at the very slowest rate of 3 min/sec, the effect of length is shown for subject R.S.; the longer the line the higher the threshold. Other subjects show a comparable result.

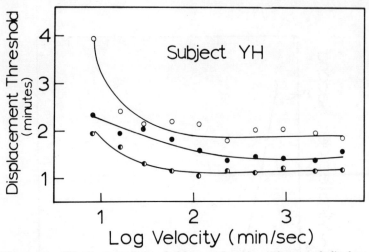

Figure 4.   Displacement threshold as a function of rate of displacement and luminance. (Threshold decreases to a minimum value at a
rate of about 2 degrees per second. Thereafter, displacement remains
unchanged. The curves are displaced on the threshold axis; the threshold curve is lowest for the high intensity condition and highest for
the low intensity.)

Some Theoretical Considerations

It is certainly true that the eye does not exhibit perfect following of a
moving target.

"Slippage" displacements i.e., errors in tracking, occur between any
arbitrarily specific object image point and a retinal point during visual
pursuit. This factor is especially important in determinations of velocity
thresholds; they are probably not very important in determinations of
velocity thresholds; they are probably not very important in the case of
displacement determinations. In any case, factors that exist during both
types of determination involve changes in retinal positions of visual excita-
tions especially at the leading and trailing edges of the object image. It
will not do now to raise uncertainly relevant questions concerning either
Sherrington-type feedback or Helmholtz-type voluntary components.

In another connection (1963) I have speculated about some of the
possible visual bases for elementary movement discrimination. It would
probably be useful to consider the coding of movement in the retina and
cells in the brain that respond to specific directions of movement. Despite

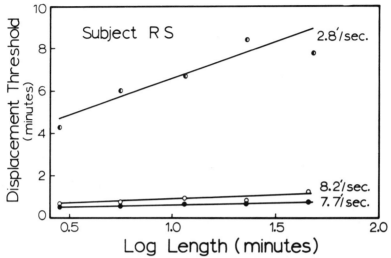

Figure 5. Displacement threshold as a function of length of stimulus line for subject R.S. (Stimulus length seems to have only a small effect at prevailing rates of displacement of 8.2 minutes or 7.1 degrees per second. However, at a rate of 3 minutes per second, the longer the line, the higher the threshold.)

the presence of cells that selectively signal movement in a restricted direction, it may be hypothesized that within overlapping receptive fields, various retinal ganglia, appropriately aligned and selective as to stimulus direction, may, in sequential combination, signal direction and distance of movement. The component effects centered at successive loci as a *point* stimulus moves in the visual field, may be thought of as constituting a contrast wave. Rodieck (1965) has presented some interesting ideas concerning movement determinants in the cells of the cat's retina.

Stimulation becomes more complex when the stimulus is not a point of light. Consider, for example, what might happen with a very narrow rectangle of light oriented in the direction of movement. Transient effects occurring at the leading and trailing edges of the stimulus may be expected to give an overall pattern different from those that occur with point sources.

Stimulation by the leading edge may, to consider a possibility, give rise to inhibitory contrast effects such as those discussed for point stimuli, and subsequent stimulation must then act on the neural background activity that is provided by the leading edge. Stimulation after the passing of the leading edge will be of a sort that maintains, due to the continued constant action of light, a relatively uniform level of brightness. Finally, to continue the example, an inhibitory background effect may be established as the trailing edge passes over the receptors.

The results on the effect of intensity, described earlier, show that as luminance increases, the displacement and movement thresholds decrease. This consideration may argue against the position of Bourdon (1902), for example, that perception of movement depends mainly on feedback mechanisms and little if at all on sensory events. The effect of luminance might be interpreted to mean that, because of factors related to magnitude of signal as it increases in each elementary sensory area with increase in intensity, the propagated movement across the visual field is seen to move faster.

The most interesting results obtained in the studies of velocity threshold and displacement threshold are concerned with the effect of size of moving object. The effect is clear for the velocity threshold where the prevailing rate is the one, in fact, that occurs at threshold. Threshold for various subjects increases by a factor of about 2 with an increase in length of stimulus line from about 17 to 52 minutes.

The story is more complicated for displacement threshold. For stimuli moving at rates of 8.2 min/sec to 7.1 degrees/sec there is little difference in threshold displacement. These rates presumably deal with the almost simultaneous perception of beginning and end of the displacement. However, at a very slow rate, 3 min/sec for one subject, threshold occurs with a displacement of about 4.5 min; thus, time of movement is about $\frac{1}{5}$ sec. In a word, the threshold rate during displacement is about 3 min/sec, a rate of the same order as the one holding for the velocity threshold. This may mean that displacement and rate thresholds are determined at nearly the same values of rate of movement, but it must be pointed out that the beginning and end conditions in the two discriminations are different. In any case, it could mean the minimum movement can be appreciated through a given, possibly constant, interval of the order of 3 minutes in a duration of 1 second on the retina. This figure is more than one minute of angular separation for nearly maximum acuity, which, in turn, is near the figure of minimum separation between adjacent cones. However, the discrimination is one of succession (not simultaneity as for acuity) and transient interaction effects, involving the leading and trailing edges; edges are undoubtedly important limiting factors, as in apparent movement.

Increasing the size of a bright object on a dark background at prevailing low rates increases both the displacement and rate threshold. This may mean that an increase in length of the moving stimulus provides, over its length and in adjacent unstimulated areas (particularly at the leading and trailing edges), an increase in the inhibitory background. The inhibitory effect at a given point is, presumably, increased by virtue of the greater number of converging units that exist with longer lengths of stimuli than with shorter. The data of Diamond (1962) on effects of area on contrast with stationary stimuli are in line with this interpretation. In consequence of this state of affairs, inhibitory resistance to the propagated contrast wave provides a slower appearance of movement for large areas than for small.

The experiments that I discussed have uniformly involved a bright line on a dimmer background. What may happen in a situation where the test stimulus is a black line on a bright background?

It may be that comparable results on movement are found with either positive or negative conditions of contrast, but we are not sure that this is the case. We have done some preliminary investigations on this problem, but I do not now wish to state a definite conclusion. It will be important, on theoretical grounds, to do further experiments on this topic, especially as related to effects of line length. The importance of the effect of contrast may be suggested when we consider the conditions prevailing at the leading and trailing edges, depending on whether or not the stimulus-object is bright on a dark background or dark on a bright background. In one situation the movement provides a darkening effect at the leading edge and a lightening effect at the trailing edge. In the other condition the reverse effect holds.

Monocular Movement Parallax

I should now like to consider some more complex effect than velocity and displacement thresholds. The discussion will consider movement in the frontal plane and some of its consequences for space perception.

Let me begin the account with a bit of personal history. Early in World War II, just before the ascendancy of radar, I was doing research on various problems of space perception. Although particular attention was paid to stereoscopic range finding, other questions arose concerning, among other things, monocular movement parallax. Monocular movement parallax is, as you know, the apparent change in the relative positions of two objects seen in the monocular field under conditions of prevailing movement. It seems strange that, although movement parallax effects have been known for centuries, precise psychophysical data on their characteristics were sparse or nonexistent. It was not until after the war that I was able to deal with the topic from a fairly general point of view, and a study of monocular movement parallax appeared in 1948 (Graham, Baker, Hecht and Lloyd, 1948).

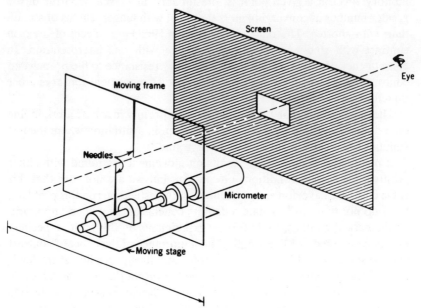

Figure 6. Apparatus for determining monocular movement parallax thresholds (from Graham, 1965).

The apparatus on which the determination of threshold was made is represented in Figure 6. The subject adjusts (in the straight-ahead direction) the lower needle attached to the micrometer until that needle appears to the subject to be in the same frontal plane as the fixed upper needle. The moving stage, containing the needles and accessory equipment, moves at constant rate in the frontal plane.

By means of this equipment one can, by computation of the parameters of the system, determine the threshold relative monocular movement as a differential angular velocity in sec of arc per sec as dependent on the prevailing rate of movement in the frontal plane. The subject is instructed, except under certain conditions, to follow with his eye the movements of the needles and to fixate the small region of separation between the two needles. Thus, his eye follows the movement of the needles and he adjusts the lower needle until the two needles appear to be in the same frontal plane. The threshold differential angular velocity is proportional to the depth difference of the needles multiplied by rate of needle movement in the frontal plane.

It turns out experimentally that the threshold movement parallax varies with background luminance, with the rate of the prevailing standard needle movement, and with visual axis of movement. One thing is sure: small differences in angular velocity may be discriminated. The thresholds are small indeed, being under optimum conditions about one min of arc per sec. Similar results were obtained by a method directed toward another objective by Aubert (1886) and in more recent experiments by Zegers (1948).

One matter is of practical and theoretical significance: the fact that for certain purposes, depth perception specified by these thresholds can be equal to and more general than depth provided by stereoscopic vision. Thresholds for movement parallax vary on different axes of movement. Thresholds are lowest on the horizontal axis, and highest on the vertical axis. Nevertheless, good depth discrimination is possible, even on the latter axis. On the other hand, with stereoscopic vision, depth difference discrimination is essentially impossible on the vertical axis.

Discussion. The discussion of monocular parallax in the particular situations that I have described is a specific topic in an extremely general context: the area of object perception (or naming) under conditions of movement. It is certainly true that movement of objects changes perception and, in fact, causes "new" phenomena, as shown by the different descriptions that may be applied to the same object before and during movement. Here one may think of Ames' trapezoidal window, to be discussed later. In other cases, the response may not change during movement and we encounter a form of "constancy." One might well, at this point, discuss constancy-producing factors, but the topic is one that must not now deter us. Let us move to the next consideration.

Depth by Kinetic Movement Perception

A demonstration of depth perception as dependent on movement and one that has very general implications is given by the kinetic depth effect studied by Miles (1929), Metzger (1934), and most recently and extensively by Wallach and co-workers (1953a, b; 1963).

Consider some changes in angular relations that are necessary conditions for seeing depth as dependent on movement.

A three-dimensional figure, e.g., a wire frame in the shape of a cube balanced on one corner, rotates in rays from a point source so that its shadow falls upon an opal glass screen. As viewed from the other side of the screen, a subject sees the shadow of the wire frame on the opal glass. However, the shadow does not appear to move in jumbled disarray in the frontal plane. Rather, due to differential angular velocities existing between

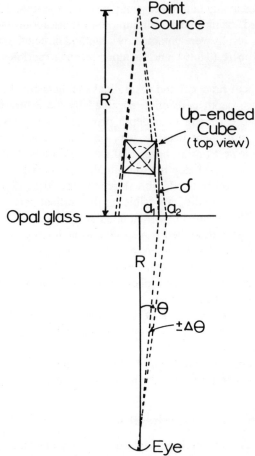

Figure 7.   A top diagrammatic view of conditions for the demonstration of depth by kinetic movement perception.

arbitrarily selected points on the frame, the silhouetted object appears to move in third dimension. The figure appears to be a rotating up-ended cubic framework.

A top diagrammatic view of the equipment used for this sort of demonstration is given in Figure 7, which shows light projected through the rotating cubic framework onto an opal glass screen. For purposes of simplicity we shall restrict our discussion to shadows a_1 and a_2 of the adjoining apices of the framework cast on the opal glass. Other parts of the cubic framework, shadowed on the screen, of course might be examined. The essential thing about the treatment is that the arbitrarily selected shadows a_1 and a_2 subtend a visual angle $\Delta\theta$ at the eye; as the cube rotates the distance between the shadows, and hence $\Delta\theta$, changes. The geometrical conditions that apply in this situation (where the point source is not too close to the framework) are represented by the expression $\delta \propto (\theta \cdot \Delta\theta)$. In Figure 7, θ is the angular position of a_1 with respect to straight-ahead regard; $\Delta\theta$ is the angle subtended at the eye by the distance between the point source and the opal screen; and δ is the "implied" depth difference, equivalent to one in real space given by an object that gives retinal projections θ and $\Delta\theta$.

It is sufficient to say that, as $\Delta\theta$ varies with time, so also do functions of θ. The essential thing is that, as in stereoscopic vision where depth differences are seen, owing to binocular differences in image shape, so the kinetic movement differences are seen because of temporal variations in a monocularly viewed image. In both cases the depth figures present views to the eye that are retinally equivalent to a view of objects in real depth.

Ames' Trapezoidal Window

Ames' window has been studied by many psychologists, both experimental and social. In the latter group, Cantril (1950) and Allport and Pettigrew (1957) have studied and discussed cultural influences on responses to rotations of the trapezoidal window. Among the experimentalists, the pioneer work and discussions of Ames (1951), Pastore (1952), and Ittelson (1952) have been influential in setting programs of research. Most recently, Zegers (1964) has studied the responses of subjects under several conditions of stimulus variation. It is a general finding that the trapezoidal window does not usually elicit responses of a final uniform sort until some time after it is initially presented to a subject. At first the subject usually reports that the window rotates, and only after a while does he say that the figure gives the oscillating back-and-forth movement which seems to characterize it. The number of presentations required to attain a uniform response varies from subject to subject but, in general, as Zegers has shown, it is possible, under stimulus conditions which he has done a great deal to specify, to prescribe a condition where subjects report a uniform effect 100% of the time.

The usual trapezoidal window (Fig. 8) consists of a flat surface cut out of wood or metal, painted to resemble a window with its panes of glass, and other characteristic features. One side of the window is larger in vertical dimension than the other. The window rotates at a constant rate about a point that is nearer the short end than the long end. As the subject looks at the window for some time he observes that, instead of

Figure 8.   The usual trapezoidal window, photographed in the fronto-parallel plane.

seeming to rotate in a circle, the window appears to oscillate back and forth through 180°, centering about a vertical axis in the frontal plane (Graham, 1963, 1965).

It should be noticed that a very favorable condition for viewing the window is as a silhouette in the frontal plane. In this situation it provides a striking example of kinetic movement. The simplified relations between environment and test object, owing to their simultaneous presentations in a frontal plane or possibly their exaggerated relations due to shape distortions attributable to a point source, may act to reduce sources of confusion. In consequence, the oscillating movement is seen more clearly in the frontal plane projection than it is in the real space situation. It is for this reason that I have demonstrated the window in the frontal plane silhouette. This means, of course, that secondary cues to space, such as shadows, etc., that are painted on the window are of no consequence and need not be used to provide the appropriate response.

The stimulus conditions providing this type of apparent movement may be discussed in terms of an analysis based on two types of cues: (1) the

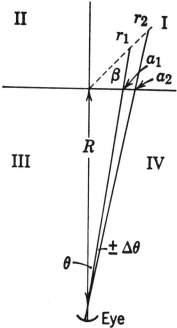

Figure 9.   Diagram of conditions in the movement of Ames' trapezoidal window (from Graham, 1965).

differential angular velocities existing between selected points on the surface of the window, and (2) other monocular cues.

(1) Let it be supposed, as represented in Figure 9, that the trapezoidal window turns about its point of rotation. Two arbitrary points, r_1 and r_2, not too far apart and lying in the same horizontal plane, occur on the surface of the window. It is supposed that the eye follows these points, i.e., θ varies so as to maintain them in central fixation. As the window rotates, the frontal plane projection (for Quadrants I and IV) of the distance between points is $a_1\, a_2$ and this distance subtends $\Delta\theta$ at any moment. (For Quadrants II and III, the distance comparable to $a_1\, a_2$ [not shown] is $a_1'\, a_2'$.) It is obvious that the conditions existing in the frontal plane very nearly represent the conditions of a projection on the retina.

(2) It is to be observed that the angle $\Delta\theta$ subtended at the eye by the points $r_1\, r_2$ changes with a differential angular velocity which, for *counterclockwise* rotation of the window, is negative for positions of the points behind the frontal plane in Quadrant I or positive for points in front of the frontal plane in Quadrant IV. (Of course, the other end of the window

which can exist in Quadrant III or II has angular relations symmetrical with those of I and IV.) Thus differential angular velocity gives a cue that is ambiguous; the subject cannot tell whether the points r_1 and r_2 on the window are approaching him in Quadrant IV or moving away from him in Quadrant I; the same sign and quantity of differential angular velocity apply in both quadrants.

(3) Under these circumstances, how does the subject see the movement of the window? A first tentative answer has suggested that he resolves the ambiguity of movement parallax cues by depending on perspective cues. The short end of the window is always seen as farther from the subject than the long end; and the ambiguous parallax cues are interpreted as in harmony with perspective cues.

With due regard to these various factors I propose the following description: conditions of angular differential velocity are such that the short side seems to move alternately toward and away from the subject behind the frontal plane with a periodic oscillating motion over an angular range of 180° in Quadrants I and II; the long end seems to move in a comparable manner through Quadrants III and IV. Thus, the oscillation seems to be a movement through a half circle and return at a rate which is the rate of physical rotation.

This account has some points in common with Ames' or Pastore's description. Guastella (1966) has presented a further version. My conclusion concerning the resolution of ambiguity by perspective cues is still tentative and depends, among other things, on further examination of some conclusions by Day and Power (1963) (which have also been discussed by Hershberger [1967]). Day and Power observed that various irregular and regular figures do, when rotating, seemingly change direction "spontaneously" and presumably irregularly. They take the fact of change of direction to be primary in the perception of any rotating configuration. From their point of view, then, the reversal is the basic, primitive influence in rotation. Oscillation is not controlled by perspective factors, but rather by "spontaneous" processes. More work is required on this analysis. Certainly we should like to learn what controls the reversal and, in the case of Ames' window, why it is so regular and "tied" to the angular relations of rotation.

These observations and speculations, will, I think, conclude my discussion. I hope that they will help to delineate some simple aspects of movement and their ramifications into perceptually more complex situations. It would probably not be wrong to say that movement variables pervade much of visual theory. Many more serious analyses will surely be needed for further understanding of this field which has major implications for all aspects of vision and visual perception.

References

Allport, G. W., and T. F. Pettigrew (1957). Cultural influence on the perception of movement: the trapezoidal illusion among Zulus. *J. of Abn. and Soc. Psychol. 55,* 104-113.

Ames, A. (1951). Visual perception and the rotating trapezoidal window. *Psychol. Monog.* No. 324.

Aubert, H. (1886). Die Bewegungsempfindung. *Arch. ges. Physiol. 39,* 347-370.

Basler, A. (1906). Uber des Sehen von Bewegungen. I. Die Wehrnehmung kleinster Bewegungen. *Arch. ges. Physiol. 115,* 582-601.

Bourdon, B. (1902). La perception visuelle de l'espace. Librairie C. Reinwald, Paris.

Brown, J. F. (1931A). The visual perception of velocity. *Psychologische Forschung 14,* 199-232.

Brown, J. F. (1931B). The thresholds for visual movement. *Psychol. Forsch. 14,* 249-268.

Cantril, H. (1950). *The "Why" of Man's Experience.* Macmillan, New York.

Day, R. H., and R. P. Power (1963). Frequency of apparent reversal of rotary motion in depth as a function of shape and pattern. *Australian J. Psychol. 15,* 162-174.

Diamond, A. L. (1955). Foveal simultaneous contrast as a function of inducing field area. *J. Exp. Psychol. 50,* 144-152.

Fisichelli, V. R. (1945). Effect of rotational axis and dimensional variations on the reversals of apparent movement in Lissajou figures. *Amer. J. Psychol. 58,* 530-539.

Graham, C. H. (1963). On some aspects of real and apparent visual movement. *J. Opt. Soc. Amer. 53,* 1015-1025.

Graham, C. H. (1965). Perception of movement, in C. H. Graham (ed.). *Vision and Visual Perception.* Wiley, New York.

Graham, C. H. (1951). Visual perception, in S. S. Stevens (ed.). *Handbook of Experimental Psychology.* Wiley, New York.

Graham, C. H., K. F. Baker, M. Hecht, and V. V. Lloyd (1948). Factors influencing thresholds for monocular movement parallax. *J. Exp. Psychol. 38,* 205-223.

Guastella, M. J. (1966). New theory on apparent movement. *J. Opt. Soc. Amer. 56,* 960-966.

Hershberger, W. A. (1967). Comment on "apparent reversal (oscillation) of rotary motion in depth." *Psychol. Rev. 74,* 235-238.

Ittelson, W. H. (1952). *The Ames Demonstration in Perception.* Princeton University Press, New Jersey.

Leibowitz, H. W. (1955A). Effect of reference lines on the discrimination of movement. *J. of Opt. Soc. Amer. 45,* 829-830.

Leibowitz, H. W. (1955B). The relation between the rate threshold for the perception of movement and luminance for various durations of exposure. *J. Exp. Psychol. 49*, 209-214.

Metzger, W. (1934). Tiefenerscheinungen in optischen Bewegungsfedlern. *Psychologische Forschung 20*, 195-260.

Miles, W. (1929). Figure for the windmill illusion. *J. of Gen. Psychol. 2*, 143-145.

Pastore, N. (1952). Some remarks on the Ames oscillatory effect. *Psychol. Rev. 59*, 319-323.

Pearce, D., and L. Matin (1966). The measurement of autokinetic speed. *Canadian J. of Psychol. 20*, 160-172.

Rodieck, R. W. (1965). Quantitative analysis of cat retinal ganglion cell response to visual stimuli. *Vision Research 5*, 585-601.

Wallach, H., M. E. Moore, and L. Davidson (1963). Modification of stereoscopic depth perception. *Amer. J. Psychol. 76*, 191-204.

Wallach, H., and D. N. O'Connell (1953). The kinetic depth effect. *J. Exp. Psychol. 45*, 205-217.

Wallach, H., D. N. O'Connell, and U. Neisser (1953). The memory effect of three-dimensional form. *J. Exp. Psychol. 45*, 360-368.

Wertheimer, M. (1912). Experimentelle Studien über das Sehen von Bewegung. *Zeitschrift für Psychologie 61*, 161-265.

Zegers, R. T. (1948). Monocular movement parallax thresholds as functions of field size, field position, and speed of stimulus movement. *J. of Psychol. 26*, 477-498.

Zegers, R. T. (1964). The reversal illusion of the Ames trapezoid. *Transactions of the New York Academy of Sciences 26*, 377-400.

Richard Held*

Vision and Movement

I should like to bring your attention to a set of problems which I believe have been at least partially neglected by the traditional psychophysical and neuropsychological approaches to vision. An increasing amount of evidence, accumulated in recent years, points to dissociation in the processing of visual stimulation according to its function: whether it serves for either brightness discrimination, hue discrimination, detection of motion, pattern recognition, or localization in space. In the case of spatially distributed stimulation, there seem to be at least two distinguishable modes of processing, one embodied in the system for the visual guidance of movement, and the other in the analysis of the spatial patterning (form) of that stimulation. This distinction has been of particular interest to me because the former system appears to be the one which is modifiable to a significant extent and which depends upon contact with the sense-stimulating environment for its earliest perfection and continued maintenance (Held, 1965, 1968; Held and Hein, 1967; Held and Freedman, 1963). The latter system appears to be relatively fixed from birth.

I propose to review some of the evidence for this dissociation of function and to conclude with a speculative interpretation of its significance. The evidence comes from two sources:

1. Neurobehavioral research, and
2. Studies of the effects of deprivation of vision and motility in neonatal animals and of visual rearrangements in man.

* Research of the author was supported by the National Institute of Mental Health (Grant MH-07642) and the National Aeronautics and Space Administration (Grant NSG-496).

Neurobehavioral Research

There are at least three lines of evidence which indicate the existence of separable neural systems for processing visuospatial information. The first of these consists of studies of split-brain animals, monkeys in particular. A normal monkey will readily learn to choose one of two forms using one eye while the other is occluded. When tested with the formerly occluded eye, he will immediately make the correct choice. The split-brain animal, of interest here, is prepared by midline sectioning of optic chiasm, corpus callosum, and other forebrain commissures joining the two cerebral hemispheres. As a consequence of this operation the cerebral projections originating from each eye are limited to the hemisphere on the side of that eye. Neurally-encoded information derived from stimulation of one eye is then accessible to the opposite hemisphere only by way of midbrain and lower level commissures. Such an animal easily learns pattern discriminations with one eye but, in general, fails to show evidence of this learning when the other eye is tested (Sperry, 1961, 1964). Apparently whatever change in learning has been effected by means of the one eye and hemisphere is simply not available to influence behavior controlled by the contralateral eye. In addition, direct tests of the animal's ability to compare two patterns when each is presented separately to one eye show great deficits when the results are compared with those obtained from normal animals (Trevarthen, 1968; Hamilton, Hillyard and Sperry, 1968). These experiments lead one to conclude that interhemispheric communication of the information derived from patterned stimulation of one retina is precluded by section of the forebrain commissures.

In contrast to the failure of interocular transfer of information about pattern in the split-brain animal, little if any loss appears in the visual control of movements of parts of the body, including the eyes, which demand cooperation of the two sides of the brain (Myers, Sperry, and McCurdy, 1962; Gazzaniga, 1966). There are few observable deficits in the control of reaching using any one of the four eye-hand pairs (Hamilton, 1967). The analogue to testing for the interocular transfer of a learned discrimination consists in testing for interocular transfer of altered eye-hand coordination. When the hand is viewed through laterally displacing wedge prisms, an alteration of the direction of reaching occurs which outlasts the presence of the prism. This sort of adaptation has been used to test for interocular transfer in split-brain monkeys (Hamilton, 1967; Bossom and Hamilton, 1963). In an extensive series of experiments, Hamilton found that interocular transfer of the adapted reach was as efficiently performed by split-brain monkeys as by normal control animals.

Comparison between the two types of experiment on split-brain animals shows that the central processing of visual stimulation designed to control movement can be partially dissociated from that involved in the analysis of pattern. The former seems to entail a vertical integration between higher and lower centers within the nervous system consistent with communication across commissures of the midbrain and brain stem, whereas the latter is carried on primarily within the cerebral hemispheres and requires the forebrain commissures for interhemispheric transfer.

The studies of split-brain monkeys are consistent with another line of investigation which has recently been reviewed. For the last few decades the dominant view of the visual performance of monkeys missing striate cortex has asserted that they can discriminate little more than differences in light flux available at the retina; that in this regard they behave as if their eyes served essentially as light meters and they are pattern blind (Kluver, 1942). However, in recent years several investigators have reported that these animals are, in fact, capable of a considerable range of visually guided behavior (Humphrey and Weiskrantz, 1967; Denny-Brown and Chambers, 1958). In particular, they are capable of reaching for an isolated object, viewed in dim background illumination, provided it has either been moving or is intermittently illuminated. These findings suggest that the control of visually guided movement, found unaffected by section of forebrain commissures, can in fact be sustained in the monkey without mediation by the striate cortex thought to be the primary cortical projection of the visual nervous system.

Some further light may be cast on the performance of the destriate monkeys by recent studies of the role of midbrain structures involved in visual performance. In a remarkable series of experiments, Gerald Schneider (1967) showed that in the golden hamster it is possible to dissociate rather completely form discrimination from orientation to visible objects by selective lesions either of cortex or of the superior colliculus. The latter nucleus is the principal visual center in the midbrain. Lesions of striate cortex that prevented the animal from discriminating between differently patterned surfaces left it quite capable of orienting its head and body to well-defined visible targets. On the other hand, removal of the superior colliculus, leaving the cortex intact, left the animal apparently incapable of orienting itself to visible objects but quite capable of discriminating between different visual patterns. This double dissociation suggests that in the hamster, at least, midbrain systems are required to control visual orientation while forebrain systems are necessarily involved in pattern discrimination. Work on higher mammals, the cat for example, suggests that such complete dissociation of function between forebrain and midbrain is unlikely. Apart from the small size of the midbrain relative to the forebrain in these animals, the two structures appear to work in a complex

synergy according to the research of Sprague and his collaborators (1966).
In any event, neurobehavioral research gives ever increasing support to
the distinction between two visual systems, one selectively concerned with
analysis of pattern, and the other primarily with control of action oriented
with regard to visible objects and events.

Early Deprivation and Rearrangement

My interest in the idea of two processing systems for vision stems from
our own work on the maintenance and development of visual-motor
coordination (Held and Hein, 1963; White and Held, 1966; Held, Bauer,
Jr., 1967). In the course of this research we have had reason to make
a distinction which appears to be related to that inferred from the neurological data.

The first bit of evidence I want to mention derives from the studies of
animals deprived of patterned visual stimulation as well as movement-
produced visual change. For a time it was supposed that absence of
normally patterned visual stimulation during the earliest development of
the neonatal mammal (cat and primate, for the most part) precluded the
development of visual performance in general. Most of the tests of pattern
vision then made involved responses that were guided by visible aspects
of the testing apparatus, thus confounding discrimination of pattern with
visually guided behavior (Riesen, 1958). More recently, the weight of
evidence incidates that by far the greater deficit occurs not in pattern
discrimination but in visually guided movement. This difference has been
most clearly demonstrated by Meyers and McCleary (1964) who tested
cats, reared from birth without pattern vision, under conditions in which
the indicant response was not itself visually guided. Under these circumstances, the animals showed themselves quite capable of discriminating
cross from circle and of transferring that discrimination from the eye
which had learned it to the contralateral eye which had been occluded
during training. However, at the same time these pattern-discriminating
cats were incapable of guiding their behavior in space in relation to visible
objects. One implication of this result was that previous investigators had
overlooked the pattern discriminating capabilities of the deprived animal
by testing for them with visually guided responses not available to the
animal.

Following up earlier work of our laboratory (Held and Hein, 1963),
we have demonstrated that both kittens and monkeys reared without
sight of their limbs and bodies will fail to develop accurate visual guidance
of their limbs although they will be quite capable of controlling movement of their eyes and heads (Held, Bauer, Jr., 1967; Hein and Held,

1967). The kittens were prevented from viewing their bodies when in illuminated surrounds by placing ruffs around their necks which occluded vision of their bodies but did not prevent them from locomoting and viewing their environments. The monkeys were reared with shields around their necks attached to body holders which allowed them considerable freedom of movement but precluded sight of their limbs. In further experiments, Alan Hein (1968) and I have shown that the two eyes of kittens may be dissociated with respect to control of the limbs by a special rearing procedure. When the kittens wore the ruff which prevented sight of their limbs and bodies, vision of one eye was occluded by an opaque contact lens while the other remained open. When the ruff was not worn (half of the time in light) the opposite eye was occluded and the formerly occluded eye was free to view the limbs. Consequently, one eye had opportunity to view the limbs while the other did not. After many hours of exposure to light, the eye which had viewed the limbs was capable of guiding them whereas the opposite eye was not. These results tell us that, unlike pattern discrimination, the visual guidance of movement appears to require exposure to the environment for its development, and, as a further distinction, allows of dissociation of the two eyes with respect to the control of movement.

Two findings from the study of adaptation to rearrangement in the human adult are also relevant here. For many years we have been studying the compensation that human adults and other primates show to displacements and distortions of spatial vision (Held, 1965; Held and Freedman, 1963). The wedge prism goggle has been a favorite tool of these investigations. Recently, Ann Graybiel, of our laboratory, discovered that adaptation of eye-hand coordination may occur under the dark adapted state of vision as readily as it does during the light adapted state. This result indicates to us that the high resolution system of foveal vision is not essential to the maintenance of visually guided movement, another bit of evidence consistent with the separation of pattern analyzing from orienting systems.

A second kind of evidence suggests that an interaction between the two systems constrains the amount of compensation that the more plastic orienting system may undergo. Rearrangement experiments which employ linear transformations, such as displacement or rotation of the visual field, are reported to allow full and exact compensation after sufficient exposure (Held and Freedman, 1963). However, nonlinear transformations, such as prism-induced curvature, show rather limited amounts of compensation. And perceived discontinuities introduced into the field of vision, such as may be produced by prisms in half-spectacle frames, do not show any reduction according to available reports. Implicit in these results is the idea that the pattern-analyzing system will constrain alterations in the perception of form generated by an altered set of directions of orientation to the parts of the form (Held and Hein, 1967; Held, 1968).

Conclusion

From this all too summary review I should like to conclude that the evidence for two systems of utilization of visual information is becoming increasingly convincing.* For those of us interested in the modifiability of visually controlled behavior, the distinction is important because the data indicate that while the system controlling visual orientation is modifiable, that concerned with analysis of pattern and form is relatively fixed. Research on modifiability is being pursued in many laboratories including our own. The results are and should continue to be of obvious relevance to practitioners of visual science. From a teleological point of view, the difference in modifiability between systems makes sense according to the following reasoning. An effective system for analysis of pattern takes little account of differences in size, orientation, and certain kinds of perspectival distortion. In short, the metrical properties of visual space are relatively unimportant. However, it is just these metrical properties which are of great importance in visually guided behaviors. Moreover, the visual stimuli which are cues to these characteristics of objects change with growth and, hence, some sort of adjustment is required. For example, as the head of a child grows, the interocular distance changes. This distance is a determinant of binocular disparities as well as the magnitude of convergence of the eyes required for binocular fixation at given distances. Both of these serve as cues to the metrical properties of space. The modifiability we observe in adult vision then represents a kind of residual capacity derived from the need of a growing animal to adjust to growth-induced changes in proximal stimulation originating from events in the environment.

Postscript

In the question period following the above address, Dr. George Wald called my attention to a study of strabismic amblyopia he and H. M. Burian (1944) performed some years ago. They had been struck by the fact that although the acuity of an amblyopic eye may be as low as 20/200

---

* A much more complete review of evidence supporting this point of view is to be found in a Symposium under the title, "Locating and Identifying: Two Modes of Visual Processing" with contributions by D. Ingle (1967), G. Schneider (1967), C. Trevarthen (1968), and R. Held (1968).

or 20/400, the amblyope's ability to fixate and localize points in space was almost normal. On the basis of their observations, they noted the resemblance between the vision of amblyopic patients and that of animals deprived of visual cortex and suggested that spatial orientation may be served by subcortical structures whereas pattern vision requires cortical mediation.* Their results and interpretations anticipated our own by twenty-five years.

References

Bossom, J., and C. R. Hamilton (1963). Interocular transfer of prism-altered coordinations in split-brain monkeys. *J. Comp. Physiol. Psychol. 56,* 769-774.

Denny-Brown, D., and R. A. Chambers (1958). Visuo-motor function in the cerebral cortex. *Trans. Am. Neuro. Assoc.* 37-39.

Gazzaniga, M. S. (1966). Visuo-motor integration in split-brain monkeys with other cerebral lesions. *Exp. Neurol. 16,* 289-298.

Hamilton, C. R. (1967). Effects of brain bisection on eye-hand coordination in monkeys wearing prisms. *J. Comp. Phys. Psych. 64,* 434-443.

Hamilton, C. R., S. A. Hillyard, and R. W. Sperry (1968). Interhemispheric comparison of color in split-brain monkeys. *Exp. Neurol.* In press.

Hein, A. (1968). Labile sensorimotor coordination. *Proceedings of the M.I.T. Conference on Malnutrition, Learning, and Behavior.* Nevin S. Scrimshaw and John E. Gordon, eds. Cambridge, M.I.T. Press.

Hein, A., and R. Held (1967). Dissociation of the visual placing response into elicited and guided components. *Science 158,* 390-392.

Held, R. (1965). Plasticity in sensory-motor systems. *Scientific American 213,* 84-94.

Held, R. (1968). Dissociation of visual functions by deprivation and rearrangement. *Psychologische Forschung 31,* 338-348.

Held, R., and J. A. Bauer, Jr. (1967). Visually guided reaching in infant monkeys after restricted rearing. *Science 155,* 718-720.

Held, R., and S. Freedman (1963). Plasticity in human sensorimotor control. *Science 142,* 455-462.

Held, R., and A. Hein (1963). Movement-produced stimulation in the development of visually guided behavior. *J. Comp. Physiol. Psychol. 56,* 872-876.

* Editorial comment: Eccentrically fixating amblyopes typically demonstrate abnormal directional localization, which tends to support Dr. Held's implications. However, non-amblyopic central fixating unilateral strabismics often show abnormal directional localization which is in direct conflict with his conclusion. (J. Pierce)

Held, R., and A. Hein (1967). On the modifiability of form perception. *Models for the Perception of Speech and Visual Form.* Weiant Wathen-Dunn, ed. Cambridge, M.I.T. Press, pp. 296-304.

Humphrey, N. H., and L. Weiskrantz (1967). Vision in monkeys after removal of the striate cortex. *Nature 215,* 595-597.

Ingle, David (1967). Two visual mechanisms underlying the behavior of fish. *Psychologische Forschung 31,* 44-51.

Kluver, H. (1942). Functional significance of the geniculo-striate system. *Biol. Symposia 7,* 253-299.

Meyers, B., and R. A. McCleary (1964). Interocular transfer of a pattern discrimination in pattern deprived cats. *J. Comp. Physiol. Psychol. 57,* 16-21.

Myers, R. E., R. W. Sperry, and N. M. McCurdy (1962). Neural mechanisms in visual guidance of limb movements. *Arch. Neurol. 7,* 195-202.

Riesen, A. H. (1958). Plasticity of behavior: Psychological aspect. *Biological and Biochemical Bases of Behavior.* H. F. Harlow and G. N. Woolsey, eds. Madison, University of Wisconsin Press, pp. 425-450.

Schneider, Gerald E. (1967). Contrasting visuomotor functions of tectum and cortex in the golden hamster. *Psychologische Forschung 31,* 52-62.

Sperry, R. W. (1961). Cerebral organization and behavior. *Science 133,* 1749-1757.

Sperry, R. W. (1964). The great cerebral commissure. *Scientific American 210,* 42-52.

Sprague, James M. (1966). Interaction of cortex and superior colliculus in mediation of visually guided behavior in the cat. *Science 153,* 1544-1547.

Trevarthen, C. B. (1968). Two mechanisms of vision in primates. *Psychologische Forschung 31,* 299-337.

Wald, G., and H. M. Burian (1944). The dissociation of form vision and light perception in strabismic amblyopia. *Amer. J. Ophth. 27,* 950-963.

White, R., and R. Held (1966). Plasticity of sensorimotor development in the human infant. *The Causes of Behavior: Readings in Child Development and Educational Psychology.* J. F. Rosenblith and W. Allinsmith, eds. Boston, Allyn and Bacon, Inc.

S. Howard Bartley

Progress in a Long-Term Study of Some Temporal Processes in Vision

For many years, one of my major concerns has been the interrelating of three sets of phenomena—the fundamental variables of stimulus input, the visual phenomena produced, and the underlying neural processes which can be isolated. Neuro-anatomical and general neurophysiological understandings have provided a starting point for doing this. Experimental answers to certain questions turned out to provide data that were guidelines to a program of further neurophysiological exploration. This meant that new information, as it was obtained, had a maximal probability of being interpretable, and after-the-fact guesses and explanations were held to a bare minimum.

Bishop and I, in beginning our studies, did not do the usual thing, namely, start at the periphery and clear up certain questions there. We were concerned with the brain activity and luckily chose the visual system, the optic pathway, as about the most accessible and highly manipulatable system to study. We began with the optic nerve, the eye removed, and supposed we could most fundamentally manipulate input by so doing. Our first publications were exploratory and if further experiments had continued in dealing with some of the initial problems, we likely would have gotten nowhere. It so happened that we became concerned with the fact that despite equal energy inputs to the optic nerve, cortical responses varied widely in amplitude. Curiosity about this led to supposing that there was something periodic involved in the determination of response amplitude. This being the case, timing was called for, to ascertain whether or not, if the right temporal separation between stimuli was found, all responses would be of the same height. This proved to be the case for responses to supra maximal inputs (Bishop, 1933). A second stage in timing was attempted by shifting the phase at which the inputs were delivered within this cycle. It was supposed that all small, all large, or all medium-sized responses could be obtained. This, too, turned out to be the case. It has been a far more complex matter to ascertain the nature of the processes responsible for the demonstration of the periodicity, but an accumulation of information has led to a comprehensive, though skeletal, picture of what must be happening.

The analysis of the components of the cortical response went hand in hand with the study of the temporal relations between input and response (Bartley and Bishop, 1933). The response was a wave complex followed by one or more repetitions of the final wave component of the complex. This train had the essential appearance of a train of the animal's alpha waves. While it is not relevent to dwell on this analysis, it can be said that the findings were positive (Bartley, 1933; Bartley, O'Leary, and Bishop, 1937A, 1937B). That is, the waves were shown to be produced by the activity of the same elements as those producing the alpha rhythm.

The connection of the elicited cortical response and alpha activity and the brightness enhancement maximum to be mentioned later was a strong cue for connecting our neurophysiological findings and brightness enhancement data.

Anatomy and the response data led us to regarding the pathway as a population of functional parallel units which could be activated in many temporal patterns. I call these units channels, although the term is unfamiliar in neurophysiology. The stimulation producing this range of response pattern varied from brief stray inputs to a series of inputs variously spaced in time.

A channel in the unmodified experimental preparation, in some respects, is similar to a retinal receptive field with its chain of neurons to the visual cortex. Just as those who isolate and study receptive fields think of these as having some functional unity, a channel is to be considered in the same manner, so that when the whole retina or the whole optic nerve (if the retina is removed) is stimulated, it involves the total population of possible channels. If a restricted retinal area or a few of the fibers in the optic nerve are activated, in other words, one can say that only some of the channels have been activated. We should expect considerable flexibility or variety in the particular grouping of channels and in the temporal sequences of the participation in the total discharge sent to the visual cortex. It seems quite reasonable that two things determine end results, namely, *which* channels are activated and in *what temporal* pattern they are activated. Thus, *which* and *when* are the two variables. Some researchers have chosen to study the *which*, or the *spatial* aspect of visual activity, and only recently has there been a growing interest in the *when* or the *temporal* aspect. In fact, it seems to me that Bishop and I, in 1931, wittingly or unwittingly, were the forerunners of the study of the temporal features of the activity of the optic pathway. Surprisingly, some have since declared this sort of study has yielded little or nothing. I believe experimental results have demonstrated the fallacy of this conclusion.

Soon this electrical stimulation without the eye in the circuit was extended to comparisons between it and brief photic pulses to the eye using the same temporal ranges of stimulus intermittency (Bartley, 1934). It turned out that the two forms of stimulation produced essentially the same cortical results. Our first photic stimulation consisted not of restricted stimuli but ones that flooded the whole eye, so as to possess characteristics which would most nearly simulate the electrical stimulation of the optic nerve. Hence, by using photic stimulation, we were intentionally in no way testing the enormous possibilities of the retina to structure complex patterns of input to the optic nerve and thus to the cortex.

Brightness

Parallel to the neural studies, I soon began to study the human observer's sensory response to intermittent photic stimulation (Bartley, 1938). Instead of extending the already ongoing analysis of the conditions for CFF (critical flicker frequency) then found in the literature, I chose to study the sensory results of using subfusional pulse frequencies. This line of work was essentially a human psychophysical parallel in terms of stimulus conditions, to those used in the neurophysiological studies on rabbits and cats. In these human studies, the factor of brightness was singled out for attention. The question was thus how did the factors of rate, intensity, and ratio of photic pulse to the null period of the intermittency cycle affect brightness. The chief finding here was that by reducing the stimulus intermittency rate from critical flicker frequency (fusion) down to rates so low that a succession of separate flashes began to replace flicker, it was found that brightness increased until a critical rate of around ten per second was reached. This increase in brightness, beyond that produced by a continuous steady stimulus was called *brightness enhancement*. Under some conditions, it was later found that a submultiple of this rate, namely, five per second, produced the maximal brightness. At first it was not apparent that rates much below this should not be counted into this series since brightness of a field that could properly be compared with a steady field was no longer being studied. For a short time, I overlooked this rule that chickens and cows could not be compared and all called cows. This same pitfall has entrapped a number of other workers who have later persisted in supposing some of my interpretations about brightness maxima are valid.

With the two sets of results at hand, one set having to do with the stimulus conditions for producing cortical responses of various amplitudes, and the other having to do with conditions for producing various levels of brightness, it appeared evident that these two sets of conditions were essentially identical. This provided a basis for a unified interpretation which dealt with both sets of phenomena, the neurophysiological and the sensory psychophysical. This interpretation was an extension of the alternation of response theory which I have been briefly referring to.

To be brief about this concept, I shall mention only some of its main features. First, it supposes that the amplitudes of the cortical response obtained with peripheral stimulation is determined by the number of separate channels that are activated relatively close together in time, brought into relative synchrony. Repeated photic pulses produce maximal brightness enhancement when their rate of repetition is the same as the periodicity of the optic pathway.

Only strong stimulation and only targets subtending more than very small visual angles should produce brightness enhancement; for only such impingements will activate sufficient channels. This expectation was verified. In using optic nerve stimulation to elicit cortical response, only strong stimulation had produced the effects displaying the periodicity of the central end of the optic pathway.

Another prediction was that short, intense photic pulses would produce more enhancement than longer ones of the same intensity (Bartley, Paczewitz, and Valsi, 1957). That is, small PCF's would be more effective than larger ones. This, too, was confirmed in a number of investigations. This expectation was based on the supposition that extended inputs destroy synchrony. No investigation of any sort has contradicted this expectation.

Another, although incidental, supposition was that greater brightness enhancement will result under conditions that avoid the usual effects of entoptic stray retinal illumination. I had demonstrated that a sizeable amount of stray (not a focused part of the retinal image) illumination is typically involved even with images of very small angular subtense.

As a result, the portion of the retina supposedly activated by the intermittent stimulus was illuminated by the stray illumination from the steady standard target. This expectation was also verified by placing the image of the standard target on one retina and the image of the intermittent on the other retina.

The theory has dealt not only with conditions for producing maximal brightness but for predicting or accounting for other levels of brightness and critical flicker frequency, and for describing the conditions for passing from flicker to steady light. It has done this by being able to specify the neural consequences in terms of various forms and degrees of asynchrony in the distribution of channel activity.

We have moved to studying other sensory phenomena in addition to brightness, as they are seen to be inherent in the picture. This has been particularly true in dealing with color. The move to study color has been simply a matter of analyzing the relative effectiveness or even modes of effectiveness of different wave-bands in the overall photic stimulus we have been using. As far as I was concerned, it was a reluctant step, for I knew very well that I was unacquainted with spectral analysis and its sensory correlates in the manner studied by color experts.

As a consequence of the publication of my earlier work on brightness enhancement, several investigators supposed that the conditions producing brightness enhancement should be productive of increased visual acuity. It is known that, at least within certain ranges of luminosity, visual acuity improves with luminosity. Of course, brightness itself improves, so the expectation of the authors alluded to were that any condition that increases brightness improves visual acuity. Brightness enhancement then, according to this, has increased visual acuity associated with it.

In fact, they seemed to go beyond this and simply rather broadly suppose that intermittent stimulation should produce greater visual acuity than steady stimulation. They found, to their surprise, that just the opposite was the case and were unable to give any reason. Their studies were not conducted in a way that made them optimally applicable of interpretation in light of what we had already learned about intermittent stimulation and brightness enhancement (Senders, 1949; Gerathewohl and Taylor, 1953). So in 1963, Nelson, Soules, and I reported a study on the effect of intermittent stimulation on visual acuity. We used Snellen E's in each of four clock positions and different PCF (pulse-to-cycle fraction) conditions. Four rates of intermittency were used, namely, 4, 10, 24, and 48 cycles per second.

For the two intermittency cycles, the luminosity of the target was compensated so as to be the same per unit time as the steady target. Thus all three conditions, i.e., the two intermittent and the one steady, were equal. The only variables were the PCF and the rate of intermittency. If rate and PCF were to have no effect, visual acuity should have been the same in all the conditions. The results showed that both rate and PCF were effective. As rate dropped from 48 to 4 cycles per second, visual acuity dropped and the .25 PCF condition impoverished visual acuity more than the .75, and these two conditions produced a lower visual acuity than the steady condition. The difference between the PCF .75 condition and the steady emerged only at the lower rates of intermittency.

The results were in line with expectations gained from previous work regarding the way the optic pathway functions. We supposed that there are two directions of activity in the system (probably the eye in this case), one is the longitudinal activity running from eye to brain. We had already been dealing with it in the studies on brightness. The other is the transverse or crosswise activity which had already been shown to involve a certain amount of time in order for borders to be perceived. I refer here to the work of Werner (1935) and some of my own in 1939. Using intermittent photic inputs synchronizes or bunches the longitudinal activity and, when in certain temporal patterns, could be expected to disrupt the transverse activity, the crucial activity in determining visual acuity. Thus although brightness is enhanced at rates of intermittency synchronizing longitudinal activity, visual acuity is impoverished.

A couple of years later, Bourassa and I (1965) studied visual acuity by a different method to relate not only to intermittency but more specifically with conditions which produce brightness enhancement. In this study, the observer's task was to report when two bars (illuminated first-surface mirrors) were seen as separate and when no interval between them was apparent.

One objective in this case was to relate visual acuity still more closely to brightness enhancement by using not only several photic rates but to classify our conditions into three categories: (1) those producing brightness enhancement; (2) those producing reduced brightness; and (3) those producing reduced visual resolution.

The results were as follows. (1) Conditions producing brightness enhancement adversely affected visual acuity. (2) For the conditions in which brightness enhancement did not occur, visual resolution depended on target size and luminosity. (3) *With small targets,* raising luminosity decreased acuity. Thus small PCF's providing less flux per unit time than larger ones provided better acuity, possibly because such targets produced less synchrony of discharge in the longitudinal activity in the pathway. As pulse rate decreased, thus increasing brightness, visual acuity became poorer. *With large targets,* there was a wide range of luminous intensities which did not affect acuity. Here neither pulse rate nor PCF appreciably affected visual resolution.

Just recently, Ball and I (1968B and Bartley and Ball, 1968) have returned to more study of visual acuity. We have used a black Landolt C on a white background which covered a visual angle of about twenty minutes of arc. The C subtended five minutes of visual angle with a one minute gap when viewed at twenty feet. The viewing distance used was thirty feet; thus the visual acuity necessary for resolution was 30/20 or 20/13. Luminosities between fifteen and 245 c/ft² and intermittency rates of 1, 5, 10, and 15 cycles per second were used. PCF's of .25, .50, and .75 were employed at each rate and at each luminosity. The observer had to determine in which of four clock positions the gap appeared in each of several runs of ten trials each.

The results were interpreted as follows:

1. Certain of the temporal manipulations produced poorer visual acuity than comparable steady targets.

2. This decrement was maximal at about five cycles per second.

3. The decrement was greatest for a PCF of .25 and progressively decreased for PCF's of .50 and .75. A PCF, you will recall, was maximal for brightness enhancement.

4. *Pulse length* was not the prime factor in reducing visual acuity.

5. Except at very low luminosities, total flux per unit time was not the crucial factor in determining visual acuity.

6. Decrements in visual acuity were not pronounced with intermittent targets at low luminosities. That is, only those luminosities which produce brightness enhancement produce impoverished acuity.

7. A black Landolt C on a white background provided for more of the temporally induced visual acuity decrement than a white C on a dark background. This is in line with the expectations from the original brightness enhancement studies, where it was shown that considerable intensities were necessary, or major portions of the retina had to be illuminated to produce it.

8. Some luminosity levels with a white C on a black background show less impoverishment of visual acuity under intermittency conditions than with steady stimulation. This impoverishment is related to timing and not directly to total flux.

9. There is a marked similarity between the intermittency conditions producing acuity impoverishment and the intermittency conditions producing the brightness enhancement, and the color effects we shall describe later.

10. Again, I would point out that it would seem that acuity impoverishment results from certain forms of interference between transverse neural processes responsible and the longitudinal processes set up by certain rates of intermittent stimulation in which neural response is synchronized into bursts and null periods between.

In a study on visual acuity as affected by intermittent stimulation presently in progress, a Landolt C was again used, part of the time a white C on a black background and part of the time a black C on a white background. In this investigation, the visual angle subtended by the C and by the gap were varied by progressively increasing the viewing distance at which the target's resolution required a visual acuity of 7.5/15 to one in which it represented 30/15, in seven steps. This is to say, a range of visual acuity from 2.0 to 0.5. There were two luminosity levels for the steady target, one was equal to the luminosity for the intermittent target and the other was ¼ the luminosity to compensate for the PCF of .25 of the intermittent target.

From this study, it can be concluded that little impairment results in any case until visual acuity demands are reduced to a critical point. For a white C on a black background, intermittency produced nearly the same effects as the steady.

A very different result occurred when the C was black on a white background. For both observers, the intermittent target produced definitely more errors than the steady. Here it is the background that contributes the illumination, and its area is many times greater than the area of the white Landolt C on the black background. This again demonstrates that for intermittency to have an effect, a sizeable amount of the retina must be involved, and this is definitely in line with the initial findings on brightness enhancement and is also in line with the expectations from the neurophysiological aspect of the alternation of response theory. It was in the neurophysiological studies that it was shown that a sizeable number of channels in the pathway had to be involved to be effective in disclosing the periodicity effect on the size of the cortical response.

As was pointed out earlier, we needed to study the effects of parts of the spectrum as well as variations of time and intensity. Using part-spectral stimuli was simply a way of selecting out portions of the input channels to study another aspect of the general understanding to which I have given the name of alternation of response theory.

So, in 1960, Bartley and Nelson first reported on certain chromatic and brightness changes produced with intermittent photic stimulation. This was a purely preliminary study and used a series of Wrattan filters instead of colors produced by a monochromator or narrow band pass filters. Not all the filters transmitted the same overall flux. The observer's task was to report changes in brightness and color as rate of intermittency was manipulated. These two sorts of reports were made in separate series. The color changes were as follows:

Filter 70 brilliant red changed to amber, then to orange, and finally back to red as intermittency rate varied from about 1 cycle per second to CFF.

Filter 71a red yellowish red to orange-amber, and back to red with an ochre or amber tinge in the red above CFF.

Filter 72   clear orange to whitish glaring yellow, back to orange. Above fusion the orange brightens.

Filter 73   greenish yellow to golden (no glare or glitter) to amber. Supra CFF color about same as original.

Filter 76   bluish violet to yellowish green to glittering whitish purple or violet, returning to yellow with greenish tinge and then back to violet.

The results were discussed in light of findings in the neurophysiology of the optic tract with part-spectrum inputs.

In 1961, Nelson and Bartley studied the role of PCF in the production of hue changes. PCF did not seem to affect the rate at which maximum desaturation occurred, although various PCF's did produce different amounts of desaturation. The phenomenon was most marked with moderately small PCF's. For example, both of the observers consistently reported desaturation of red for PCF's of .12 and .25, blue desaturation at .12, .25, and .50. Frequency of reports of desaturation declined at PCF .50 for the red and at .97 for the blue. When desaturation occurred, it began at once, i.e., with intermittency rates of about one per second. Hand in hand with desaturation went the previously reported hue changes. Desaturation was maximal in the frequency range at which maximal hue shift occurred.

Both of these studies were definite demonstrations of the fact that intermittency within certain rate ranges are definite factors in determining color change, and thus that a temporal factor enters into the production of color.

Nelson, Bartley, and Ford in 1963, using part-spectrum targets, showed that a PCF produced maximal brightness enhancement and that all portions of the spectrum did not produce equal amounts of brightness enhancement. Of the five filters used, a peak of brightness enhancement was shown in the region of 490 mμ and at 570 to 600 mμ. A PCF of .75 produced no brightness enhancement for the short wavelengths; a peak in the region of 570 mμ with brightness enhancement disappears for longer wavelengths.

Nelson, Bartley, and Bleck, in 1964, showed that CFF was varied by PCF, spectral purity, and wavelength.

Ball, in 1964, made extended study of brightness enhancement by using a modified Fry prism monochromator instead of color filters, thus introducing far greater analytical precision than Nelson and I had used.

He confirmed an earlier finding that a PCF of approximately .25 provides for maximal brightness enhancement. Maximal enhancement was obtained at approximately ten cycles per second, although rates just below this produced almost as much. Maximal enhancement effects are obtained with high luminosities. Brightness was extremely dependent upon wavelength. A narrow range around 500 mμ manifested brightness enhancement when luminosity and temporal factors were properly adjusted. A target with an illuminated target surround provides for less brightness enhancement than a target with an unilluminated one. As intensity of the surround was increased, brightness enhancement decreased.

Surprisingly, a surround of the same wavelength of the target provided for more brightness enhancement than one of a widely different wavelength. This is as yet a finding that has not been resolved. Larger targets gave more brightness enhancement than those of lesser area.

The study showed that all three aspects of color, namely, hue, saturation, and brightness, were highly dependent upon the temporal features of the photic input. Wavelengths outside the neighborhood of 500 mμ produce either reduced brightness enhancement or none at all.

In 1965, Ball and Bartley varied PCF from $\frac{1}{32}$ to $\frac{3}{4}$. The two highest PCF's ($\frac{1}{2}$ and $\frac{3}{4}$) produced hardly any brightness enhancement. Previous research had shown that mere duration of the pulses in the intermittent train was not the critical factor in producing enhancement. Here, for example, a PCF of $\frac{1}{4}$ and a rate of 10 per second and a PCF of $\frac{1}{2}$ and a rate of 20 per second involved equally long pulses; the one produced great enhancement and the other produced virtually none. In fact, for some wavelengths, the effect was below the Talbot level. In the second slide the pulse durations were compensated for by adjustment for luminance level so that the total flux per cycle was the same for all wavelengths.

This is analogous to the miles per gallon of gasoline used by a car. By properly distributing the photic input in time and in selecting intensity and wavelength, we can create situations of greater or lesser efficiency in utilization of photic energy. It seems that proper programming will provide more efficient use, or, so to speak, more miles per gallon, out of the incident energy.

Ball and Bartley (1966A) studied the changes in brightness index, saturation, and hue produced by luminance, wavelength, and temporal interaction. They found that at the higher luminances used the 500 mμ waveband produced the greatest brightness index, while the 660 mμ band produced the lowest. The 580 mμ band was intermediate.

As luminance was lowered, the brightness index of the 500 mμ band dropped while the indices of the 580 and the 660 mμ bands rose. All three converged at a luminance level of 30 c/ft². Additional lowering of luminance produces further decline in brightness index, but wavelength does not seem to be a differentiating factor. A luminance within a range of from 200 down to 30 c/ft² produces maximal index while shifts up or down beyond this range decrease it.

Brightness indices of less than 100 (the Talbot level) occur at the higher luminances for the 660 mμ waveband, and in some cases, the 580 mμ waveband.

The 500 mμ band was the only one yielding large amounts of temporally produced desaturation. At the higher luminances, all four of the observers reported total desaturation with this waveband.

At about 20 c/ft² desaturation vanished. Temporally induced hue shifts were found for the 660 mμ waveband and these shifts were toward colors ascribed to shorter wavelengths. Only one observer found any hue shifts for the 580 mμ waveband and it was slight. Three of the four observers reported hue shifts for the 500 mμ waveband. For two of them, it was toward color ascribed to shorter wavelengths and then toward longer. The other observer found only a shift toward the longer, but it began at the same luminance as for the other two.

Ball and Bartley (1966B) studied effects produced by the use of wavelengths in the range of 650 mμ to 710 mμ. Both hue and brightness shifts were found as this range was traversed. Manipulation of target area from 6 degrees down to 1½ degrees did not change brightness. This was interpreted as showing that brightness was not differently dependent upon rods and cones. Desaturation declined as the range was traversed. Luminosity was not controlled. A PCF of ¾ produced only slight desaturation anywhere in the wavelength range. With wavebands centered at 680 and at 690 mμ, the two phases of the intermittency cycle yielded different hues. These hues lay in different planes for the observer. This is a phenomenon earlier reported by Bartley with achromatic targets.

In 1967, Ball and Bartley began a further study of color vision by the use of the American Optical H-R-R pseudoisochromatic plates under intermittent tungsten illumination of 60 footcandles with the recommended Corning filter. Nelson, Bartley, and Mackavey (1961) had previously showed that certain rates of intermittent illumination impoverished the performance of normal observers. The attempt at this time was to make a more detailed investigation of the perceptual results of viewing pseudoisochromatic plates under intermittent illumination.

The study confirmed the fact that color deficient performance can be produced by temporal manipulation of stimulus input. This is shown in Figure 1. No errors in viewing the plates are made with an intermittency rate of 1 per second and a very few at 21 cycles per second. The maximal

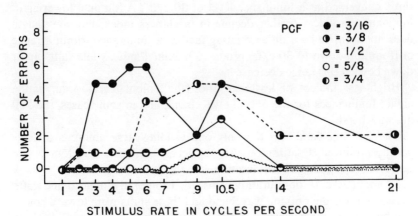

Figure 1. Performance of a color-normal observer as a function of photic intermittency. Each curve is for a different pulse-to-cycle fraction

deficit is produced with rates lying in a range of 3 to 10 cycles per second. PCF's of less than ½ produce the deficit while greater PCF's do not. Figure 2 shows the data for the same observer as the previous slide but errors here are plotted against PCF instead of rate. Each curve is for a separate rate.

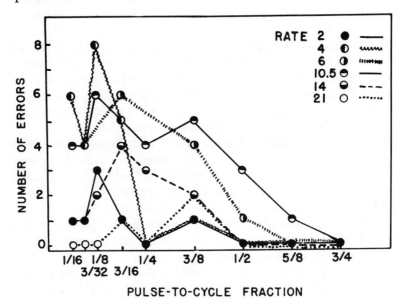

Figure 2. Performance of a color-normal observer. Each curve is for a different rate of photic intermittency

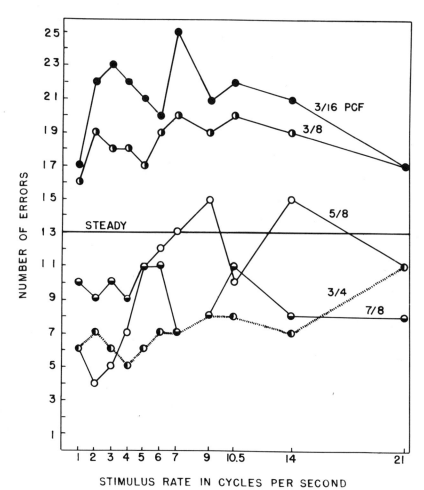

Figure 3.   Errors made by a color-deficient observer (deutan). With steady illumination, 13 errors were made. At some rates more and other rates fewer errors were made than with steady. Each curve represents a different pulse-to-cycle fraction

Figure 3 shows the performance of a deutan color deficient individual. While the normal made no errors, the deutan made 13 errors with steady illumination. Now with intermittent illumination the deutan made still more errors with PCF's of ³⁄₁₆ and ⅜ and a highly varied number of errors with a PCF of ⅝ ranging from only 4 at 2 per second to 15 at 9 and 14 cycles per second. With PCF's of ¾ and ⅞ fewer errors for all rates were made with intermittent stimulation than with steady. This means that this Deutan color deficient observer greatly improved in his performance, a result in no way predicted by conventional color vision theories which suppose that color deficiency results from a lack of one of three kinds of cones. It is difficult to conceive of how controlled timing of photic input could compensate for an anatomical lack of proper receptors.

Figure 4 shows the errors made by the deutan plotted against PCF and shows that PCF is a prime factor in determining the observer's performance. Figure 5 shows the same thing in a different way.

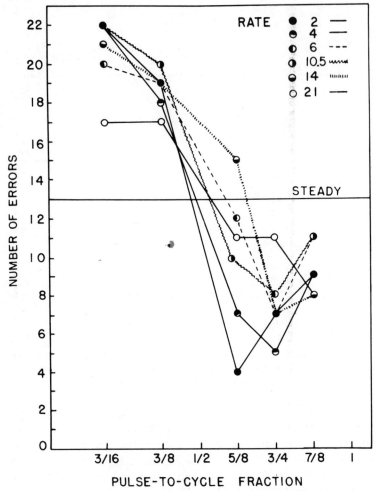

Figure 4.   Errors made by a color-deficient observer (deutan). Pulse-to-cycle fraction is a prominent factor in determining performance

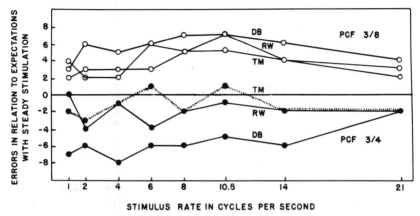

Figure 5. Errors made by three color-deficient observers with two different pulse-to-cycle fractions as stimulus rate was varied. All three types of color-deficient observers made fewer errors than with steady stimulation, when the pulse-to-cycle fraction was ¾

In 1968, Ball and Bartley (1968A) tested four observers with American Optical H-R-R pseudoisochromatic plates. PCF and intermittency rate were manipulated while the illumination of the targets was held constant. Twelve wavebands between 480 and 680 mμ were tested. Desaturation was the end result tested. This was reported upon the basis of a nine-point scale.

For all PCF's except ¾, desaturation was marked and peaked at about 500 mμ where achromaticity becomes total. A second desaturation peak occurs in the neighborhood of 640 mμ, decreasing at PCF increases. PCF of ¾ produces no desaturation for any part of the spectrum.

For the deutan, the trend was for desaturation to decline as wavelength increased. Total desaturation occurred at 500 mμ but no second desaturation region emerged at 640 mμ. The greatest desaturation occurred with PCF's ½ and ¾.

For the protan, there was a sharp peak of desaturation at 500 mμ, little at 480 mμ, and none at all above 520 mμ. Intermediate PCF's produce maximal desaturation.

For the mixed observer, desaturation declined as wavelength increased, and intermediate PCF's were the more effective. There was only a slight peaking tendency around 500 mμ. Thus it can be said that the shapes of the brightness curves were very much alike while there was a lesser similarity between the observers for desaturation.

Summary and Conclusions

The foregoing paper is an account of a long-term study of vision and some of its mechanisms, beginning with a study of the cortical effects of manipulating the fundamental variables of stimulus input; first electrical stimulation of the optic nerve with eye removed, then brief photic pulses to the eye of the experimental animal using the same or similar variables, and finally to the use of some of the same variables in human psychophysical experiments. Fortunately, not only was it possible to use similar photic inputs in the neurophysiological and the psychophysical experiments, but the resulting sensory outcomes were relatable to certain features of the neural records. This has enabled the building of a meaningful picture of relation between cause and effect from external stimulus variables to such sensory end results as brightness, flicker, hue, saturation, and visual acuity, as well as to demonstrate the possibility of fruitfully studying other sensory qualities as well.

References

Ball, R. J. (1964). An investigation of chromatic brightness enhancement tendencies. *Amer. J. Optom. Arch. Amer. Acad. Optom. 41*, 333-361.

Ball, R. J., and S. H. Bartley (1965). Effects of temporal manipulation of photic stimulation on perceived brightness, hue, and saturation. *Amer. J. Optom. Arch. Amer. Acad. Optom. 42*, 513-581.

Ball, R. J., and S. H. Bartley (1966A). Changes in brightness index, saturation, and hue produced by luminance-wavelength-temporal interactions. *J.O.S.A. 56*, 695-699.

Ball, R. J., and S. H. Bartley (1966B). Further investigations of color perception under temporal manipulation of photic stimulation. *Amer. J. Optom. Arch. Amer. Acad. Optom. 43*, 419-430.

Ball, R. J., and S. H. Bartley (1967). The induction and reduction of color deficiency by manipulation of temporal aspects of photic input. *Amer. J. Optom. Arch. Amer. Acad. Optom. 44*, 411-418.

Ball, R. J., and S. H. Bartley (1968A). Brightness and saturation changes resultant from temporal manipulation of spectral photic input for normal and color deficient observers. *J. Psychol. 68*, 55-61.

Ball, R. J., and S. H. Bartley (1968B). Visual acuity decrement for a subfusional intermittency with two directions of border contrast. *J. Psychol. 71*, 185-189.

Bartley, S. H. (1933). Action potentials of the optic cortex under the influence of strychnine. *Amer. J. Physiol. 103*, 203-212.

Bartley, S. H. (1934). Relation of intensity and duration of brief retinal stimulation by light to the electrical response of the optic cortex of the rabbit. *Amer. J. Physiol. 108*, 397-408.

Barley, S. H. (1938). Subjective brightness in relation to flash rate and the light-dark ratio. *J. Exp. Psychol. 23*, 313-319.

Bartley, S. H. (1939). Some effects of intermittent photic stimulation. *J. Exp. Psychol. 25*, 462-480.

Bartley, S. H., and R. J. Ball (1966). The study of target size in the production of brightness, hue, and saturation for intermittent photic stimulation in the red end of the visible spectrum. *J. Psychol. 64*, 241-247.

Bartley, S. H., and R. J. Ball (1968). Effects of intermittent illumination on visual acuity. *Amer. J. Optom. Arch. Amer. Acad. Optom. 45*, 458-464.

Bartley, S. H., and G. H. Bishop (1933). Factors determining the form of the electrical response from the optic cortex of the rabbit. *Amer. J. Physiol. 103*, 173-184.

Bartley, S. H., and T. M. Nelson (1960). Certain chromatic and brightness changes associated with rate of intermittency of photic stimulation. *J. Psychol. 50*, 323-332.

Bartley, S. H., T. M. Nelson, and E. M. Soules (1963). Visual acuity under conditions of intermittent illumination productive of paradoxical brightness. *J. Psychol. 55*, 153-163.

Bartley, S. H., J. O'Leary, and G. H. Bishop (1937A). Differentiation by strychnine of the visual from the integrating mechanisms of optic cortex in the rabbit. *Amer. J. Physiol. 120*, 604-618.

Bartley, S. H., J. O'Leary, and G. H. Bishop (1937B). Modification by strychnine of responses of the optic cortex. *Proc. Soc. Exp. Biol. Med. 36*, 248-250.

Bartley, S. H., G. Paczewitz, and E. Valsi (1957). Brightness enhancement and the stimulus cycle. *J. Psychol. 43*, 187-192.

Bartley, S. H., and F. R. Wilkinson (1952). Brightness comparisons when one eye is stimulated intermittently and the other eye steadily. *J. Psychol. 34*, 165-167.

Bishop, G. H. (1933). Cyclic changes in excitability of the optic pathway of the rabbit. *Amer. J. Physiol. 103*, 213-224.

Bourassa, C. M., and S. H. Bartley (1965). Some observations on the manipulation of visual acuity by varying the rate of intermittent stimulation. *J. Psychol. 59*, 319-328.

Gerathewohl, S. J., and W. Taylor (1953). Effect of intermittent light on the readability of printed matter under conditions of decreasing contrast. *J. Exp. Psychol. 46*, 278-282.

Nelson, T. M., and S. H. Bartley (1961). The role of PCF in temporal manipulations of color. *J. Psychol. 52*, 457-477.

Nelson, T. M., S. H. Bartley, and F. C. Bleck (1964). The effects of cycling upon CFF when stimulus purity is varied. *J. Psychol. 58*, 343-352.

Nelson, T. M., S. H. Bartley, and Z. Ford (1963). The brightness of part spectrum targets stimulating the eye parafoveally at intermittent rates below fusion. *J. Psychol. 55,* 387-396.

Nelson, T. M., S. H. Bartley, and W. R. Mackavey (1961). Responses to certain pseudoisochromatic charts viewed in intermittent illuminance. *Percept. Mot. Skills 13,* 227-231.

Senders, V. L. (1954). Visual resolution with periodically interrupted light. *J. Exp. Psychol. 47,* 453-465.

Werner, H. (1935). Studies in contour: I. Qualitative analysis. *Amer. J. Psychol. 47,* 40-64.

Raymond Crouzy

Signal, Noise, and Decision
in Visual Detection

It is rather disappointing that today, in spite of the progress achieved in neurology, electrophysiology, and biochemistry, and despite the various mathematical models elaborated for the nervous activity and the bulk of data accumulated by psychophysics, we are still unable to propose a complete model of light detection by the visual system, even in the simplest case: the one of binary experiment with a stimulus defined by only one variable.

The reasons for this delay seem to be twofold: first, a great deal of time and effort have been devoted to performing measures in very complicated cases, which can lead to nothing other than measurements, as it is far beyond our present possibilities to draw an explanation of what happens in such intricate experiments. Second, some misunderstanding seems to exist between theoreticians and laboratory searchers that prevents synthesis of the knowledge acquired in the different branches of research. Those misunderstandings concern mainly the notion of signal and the problems of background noise. The present paper is nothing more than an attempt to analyze and clarify the basic concepts in view of a general synthesis.

Notion of Threshold

Since early times the study of sensitivity has rested on the concept of threshold, and it is important to investigate what, in fact, the word threshold means.

It is usual to define threshold in ways similar to the following: "The visual threshold is the least amount of light energy required to give rise to a visual sensation" (Wyburn et al., 1964). Such a definition may look clear, but in fact it is inapplicable in psychophysics. It is inapplicable because it assumes that in all circumstances it is possible to decide whether or not a given amount of light energy gives rise to a visual sensation. This assumption is false.

Let us consider a constant stimuli experiment. If, to ascertain the least amount of energy, the stimulus intensity is progressively weakened, one finds that under a certain level a range exists where the visual apparatus no longer operates as a reliable detector, which means that if the trial is repeated—taking care to keep the conditions of the experiment as constant as possible—the subject sometimes answers "seen" and sometimes "not seen."

Experimentally the only thing that can be done is to link a certain frequency of "seen," which we will call f, to each value of the stimulus, and as one proves that f increases with the stimulation intensity s, then the result is the psychometric curve $f(s)$.

In these conditions, if one wants to determine a threshold, one has to choose an arbitrary value f_o of f and to stipulate that the stimulation which incites the frequency f_o of responses "seen" will be considered as the threshold.

Such a definition is still far from being satisfactory, as it does not take into account a fundamental point. All psychophysical experiments, if they are correctly carried out, require the use of "traps" or "blanks" or "null stimuli," that is to say, of assorted presentations of no physical stimulation (or, when the differential threshold is concerned, of no difference between the stimulations).

When the answers become uncertain, one can see that from time to time the subject answers "seen" in the absence of any stimulus. These "mistakes on the trap" or "false detections" have been a cause of perplexity in psychophysics and considered as errors to be avoided. When it was not possible to eliminate them completely, they were swept away with the help of so-called "correction formulas" which generally contain an implicit assumption about the response mechanism. This attitude seems to be unjustified, since "false detections" are experimental results, like others. Even more, they may be the most significant data.

However that may be, it appears that the false detection frequency is not independent of the frequency f of "seen" and that they increase at the same time. So, a lowering of the threshold f_o can only be obtained at the cost of an increase of the false detection frequency c. Such are the experimental data for which all theories of detection must account. In the end, we see that if one tries to conceive a definition of threshold, it should be said, "the threshold is the stimulus that gives an arbitrarily chosen

frequency f_0 of 'seen' with an arbitrary frequency c_0 of false detections."
Considering these two arbitrary choices, one can wonder if the notion of
threshold is anything more than a reference mark of sensitivity, and perhaps not the clearest one.

Views are still very different on this point and they can be classified
roughly into three categories:

1. The theories of high threshold according to which the visual detector
requires a certain minimum of energy to commence operation. This
minimum sets a limit to its sensitivity.

2. The theories of low threshold which claim that long before this natural
working threshold is reached, the visual performances are limited by the
raising of the false detection rate, which makes the result exceedingly
unreliable.

3. Recent theories, which state that when looked at from near enough,
any detector is unstable (especially owing to the unavoidable presence of
background noise) and, also, that the intensity of a stimulus can never be
specified with certitude. It follows then, that the very notion of threshold,
which rests on the fiction of a reliable detector, does not correspond to
reality.*

Classification of Constant Stimuli Experiments

Whatever may be the favorite theory of the specialist, the experimental
results keep their specific value.

An important number of different types of measurements have been
thought of, which results in confusion. It seems imperative to analyze
their principles from a logical point of view.

We will consider only those belonging to the so-called constant stimuli
method. This is in favor among the psychophysicists because it gives more
precise determinations and permits relatively easy interpretation.

It seems that they could be classified in the following way, and as shown
in Table 1.

1. Nature of the trial. The trial (this word being interpreted with the
same meaning it has in the study of probabilities) will be of binary type,
if it involves only two possible outcomes, that is, if the subject has to
choose between two responses, "yes" or "no." In other words, the result
admits one bit of uncertainty per trial.

In the non-binary cases, either the subject will have to select one of the

---

* In our opinion it would not even be pertinent to say "the threshold must be at
least equal to one quantum" as we have seen that a threshold can be determined only
in terms of statistical values, and the *mean* energy of a stimulation can be less than
one quantum.

Table 1. Classification of the Different Types of Measurements Using Constant Stimuli Method

CONSTANT STIMULI METHOD

Type of Trial	Number of Constant Stimuli Used in the Experiment	Number of Degrees of Freedom Involved in the Nature of the Stimulation
Binary	Only One Model of Stimulus Used	The Stimulation is Completely Determined by only One Parameter
		More than One Parameter are Necessary to Determine the Stimulus
	More than One Model of Stimulus	Only One Parameter
		More than One Parameter
Non-binary	Only One Model of Stimulus	Only One Parameter
		More than One Parameter
	More than One Model of Stimulus	Only One Parameter
		More than One Parameter

presentations which he will generally point out by its position in space or time (experiments of "forced choice" which, perhaps, it would be preferable to call "forced localization") or he must carry out a classification (differential measures of "rating" [ranking] experiments). Naturally, it is possible to complicate, endlessly, these fundamental types, for example, by letting the subject interfere in the succession of presentations as in the so-called sequential experiments.

2. *Number of stimuli used.* With a method of constant stimuli of binary type it is of course absolutely necessary to present "blanks" to the subject. So there will be at least one stimulus and a blank. In such a case the subject will have to answer "yes" (stimulus present) or "no" (stimulus absent). This is precisely the type of so-called yes-no detection experiments.

Table 1. *(Continued)*

Number of Parameters Which Are Modified Simultaneously (in the Case of Several Constant Stimuli)	Nature of the Experiment
(No Variation Possible)	Detection Without Redundancy
(No Variation Possible)	Detection With Redundancy
(As Many Distinct Values of the Parameter as Different Stimuli	Differential Detection
Only One Parameter is Modified	Differential Detection
More than One Parameter are Modified Simultaneously	Identification
(No Variation Possible)	Forced-choice Detection Without Redundancy
(No Variation Possible)	Forced-choice Detection With Redundancy
(As many Distinct Values of the Parameter as Different Stimuli)	Measure of Differential Sensitivity by Forced-choice or Rating (Ranking) Experiment
Only One Parameter is Modified	Measure of Differential Sensitivity by Forced Choice or Rating (Ranking) Experiment
More than One Parameter are Modified Simultaneously	Identification by Forced-choice

It is conceivable also that several different stimuli, A, B, C, . . . will be presented. If the question asked of the subject is merely "is there a difference between those stimuli?" (a circumstance which is rarely encountered in practice) or "does the stimulus A appear among the presented stimuli?" the case still belongs to the binary type.

On the other hand, if it is asked of the subject, "what is the more intense stimulus?" or, "what are the presented stimuli?" then the experiment is non-binary.

3. Number of degrees of freedom involved in the stimulation. A great number of independent parameters are generally required to define completely a luminous stimulation; a certain number of geometric sizes determines its shape and dimensions, but its luminance must still be specified (and it may not be uniform), its duration, and the spectral composition of its light. Eventually, parameters of position will be necessary if the position of the stimulus plays a part in the experiment.

The simplest case is the one where the stimulation is completely settled by the knowledge of a single parameter.

Can this case correspond to reality? For vision, the answer is yes, and this is an important point which we will cover when discussing redundancy.

4. *Number of modified parameters.* If there are several distinct stimuli they must differ by at least one of their parameters (it will often be the luminance or the duration), but several parameters at the same time can also be varied.

If only one model of stimulus exists, it will be said to be a *detection* experiment.

In the case where there is more than one, if they differ only by the value of a given parameter (all the other characteristics being left invariable) the matter must be to detect this single difference, in which case the experiment is a *differential* or a *rating* one.

If several parameters vary together, it will be an *identification* (or *recognition*) experiment.

The distinctions between those different types of experiments can seem rather artificial. They appear still more so if the notion of background noise is brought in.

On the' other hand, the notion of redundant information, which is generally ignored, seems to us essential, and we will study in particular its influence on false detection rate.

Transmission of Information

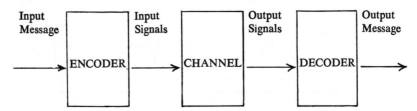

Figure 1.   General schema of an information transmission

Figure 1 shows the general schema of all transmission of information. Information is contained in an *input message* that an *encoder* transforms into signals fit to convey through the following medium, denoted under the name of *channel*. The signals reach the decoder, which, by a new transduction, supplies an *output message*.

In practice, this output message is never a faithful image of the input message for two reasons: on the one hand, codage or decodage can be altered by mistakes, and on the other hand, in the course of traveling through the channel, the two following events may happen simultaneously: (1) the fading of certain signals whose energy is dissipated, and (2) the spontaneous appearance of new signals which mix with those originated from the message. These two events are due to the fact that the channel is permanently in an unstable state. The whole of these effects is covered under the expression of "background noise."

According to the unavoidable presence of noise, there can never be a message transmission without the loss of a certain quantity of information. This is one of the fundamental laws of the universe, of which the second principle of thermodynamics seems to be just one particular aspect (Brillouin, 1959).

If we try to apply such a schema to an experiment of visual perception, we realize that there is in fact a sequence of codage, with the signal changing its physical nature several times.

There is of course a great deal of arbitrariness in the choice of sections for the channel diagram, so we will use, for the sake of our illustration, the schema of Figure 2. If we have brought together in one block the whole of the nervous tracks, it is not that they are less important, but because there is no transformation of the physical nature of the signal. Also, we know very few details about these transmissions.

From what has been said about background noise, all the events that take place between the coder and the decoder are very uncertain. The only two events that are certain are: at one end, the input message exists, and at the other end, the answer of the subject exists. The human decoder is able, from the output signals which show a random character, to make a choice in order to supply an irreversible answer to some preliminary question. That is why we have called it elsewhere a "decision center" (Crouzy, 1963).

The threshold theories which we alluded to are in fact hypothetical schemas on the operating of the decision center. It is important to specify what must be understood here by "signal" and by "background noise." A signal is any modification induced in the channel by the input message and able to affect the choice of the answer. There exists a natural leaning to mix "signal" with "excitation" or "impulse." In fact, the word signal

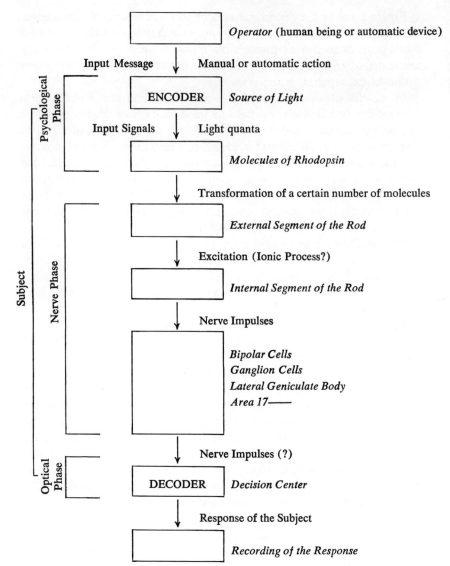

Figure 2. Schema of the information transmission in a visual detection experiment

is more general and an inhibition or a blockage may represent a signal as well as an excitation. In the same way, a group of impulses may form a single output signal if it is to be considered as indivisible for the decision center.

On the other hand, every modification appearing spontaneously in the channel, and which, for the subject, is indistinguishable from a signal will be considered belonging to the background noise. There exists at least one origin of noise whose existence is beyond question; it is the one linked to the quantal nature of radiation. It leads to random fluctuations in the number of quanta absorbed by the visual pigment. Considering the small number of quanta needed to obtain responses "seen," these fluctuations are certainly of practical importance.

Now we have to make clear the possibility of visual stimuli totally defined by only one parameter, which we mentioned on the first line of Table 1.

Let us consider the case of scotopic vision. We know that if the angular diameter u of the stimulus is reduced under a certain limit value u_o, the physical size which settles the response of the subject is not the retinal illumination but the luminous flux received by the retina, that is to say, the product of the illumination by the image area.

In a similar way, if we now reduce the duration t of the stimulation under the critical value t_o, the photometric size upon which the response depends is no longer the flux but the amount of light (product of the flux by t), that is to say, in fact, the total quantity of energy absorbed by the retina. One could object, perhaps, that beyond the amount of energy, it is still necessary to specify the wavelength used, because of the differences in luminous efficiency, which leads to a second parameter.

This last difficulty vanishes if we consider not the energy but the number of quanta, q, absorbed by the rhodopsin, for equal sensations correspond to equal numbers of absorbed quanta.

Finally, under the two conditions

$$u < u_o$$
$$\text{and } t < t_o$$

we are in the case of "total summation" and the sensation of the subject depends uniquely upon the variable q. This is a fact which deserves attention. If only one parameter is needed for the definition of the stimulation this means apparently that only one parameter is enough for determining the phenomenon which brings the information to the decision center. This one-dimension message may be defined as a monosignal, isolated in space and time, and totally described by one number which we could, for instance, call its "energy."

It is easily understood that this stimulation, whose area is not null, can be condensed to a monosignal. This means that at least for part of its way, the transmission must be "monochannel" as a result of a convergence effect. What is known about the organization of the retina allows us to think that the "bottle neck" should take place at the level of the ganglion · cell. Let us notice that if the transmission is monochannel in one point it does not matter if, later on, the signal is divided in several branches by the effect of a divergence. It is obvious that there cannot exist more information downstream than there is at the level of the "neck" and the "one dimension" character is kept up to the end.

It is not so easy to image how a stimulation of definite duration may lead to the propagation of a monosignal without any repetition in time. The reason should lie apparently in the physiology of neurones or synapses, but we do not know any work referring precisely to this problem.

Statistical Decision Theory

In any case, having made clear the notion of background noise and that of signal with one parameter, it will now be easier to mention the statistical decision theory (Green and Swets, 1966) whose basic idea we will merely refer to.

We will consider the case of an experiment of binary detection and look at some variable x which is called "observation" or sometimes "sensory excitation." Owing to the permanent presence of the background noise, this quantity can never be null and forms a random variable whose probability density function, in the absence of any external stimulation, is shown by the curve F_1 in Figure 3, and whose average is x_1. If the bringing in of an input message increases the quantity Δx all the values of x, we will obtain a new function of probability density represented by the curve F_2, deduced from F_1 by the translation:

$$\overline{M_1 M_2} = \Delta x$$

The subject has no other indication for choosing his response than the value of x, and it is easy to see that he never can be sure of the presence of the stimulus, as a given value of x may result either from noise alone or from an input message added to the noise, but with different probabilities according to x value.

Practically, the subject will choose some value x_o of x, forming a sort of internal criterion M_o, and he will give the answer "yes" each time x is above x_o and "no" in the opposite case. The frequency of correct detections (that is to say, answers "yes" in response to a presentation of the

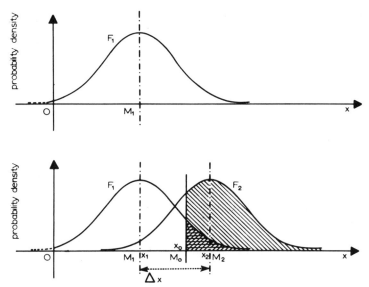

Figure 3. Basic concept of statistical decision theory applied to a threshold measure

stimulus) will be proportional to the hatched area under curve F_2 and the frequency of "false detections" (answers "yes" on the blank, that is to say, in the presence of noise alone) will be proportional to the hatched area under curve F_1, both quantities being linked through the value x_0 adopted as internal criterion.

On the schema it is possible to find all the features of a threshold measure. After sufficient training, the subject acknowledges that if he will not exceed the allowed percentage of false detections, he has to set his internal criterion at the value x_0, from which the stimulation Δx shown in Figure 3 provides, say, 75% or "yes." The x value which would give exactly 50% of "yes" would be the one which brings M_2 to coincide with M_0 (i.e., $\Delta x = x_0 - x_1$).

This schema, then, accounts in a clear way for all the experimental circumstances related before, and it agrees remarkably well with the feeling of the subject during the measures of a threshold in the elementary case considered (binary decision with stimulus settled by only one parameter).

It is to be noticed that the answer "seen" has not the same meaning according to the classical threshold theory and the statistical decision theory. In the first case, "seen" means "stimulus perceived," but for the statistical theory, it means "having to give a response anyway, I think that the best choice is yes." A special feature of this conception is that it rests entirely on that value called "observation" which is, in short, the overall estimation of a probability but whose connections with concrete events is not at all explicit. Green and Swets (1966) say themselves, "The formal theory does not treat the substantial aspects of the detection situation" (p. 235). On one hand, this gives a very broad power of explanation to the theory, as it may be applied to any sensory detection, and even to any sort of decision. On the other hand, if we are interested in a physical problem (such as visual perception) we cannot help feeling frustrated as long as we are not able to relate that mysterious value x to some biophysical (and if possible, measurable) event.

The consideration of monosignal information could bring the beginning of a solution to that problem. If summation conditions are satisfied, only one sensory unit of the retina is activated during only one elementary duration of perception, which results in a monosignal. In such a case, the interpretation of statistical decision theory looks clear; the "observation" may be considered identical to the monosignal energy (at least to a quantity proportional to this energy).

Redundancy and Detection

If we are in presence of two or more distinct signals the problem becomes at once more complicated as the information about the presence of the stimulus may be redundant. This will be the case, for instance, if more than one summation area is stimulated by the same flash or if only one summation area is stimulated during a time longer than the summation duration. Then two questions arise: (1) can the signal transmissions through the channel be considered as independent events or are there mutual interactions between them, and (2) in which way will the center of decision take advantage of this repetition of information?

Concerning the first problem, owing to the lack of knowledge and for the sake of simplicity, we will admit that the signals are independent. This is certainly not true at usual levels of luminance, but may be accepted near the absolute threshold. As regards the second point, we have to make assumptions and we will meet once more the difference between classical threshold and statistical decision concepts. On this ground, we will follow Green and Swets (1966) who call the first case "decision threshold model" and the second "integration model." The decision threshold model assumes that each monosignal elicits a particular decision and that from the "indicator board" formed by all these elementary "yes" and "no," the decision center draws its conclusion.

The two simplest rules that can be imagined are, on the one hand, that the subject answers "seen" as soon as he has recorded a single elementary response "yes." In such a case, the number of monosignals involved in the experiment being n and assuming that the probability of an individual response "yes" is the same for all the monosignals and equal to p, the probability of an overall answer "seen" from the subject is

$$p_n = 1 - (1 - p)^n.$$

On the other hand, that the subject answers "seen" only if all the n elementary responses are "yes." Then, with each individual probability equal to p, the probability of an overall answer "seen" is

$$p_n = p^n$$

Between those extreme behaviors, one may assume that the response will be "seen" if, among the n elementary answers a certain number m, or a certain fraction m/n, is "yes."

According to the integration model, the mechanism is quite different. Each monosignal provides a certain probability that the stimulus is present. The combination of all those partial probabilities results in an overall probability which determines the decision. Under the new assumptions:

1. that noise alone and signal plus noise fluctuation fit the same Laplace-Gauss distribution,
2. that the individual observations are combined through a linear law,
3. that this combination is achieved with no loss of information (even when memory plays a role),

it has been established that the result is the same for n elementary observations of value x as that for a single observation of value $xn^{1/2}$. This means that a threshold measured with n monosignals will be $n^{1/2}$ times lower than if measured with a single monosignal.

The result to be noticed is that, as one knows, taking into account the quantal fluctuations of the stimulus leads to the conclusion that there exists an absolute limit for the differential threshold which is proportional to the square root of the mean number q of quanta absorbed by the visual pigment. Attempts have been made (Rose, 1948) to check whether this lower limit (whose reality is sure) was only a theoretical view, or was of practical importance. As we have pointed to elsewhere (Crouzy, 1963), this checking should be carried out only in the case of monosignal; otherwise there is redundancy and we meet the above mentioned difficulties Number 1 and Number 2. We can see now the risk encountered in the last case. If we use stimuli of various areas (and this was the case in Rose's measures) the total number of absorbed quanta is proportional to the area, but the number of retinal sensory units is also roughly proportional to the area, and so is the number n of monosignals. Thus, if we find in the experimental results a factor proportional to the square root of the stimulus area, we could ascribe this square root law to the quantal fluctuation when it is to be attributed to the operating rule of the decision center.

Noise and Redundancy

We have seen that, in a monosignal experiment, when noise alone is present, the subject's answer will be "seen" each time the observation x exceeds the criterion value x_0. If the criterion is relatively low, this event is not necessarily exceptional. But if the experiment involves a high number n of monosignals it is very unlikely that this event will happen in the same trial at all, at the needed time and place to elicit a response of "seen."

Because of this, a researcher working with short flashes and a pinpoint source will have trouble with the dark noise, while another one using large sources and long exposures will find a very small rate of false detections and will be inclined to deny the existence of noise in the visual system. In order to bring some light to these discrepancies, we have calculated some probabilities of false detection for the two models considered by Green and Swets.

Decision Threshold Model (Table 2)

It is assumed that the subject's response is "seen" when *at least* m elementary answers are "yes" among the n monosignals used. Under the assumptions stated in the preceeding pages, this leads to a sum of terms of a binomial distribution (Bernoulli's Law). Let us call p the probability of false detection for a monosignal; the probability of false detection for the set of n monosignals will be

$$p_n(m) = \sum_{z=m}^{z=n} \frac{n!}{Z!\,(n-Z)!} p^z (1-p)^{n-z}$$

Table 2. Probability of False Detection, in Case of Redundant Information,
According to the "Decision Threshold Model"

Value of $p_n(m)$

Case C: $m/n = \frac{1}{2}$

n	m	p=0.01	p=0.05	p=0.08	p=0.10	p=0.12	p=0.15	p=0.18	p=0.20	p=0.25
1	1	.01000	.05000	.08000	.10000	.12000	.15000	.18000	.20000	.25000
2	1	.01990	.09750	.15360	.19000	.22560	.27750	.32760	.36000	.43750
4	2	.00059	.01402	.03443	.05330	.07319	.10952	.15089	.18080	.26172
6	3	.00002	.00223	.00851	.01585	.02609	.04734	.07586	.09888	.16943
8	4	.00000	.00037	.00220	.00502	.00972	.02155	.03973	.05628	.11381
10	5		.00006	.00058	.00163	.00371	.00987	.02132	.03279	.07813
12	6		.00001	.00016	.00054	.00144	.00464	.01162	.01940	.05440
14	7		.00000	.00004	.00018	.00057	.00221	.00641	.01161	.03827
16	8			.00001	.00006	.00022	.00105	.00356	.00700	.02713
18	9			.00000	.00002	.00009	.00051	.00199	.00425	.01935
20	10				.00001	.00003	.00024	.00112	.00259	.01386
22	11				.00000	.00001	.00012	.00063	.00159	.00997
24	12					.00000	.00006	.00034	.00099	.00719
26	13						.00003	.00020	.00060	.00521

Value of $p_n(m)$

Case A: m=1 (whatever n may be)

n	m	p=0.01	p=0.05	p=0.08	p=0.10	p=0.12	p=0.15	p=0.18	p=0.20	p=0.25
1	1	.01000	.05000	.08000	.10000	.12000	.15000	.18000	.20000	.25000
2	1	.01990	.09750	.15360	.19000	.22560	.27750	.32760	.36000	.43750
4	1	.03940	.18549	.28361	.34390	.40030	.47799	.54787	.59040	.68359
6	1	.05852	.26491	.39364	.46856	.53559	.62285	.69599	.73786	.82202
8	1	.07725	.33657	.48678	.56953	.64036	.72751	.79558	.83223	.89989
10	1	.09562	.40126	.56561	.65132	.72149	.80312	.86255	.89262	.94369
12	1	.11361	.45964	.63233	.71757	.78432	.85776	.90758	.93128	.96832
14	1	.13125	.51232	.68880	.77123	.83298	.89723	.93785	.95602	.98218
16	1	.14854	.55987	.73660	.81469	.87066	.92575	.95821	.97185	.98998
18	1	.16548	.60278	.77706	.84990	.89984	.94635	.97190	.98198	.99436
20	1	.18209	.64151	.81131	.87842	.92244	.96124	.98111	.98847	.99683
22	1	.19837	.67646	.84029	.90152	.93993	.97199	.98729	.99262	.99822
24	1	.21432	.70801	.86482	.92023	.95348	.97976	.99146	.99528	.99899
26	1	.22996	.73648	.88558	.93539	.96396	.98638	.99425	.99697	.99943

Table 2. Probability of False Detection, in Case of Redundant Information,
According to the "Decision Threshold Model" (continued)

Value of $p_n(m)$

Case B: m/n=⅓

n	m	p=0.01	p=0.05	p=0.08	p=0.10	p=0.12	p=0.15	p=0.18	p=0.20	p=0.25
1	1	.01000	.05000	.08000	.10000	.12000	.15000	.18000	.20000	.25000
3	1	.02970	.14262	.22131	.27100	.31853	.38587	.44863	.48800	.57812
6	2	.05852	.26491	.39364	.46856	.53559	.62285	.69599	.73785	.82202
9	3	.00008	.00836	.02979	.05297	.08326	.14085	.21046	.26180	.39932
12	4	.00000	.00223	.01201	.02564	.04641	.09220	.15515	.20543	.35122
15	5		.00061	.00497	.01272	.02649	.06170	.11669	.16423	.31351
18	6		.00017	.00209	.00641	.01536	.04189	.08893	.13291	.28255
21	7		.00005	.00089	.00327	.00900	.02874	.06842	.10851	.25637
24	8		.00001	.00038	.00168	.00532	.01987	.05300	.08917	.23379
27	9		.00000	.00016	.00087	.00316	.01381	.04129	.07365	.21405

Value of $p_n(m)$

Case D: m/n=⅔

n	m	p=0.01	p=0.05	p=0.08	p=0.10	p=0.12	p=0.15	p=0.18	p=0.20	p=0.25
1	1	.01000	.05000	.08000	.10000	.12000	.15000	.18000	.20000	.25000
3	2	.00030	.00725	.01817	.02800	.03974	.06075	.08553	.10400	.15625
6	4	.00000	.00009	.00054	.00127	.00254	.00588	.01155	.01696	.03759
9	6		.00000	.00002	.00006	.00018	.00063	.00173	.00307	.00999
12	8			.00000	.00000	.00001	.00007	.00027	.00058	.00278
15	10					.00000	.00001	.00004	.00011	.00079
18	12						.00000	.00001	.00002	.00023
21	14							.00000	.00000	.00007
24	16									.00002
27	18									.00001
										.00000

Value of $p_n(m)$

Case E: $m/n=1$

n	m	p=0.01	p=0.05	p=0.08	p=0.10	p=0.12	p=0.15	p=0.18	p=0.20	p=0.25
1	1	.01000	.05000	.08000	.10000	.12000	.15000	.18000	.20000	.25000
2	2	.00010	.00250	.00640	.01000	.01440	.02250	.03240	.04000	.06250
3	3	.00000	.00012	.00051	.00100	.00173	.00337	.00583	.00800	.01562
4	4		.00001	.00004	.00010	.00021	.00051	.00105	.00160	.00390
5	5		.00000	.00000	.00001	.00002	.00007	.00019	.00032	.00098
6	6				.00000	.00000	.00001	.00003	.00006	.00024
7	7					.00000	.00000	.00001	.00001	.00006
8	8							.00000	.00000	.00001
9	9									.00000

The calculus has been achieved for five different cases:

Case A: $m = 1$ (whatever n may be)

Case B: $\dfrac{m}{n} = \frac{1}{3}$

Case C: $\dfrac{m}{n} = \frac{1}{2}$

Case D: $\dfrac{m}{n} = \frac{2}{3}$

Case E: $\dfrac{m}{n} = 1$

In each case, the values are given for nine different values of p: $p = 0.01$; $p = 0.05$; $p = 0.08$; $p = 0.10$; $p = 0.12$; $p = 0.15$; $p = 0.18$; $p = 0.20$; $p = 0.25$. The highest value of n used is 27, which means approximately, for instance, a peripheral test whose diameter varies from $1°$ to $5°$.

Table 3. Probability of False Detection, in Case of Redundant Information, According to the "Integration Model"

Value of P_n

n	p=0.01	p=0.05	p=0.08	p=0.10	p=0.12	p=0.15	p=0.18	p=0.20	p=0.25
1	.01000	.05000	.08000	.10000	.12000	.15000	.18000	.20000	.25000
2	.00050	.00990	.02330	.03770	.04750	.07080	.09680	.11700	.16850
3	.00000	.00120	.00730	.01320	.02070	.03590	.05590	.07350	.11900
4		.00050	.00250	.00520	.00910	.01880	.03290	.04650	.08690
5		.00010	.00080	.00210	.00410	.01020	.01970	.03010	.06430
6		.00000	.00030	.00080	.00190	.00540	.01220	.01920	.04850
7			.00010	.00030	.00090	.00290	.00750	.01320	.03590
8			.00000	.00010	.00040	.00160	.00470	.00820	.02740
9				.00008	.00020	.00090	.00290	.00570	.02070
10				.00000	.00010	.00050	.00190	.00360	.01580
11					.00000	.00030	.00110	.00240	.01220
12						.00020	.00070	.00160	.00940
13						.00010	.00050	.00110	.00710
14						.00010	.00030	.00070	.00550
15						.00000	.00020	.00050	.00430
16							.00014	.00030	.00330
17							.00010	.00020	.00260
18							.00000	.00020	.00200
19								.00010	.00150
20								.00010	.00118
21								.00000	.00090
22									.00070
23									.00060
24									.00050
25									.00030
26									.00030

Integration Model (Table 3)

Admitting the particular assumptions of this model, we have to deal with a Gaussian distribution. Being given the individual probability p of false detection, we may define an auxiliary variable y by the relation

$$1 - \int_{-\infty}^{y} \frac{1}{\sqrt{2\pi}} \exp\left(-\tfrac{1}{2}t^2\right) dt = p$$

The probability of false detection for a set of n monosignals will be

$$P_n = 1 - \int_{-\infty}^{y\sqrt{n}} \frac{1}{\sqrt{2\pi}} \exp\left(-\tfrac{1}{2}t^2\right) dt$$

Three remarks need to be made about these numerical results. (1) In decision threshold models the extreme cases A and E are equally hard to believe for opposite reasons. (2) With a ratio m:n around ½, the false detection rate would fall from 0.10 to 0.001 when the diameter of a peripheral test is changed from 1° to 3°, which seems reasonable. According to the integration model, the fall of the false detection rate would be from 0.10 to 0.001, which is also acceptable. (3) It is interesting to notice that while the integration model gives an almost exponential decrease of p_n, with the decision threshold model, in several cases $p_n(m)$ begins to increase for small values of n and then drops rapidly. This could provide a way of discriminating experimentally the two hypotheses.

The Discontinuous Structure of Signals

We think that all the proceeding reflections should be considered only as a first approach to the actual detection mechanism.

The observation has been treated as a continuous variable. In fact, we know that the input signals consist in discrete events, namely the absorption of light quanta by molecules of pigments. Also, it is tempting to consider, at least at low levels, the output signals have a digital nature, too, made up of nerve impulses, all equal in size and whose number only is significant. If this is true, the statistical detection model should be modified in three main respects (Crouzy, 1965). This could allow us to link more safely the physiological behavior to the theoretical model, and, perhaps, along the lines drawn by Blachman (1966) regarding information theory in the discrete case, to reach an evaluation of the entropy and channel capacity involved in a given visual process. The consequence could be not only the explanation of the experimental laws of response, but the possibility of predicting, from a small number of fundamental data, the behavior of a given subject in given circumstances. Up to now, such problems are scarcely outlined.

References

Blachman, N. M. (1966). *Noise and Its Effect on Communication*. McGraw-Hill, New York.

Brillouin, L. (1959). *La Science et la théorie de l'Information*. Masson et Cie, Paris.

Crouzy, R. (1963). La Structure quantique de la lumière et la sensibilité différentielle de l'appareil visuel considéré comme un détecteur de signaux. Thesis. Faculté des Sciences, Paris.

Crouzy, R. (1965). Sur la possibilité d'appliquer la théorie de la détectabilité d'un signal au calcul de certaines fonctions psychométriques. *C. R. Acad. Sci.* (Paris) *260,* 1773.

Green, D. M., and S. A. Swets (1966). *Signal Detection Theory and Psychophysics.* John Wiley and Sons, New York.

Rose, A. (1948). The sensitivity performance of the human eye on an absolute scale. *J. Opt. Soc. of Amer. 38,* 196-208.

Wyburn, G. M., R. W. Pickford, and R. J. Hirst (1964). *Human Senses and Perception.* Oliver and Boyd, London.

John Lott Brown

Problems in the Specification of Luminous Efficiency

Introduction

The spectral response characteristic of the human eye is usually represented by two functions, one of which represents the photopic or cone process and the other of which represents the scotopic or rod process. An illustration of these functions is presented in Figure 1. It is customary to

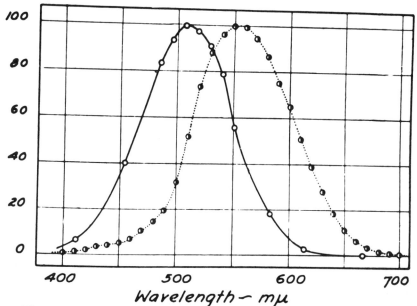

Figure 1. Spectral sensitivity for photopic (dotted curve) and scotopic (solid curve) vision

243

present these distributions of luminous efficiency with maxima set equal to one or to one hundred as in Figure 1. Ordinate values of the two curves thus may not be compared in an illustration such as this. Under most circumstances, the eye is appreciably more sensitive under scotopic conditions than under photopic conditions. For a comparison of the two processes it is more useful to compare distributions of the kind of data from which the curves in Figure 1 have been derived. Two such distributions are presented in Figure 2. Here, threshold values are presented as a

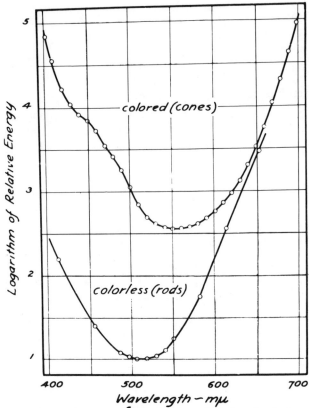

Figure 2.　Spectral energy thresholds for colored and colorless criteria of response

function of wavelength for conditions where the color of the test light is identifiable and for the simple detection of light. It is clear that the identification of color requires significantly higher energy levels than does light detection. The difference in the spectral values at which minimum energy is required corresponds to the difference in spectral values for maximal sensitivity as illustrated in Figure 1. The elevation of the energy values for the detection of color is sufficiently great, however, that all of these thresholds are higher than those for detection of light, even at the long-wavelength end of the spectrum.

The curves in Figure 2 perhaps provide a better comparison between photopic and scotopic processes, but it must be recognized that the relative positions of the two curves in this figure are themselves somewhat arbitrary. The curves will fall in different relative positions with changes in a variety of conditions. For example, if the size of the test field used in the measurement of the scotopic process is reduced, the position of the curve will be elevated significantly. The position of the photopic process is much less sensitive to this size. Test flash duration and the nature of the threshold criteria may also alter the relative positions of the curves.

The complete specification of luminous efficiency under all possible conditions requires the consideration of both photopic and scotopic processes as well as interactions between these processes. Specifications of luminous efficiency are usually based on the photopic process as this is of greatest concern under most conditions. Problems and inaccuracies associated with luminance specification are well known, but the specification system has proven useful nonetheless. The problems which I wish to treat relate primarily to circumstances in which the scotopic process must be considered as well as the photopic process .

Adapting Effects

The luminous efficiency of any light source is of significance not just in relation to the effectiveness of the light for seeing, but also in relation to the effectiveness of the light as an adapting stimulus or desensitizer (Brown, 1956). The differences between the photopic and scotopic spectral response processes have been exploited for the preservation of low luminance sensitivity under circumstances when the eye must be stimulated with light for the performance of visual tasks which require photopic vision. Years ago, roentgenologists recognized that restriction of light to the longer wavelengths for a period prior to examination of X-ray plates at very low luminances resulted in more rapid adaptation of the eyes to the low luminances. Most normal visual tasks could be performed with the red light and so the saving in time was achieved with minimal sacrifice. The use of red illumination in military situations is widespread, although there continues to be controversy over the exact value of this procedure.

Two light distributions are illustrated in Figure 3 in terms of radiant flux as a function of wavelength. One of these is an equal-energy spectrum

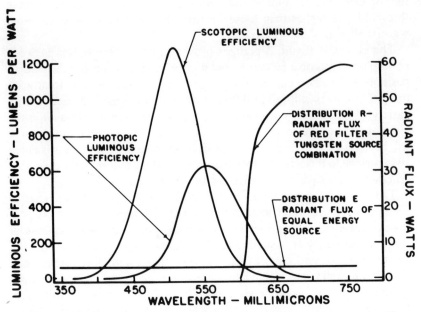

Figure 3.   Photopic and scotopic luminous efficiency curves and radiant flux as a function of wavelength for two hypothetical sources

and the other is a long wavelength distribution which includes no energy at wavelengths shorter than 600 nanometers. The spectral distributions of photopic and scotopic luminous efficiency are also illustrated in Figure 3. Here, the ordinate scale is based on the standard C.I.E. values. The data presented in Figure 3 permit interesting comparisons between the photopic and scotopic luminous flux distributions of the two light distributions. These are presented in Figure 4. The photopic luminous flux distributions are presented with solid lines. The areas under these two distributions are equivalent and hence they may be considered equally effective for vision under photopic conditions. The scotopic luminous flux distributions, presented with dashed lines, illustrate the very much higher scotopic luminosity of the equal energy source, as compared to that of the rod light distribution. The equal energy source will therefore light adapt, or desensitize, the receptors which subserve scotopic vision to a far greater extent than will the red light. Thus, performance of some visual task at a photopic level with an adequate amount of red light will have far less effect on dark adaptation than will performance of the same task with an adequate amount of light from an equal energy spectrum.

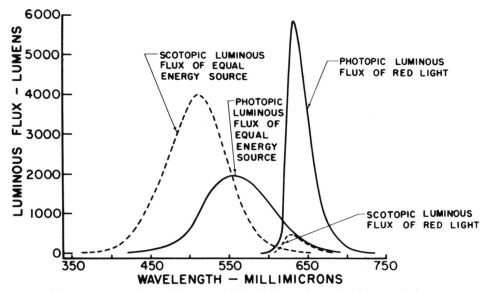

Figure 4. Luminous flux as calculated for each of the four possible combinations in Figure 3

Cone-to-Rod Ratio

It has been suggested that this advantage of long wavelength distributions of light for the preservation of scotopic vision might be subject to quantification (Chapanis, 1949). If an equal energy source is used as a reference, then any other spectral distribution might be specified in terms of the ratio between the summated photopic luminous flux and the scotopic luminous flux for that distribution. It is clear that this ratio will have higher values as the light energy is restricted to longer wavelengths. The important question is whether the quantitative value of the cone-to-rod ratio bears any relation to the savings afforded in recovery of scotopic vision following adaptation to the illumination at photopic levels.

An experiment was performed in an effort to examine this question (Brown, Adler, and Kuhns, in preparation). Curves of dark adaptation following light adaptation in each of four luminances were measured for the detection of a white test light. Light adaptation luminances were 100, 10, 1, and 0.1 millilamberts. The wavelength distribution of the adapting light was varied with the aid of Corning color filters. Nine different adapting colors were used at all except the highest adapting luminance. The deep blue filter (Corning 5543) so reduced the available light that a

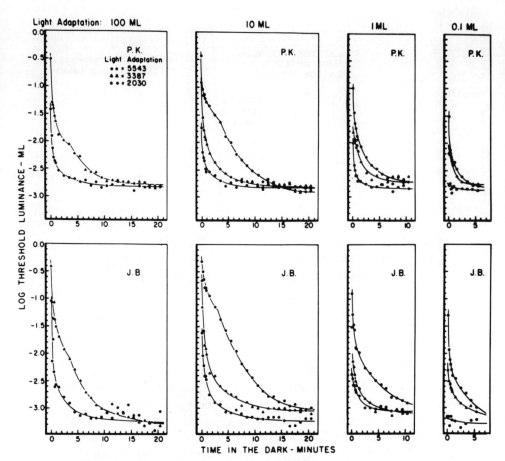

Figure 5. Dark adaptation curves following each of four adapting luminances. Wavelength distributions of the adapting light were varied with corning filters (5543, 3387, and 2030). Threshold measurements were made with a white test light. Observers P.K., J.B.

luminance of 100 millilamberts could not be achieved with it in the system. The results for three of the color filters are presented in Figure 5 for each of two subjects. Color filters represent a deep blue (5543), a yellow (3387), and a deep red (2030). For all conditions of light adaptation, dark adaptation is very nearly complete after fifteen minutes. The important generalization from data presented in Figure 5 is that, for all conditions of light adaptation and for both observers, the minimum threshold in dark adaptation is achieved much more quickly following exposure to the red adapting light than is the case with either of the other two adapting lights. Recovery is slowest following exposure to the blue adapting light.

The important question to be raised about these results is whether or not some index of recovery of dark adaptation can be found which will show a useful relation to the cone-rod ratio of the adapting light. A variety of indices of recovery were selected, from threshold after one second of dark adaptation to threshold after five or six minutes of dark adaptation. It is too early for the level of the threshold after one second of dark adaptation to be of very much practical significance while the thresholds after five and six minutes of dark adaptation occur sufficiently late so that important differences may be concealed. All of the indices investigated showed a decrease in threshold with an increase in the logarithm of cone-to-rod ratio, although the amount of decrease was somewhat reduced after five to six minutes of dark adaptation.

The relation between a log threshold luminance and the logarithm of cone-rod ratio of the adapting light are presented in Figure 6 for mean

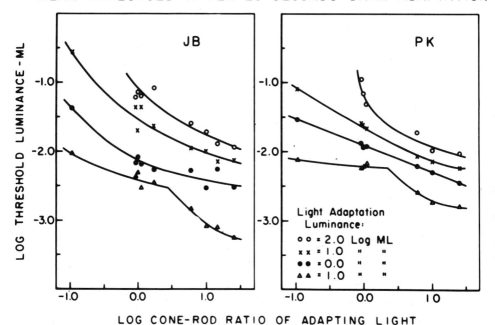

Figure 6. Mean threshold after 20 seconds of dark adaptation plotted as a function of the logarithm of the cone-to-rod ratio of the adapting light. Observers P.K., J.B.

threshold values after 20 seconds of dark adaptation for each of the two observers. The four curves represent four luminances of light adaptation. With an increase in the logarithm of cone-rod ratio of the adapting light, there is a decrease in threshold. The implication of this is that the threshold for a scotopic process is lower as cone-rod ratio is increased. Since the photopic luminance is a constant at all cone-rod ratios, any decrease in threshold should reflect reduced light adaptation of the rods. Increase in cone-rod ratio actually represents a decrease in scotopic adaptation luminance. The points on all four of the individual curves in Figure 6 could be replotted on a graph of log threshold luminance versus log scotopic adapting luminance. All that is necessary to accomplish this is to shift the upper curves to the left along the log cone-rod ratio axis by an amount equal to the difference in log adapting luminance for these curves and the log adapting luminance for the lowest curve. Thus the uppermost curve is shifted to the left by three log units, the next by two log units, and the third by one log unit. When this is done, all of the points fall to a reasonable approximation on one smooth curve which drops at a decelerated rate. When this transformation has been made the coordinates of the graph are log threshold luminance versus log scotopic adapting luminance with scotopic adapting luminance decreasing from left to right. At the end of the curve, corresponding to the end of the lowest curves in Figure 6 at the highest log cone-rod ratio values, there is a break in the curve and a drop to lower log threshold luminances. This corresponds to the lowest adapting luminance with orange or reddish light. The threshold values are lower than those which would be expected after 20 seconds of dark adaptation following light adaptation at equivalent scotopic luminances with a white light. This result is suggestive of the possibility that the advantage of red light may be greater than that which is to be expected simply on the basis of the scotopic luminosity of the adaptation.

Retinal Sensitization

Several investigators have suggested that red light may have a sensitizing effect on the retina. The first such suggestion was probably made by British investigators who worked on the problem of interior lighting and problems of dark adaptation during the Second World War (Admiralty Research Laboratory, 1942A, 1942B). The possibility of this effect has also been studied in this country (McLaughlin, 1952). The particular conditions in which an effect was found in the experiment reported above provided a basis for further exploration of this possibility (Kuhns, Brown, and Adler, unpublished report). Observers were adapted for five minutes at a luminance of 100 millilamberts to a narrow band of light with the peak value at 495 nanometers. The adapting field was 35 degrees in diameter and was centered at a point 10 degrees to the temporal side of the fovea in the right retina. Light adaptation was followed immediately by one minute of darkness. The observer then adapted to red light at a luminance of .12 millilamberts for two minutes or to darkness for two additional minutes. A Corning 2030 filter was used for the red light adaptation field. After the termination of the red adapting light or the three minutes of dark adaptation, test flashes of predetermined luminance were presented every 20 seconds for a period of seven minutes. The test field was a 1° circular field centered 10° from the fovea.

Sufficient data were obtained in a number of sessions to determine frequency of seeing measured for each of the luminances presented during dark adaptation. Results of this experiment are illustrated in Figure 7.

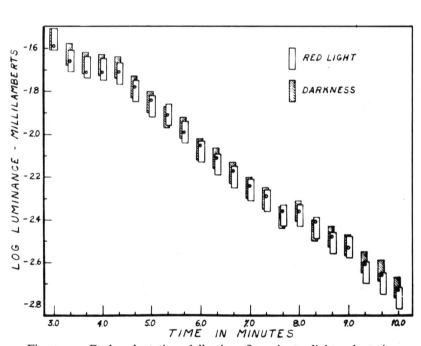

Figure 7. Dark adaptation following five-minute light adaptation with a green light at 100 millilamberts, one minute of darkness, and two minutes of red light (open rectangles) or two additional minutes of darkness (shaded rectangles). Rectangles represent the estimated range of luminances from 20% to 80% seeing. Open circles represent the luminance employed in the experiment

Thresholds following the interval during which the eye was exposed to the red light or darkness are presented as a function of time in the dark. A single luminance value was presented at each time interval during the dark adaptation process and the frequency of seeing function was determined for that luminance level. The rectangles represent an estimate of the range of luminances from 20% detection to 80% detection of the test flash. Frequency of seeing values for the test luminances employed all fell within this range and the frequencies associated with each luminance for the conditions employed are indicated by the location of the dot along the vertical extent of the rectangle in the figure. It is clear that frequency of seeing is typically higher following exposure to the red light than following exposure to darkness in the interval after light adaptation. The difference is significant at the 1% level. Exposure to a red light at a low luminance would appear to hasten dark adaptation.

The opponent processes in the human visual system may seem to provide a possible basis for the explanation of this sort of phenomenon. There is both psychophysical and electrophysiological evidence which indicates that the effect of a red light may be opposite to the effect of a green light on retinal processes. This is undeniably the case, but the opposite effects which are demonstrable psychophysically relate to color perception and are observed at levels of light substantially above threshold. It is difficult to explain any effect which consists of the lowering of an absolute light detection threshold following the exposure of the eye to additional radiant energy which may bleach photopigment. Our knowledge of energy transformation within the eye from the point where the quanta of light energy are absorbed by color-bearing molecules to the initiation of neural activity is, as yet, poorly understood. Recent evidence indicates that solid state processes may be involved (Crawford, Gage, and Brown, 1967). Red light could conceivably serve in expediting recovery of the system in some sort of spectrally dependent process (Cone, 1967). Whatever the explanation of the effect reported, it is clear that it affords luminous efficiency.

Rod-Cone Interaction

The term "mesopic" is used to refer to those levels of adaptation between the photopic range and the lower scotopic range. The explanation of the existence of a so-called mesopic range is usually given in terms of the combined function of rods and cones. The level of adaptation is certainly a factor in the determination of the nature of visual function, but it is not necessarily a limiting condition. For example, in the completely dark adapted eye, in the region of the retina which includes both rods and cones, threshold measurements may depend on cones, or photopic vision, if the appropriate criterion of threshold is selected. If luminance thresholds are measured for the identification of acuity test targets with short flashes of light, then photopic processes will be required where spatial resolution of a relatively high order is required for achievement of the criterion.

An experiment was performed to examine the question of possible interaction between rod and cone processes in the completely dark-adapted eye (Brown and Woodward, 1957). Observers were presented with a one-degree test field the center of which was located at a distance of $1\frac{1}{2}°$ away from the central fovea along a 45° meridian in the lower quadrant of the temporal retina. A fixation cross provided for appropriate localization of the direction of regard and also served as a cue for accommodation. Grating targets were presented in 0.015 second flashes in either vertical or horizontal orientation. Luminance thresholds were determined for correct identification of grating orientation. Thresholds were measured for a series of grating acuities from .054 to .490. Wavelength distribution of the test flash was varied with the aid of Corning color filters and a total of 7 different wavelength distributions was employed. Results were examined in the form of graphs of logarithm of threshold luminance presented as a function of the logarithm of visual acuity. The kind of result which might be predicted is illustrated in Figure 8. Luminance is based on the photopic luminous efficiency of the eye and hence is an inappropriate specification of the effectiveness of light under scotopic conditions. Luminance thresholds with blue light are lowered while those obtained with red light are elevated relative to thresholds obtained with white light. The hypothetical curves of Figure 8 show an increase in the logarithm of visual acuity. The function is shown to include two branches, an early branch which presumably represents a scotopic process and a later branch which represents a photopic process. It is evident that as visual acuity is increased a limiting value will be reached beyond which resolution is impossible. The limiting value for rod vision will be lower than the limiting value for cone or photopic vision. If the scotopic portion of the curve, that portion found at lower acuities and lower threshold luminance levels, is representative of a single class of receptors, all of which contain the same photosensitive substance, then the form of all of the scotopic curves should be identical, independent of the wavelength distribution of the test light employed. These portions of the curve will be distributed along the luminance axis of the graph but they will be of identical form. That is, it will be possible to superimpose any one of the branches on top of any other by simple translation along the log luminance axis. With a transition from a blue test light to a red test light the curve will be translated upwards, and for some spectral distributions will fall above the threshold for the photopic

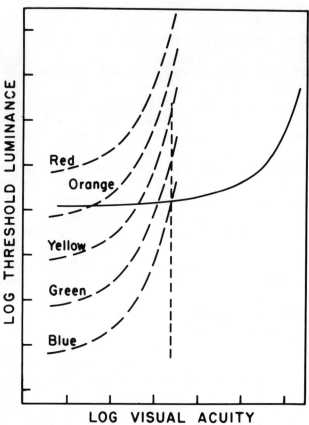

Figure 8.   Hypothetical distributions of log threshold luminance as a function of the log visual acuity of a threshold criterion for the dark-adapted eye. Solid curve represents the photopic function. Dashed curves represent the scotopic function for each of five wavelength distributions of the test light

process. The position of the photopic branch of the curve should be invariant on a luminance axis, because the calculation of luminance is based on the spectral sensitivity of the photopic receptors.

The results obtained experimentally are presented in Figure 9. Four of the curves have been displaced upwards by varying amounts above the curve obtained with Wratten neutral density filters and two have been displacd below the neutral density curve. The purpose of this shift is to afford more ready comparison among the various functions. The shapes of the portions of the curve associated with relatively high levels of visual

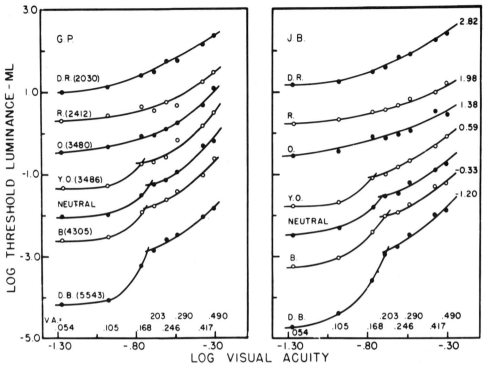

Figure 9. Log threshold luminance as a function of the visual acuity requirements of the test pattern for each of seven wavelength distributions. Curves have been shifted vertically above or below the "neutral" curve by amounts indicated at their right hand extremes. Observers G.P., J.B.

acuity are not all the same but they do not vary in a systematic way from the short wavelength distribution of illumination up through the longest wavelength distribution of illumination. The scotopic branches of the four lowest curves appear to be displaced upwards with a change from short-wavelength distribution to long-wavelength distribution of the test lights, but there is also a change in the form of the curves. They cannot be superimposed by simple translation along the log threshold luminance axis. This difference in form of the curves suggests that they represent processes having different spectral luminous efficiency functions. There may be an increase in the contribution of cones with transition toward longer wavelength of the test light, even before the "photopic" curve emerges. Thresholds obtained in the range of visual acuities from .105 to .203 may depend upon combined function of rods and cones.

A further exploration of this possibility was undertaken with mono-chromatic lights (Brown, Phares, and Fletcher, 1960). Relative energy thresholds were measured in 10 nanometer steps from 400 nanometers up through 710 nanometers. Criteria of theshold consisted of detection of light and the identification of the orientation of each of three acuity gratings requiring visual acuities of .11, 0.20, and 0.33. The results are presented in Figure 10 for each of two observers. The logarithm of relative

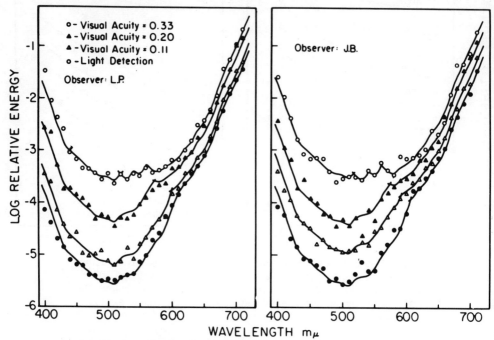

Figure 10.   Spectral threshold energy distributions for three visual acuity criteria and for light detection

energy is plotted as a function of wavelength for each of the four threshold criteria. Results for the two observers are strikingly similar.

The apparent similarity in the distributions of data points at spectral extremes prompted an investigation of the possibility of developing fitting curves based on an averaging procedure. It was assumed that all of the data could be fitted by two functions, a short-wavelength function and a long-wavelength function. The technique of fitting would be to shift the position of the short-wavelength function upward relative to the position of the long-wavelength function with change in the criterion of threshold toward higher acuity. Fitting functions were developed by averaging short-wavelength distributions of points and long-wavelength distributions of points, but the distributions of data points were broken at different wavelengths for each of the four criteria of threshold. They were then shifted for each observer by a fixed amount for each criterion so that all of the data distributions conformed approximately to the distribution for the light-detection threshold criteria. Averages of the short-wavelength and the long-wavelength distributions were computed. The results are illustrated in Figure 11 for each of the two observers. Note that in the range

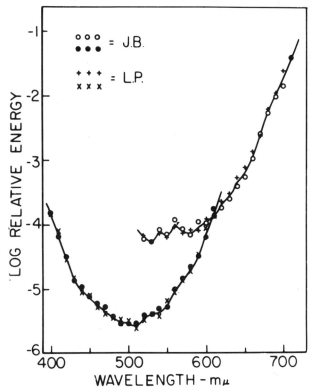

Figure 11. Average threshold data for all criteria in Figure 10. See text for explanation of the method of computation

between 520 and 610 nanometers (millimicrons) the points for the several threshold criteria are selectively assigned either to the long-wavelength function or the short-wavelength function on the basis of the author's discretion.

The interesting point to note in Figure 11 concerns the very close similarity of the distributions of averaged data points for each of the two observers. The curves in Figure 11 suggest a possible explanation for the results presented in Figure 10. The distributions of points in Figure 10 may be the result of a combination of scotopic spectral sensitivity and photopic spectral sensitivity. As the criterion of threshold demands a higher level of acuity, the scotopic threshold function is simply shifted upward relative to the photopic threshold function. The distribution of thresholds for a given criterion represents a combination of the scotopic and photopic processes with a unique relative position of each which is determined by the criterion itself.

The interpretation is subject to test. Data were available for one of the observers for the scotopic spectral relative energy function and the photopic spectral relative energy function. These were measured with the aid of a 42-minute-diameter test field located either 10 degrees away from the fovea or centered on the fovea. The distributions of data points for each of these conditions are illustrated in Figure 12. The positions of these data point distributions on the log relative energy axis are based on the actual threshold measurements themselves. The fitting curves of Figure 11 have been positioned along the log relative energy axis to provide the best possible fit of the data points. The short-wavelength fitting curve provides a very reasonable fit of the curve for relative energy thresholds. The long-wavelength curve provides a reasonable fit of the long-wavelength portion of the foveal relative energy threshold data but is much too low for wavelengths shorter than approximately 610 nanometers. Thus, although the short-wavelength fitting curve may present rod function, the long-wavelength curve cannot be explained adequately in terms of foveal cone spectral threshold data. It is possible that the region between 520 nanometers and 610 nanometers is a region in which thresholds are dependent upon a variety of spectral response functions which in turn are determined by the spatial resolution requirement of the threshold criterion.

The possibility of combining the photopic spectral response function and the scotopic spectral response function (represented by the distributions of data points in Figure 12) in some simple fashion which would provide a fit of the experimental data of Figure 10 was investigated. Resolution increases under both scotopic and photopic functions with increase in available luminance. Thus, it may be said that with an increase in the spatial resolution required by a threshold criterion there must be an increase in the level of illumination. The "sensitivity" may therefore be said to decrease for either the scotopic or photopic process as resolution requirement is increased. A simple assumption is that sensitivities for the photopic and scotopic processes are additive after they have been

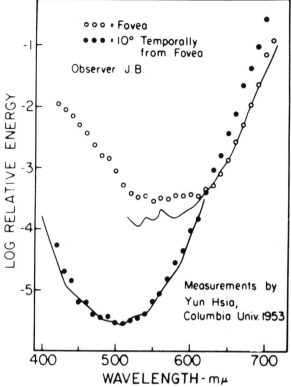

Figure 12. Data for light detection in the dark-adapted eye measured foveally and temporally with a 42-minute test field. Solid curves are those presented in Figure 11

weighted for the particular criteria under investigation. An equation for the summation of photopic and scotopic sensitivity is as follows:

$$S_{i\lambda} = c_i S_{p\lambda} + r_i S_{s\lambda}$$

In this equation $S_{i\lambda}$ represents the sensitivity of the visual system for a selected criterion of threshold designated by the letter i at a specific wavelength, λ. The sensitivities of photopic and scotopic processes, $S_{p\lambda}$ and $S_{s\lambda}$ are multiplied by appropriate weighting factors, c_i and r_i, and then added. It is assumed that the constants c_i and r_i will be invariant at all wavelengths for a given threshold criterion, i. Values of $S_{p\lambda}$ and $S_{s\lambda}$ may be calculated from the data presented in Figure 12. Values of $S_{i\lambda}$

are available by computation from the results presented in Figure 10. The above equation may be written for each of two different wavelength values for any of the three visual acuity criteria employed. The two equations can be solved for values of c_i and r_i. Several solutions were calculated for the constants, c_i and r_i, and all yielded approximately the same results. The equation presented above provided a reasonable fit of the data presented in Figure 10. The fits achieved are illustrated in Figure 13 for each

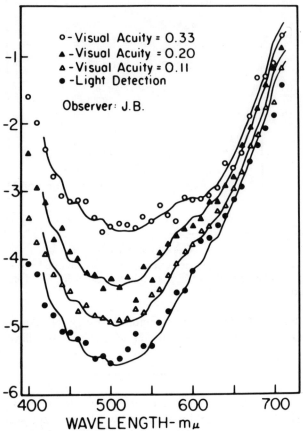

Figure 13. Data of Figure 10 for observer J.B. fitted by summing photopic and scotopic sensitivities as described in text

of the four criterion thresholds for one of the observers. The fits are reasonably good even though they are based on data which was measured a number of years prior to the data illustrated in Figure 10.

This result encouraged us to consider the possibility of applying data of the present experiment to the computation of luminance thresholds for the resolution of acuity targets with broad band distributions of colored light. If energy threshold values are available over a range of visual acuities for each of two wavelengths, one of which is in the long wavelength region at 610 nanometers or longer, and the other in the short wavelength region in the vicinity of 500 nanometers, then a spectral sensitivity function may be computed for the resolution of any acuity criterion in that range by the dark adapted eye. The relation between the logarithm of threshold energy and the logarithm of the visual acuity of the criterion task with a grating test object is shown in Figure 14 for each of two wave-

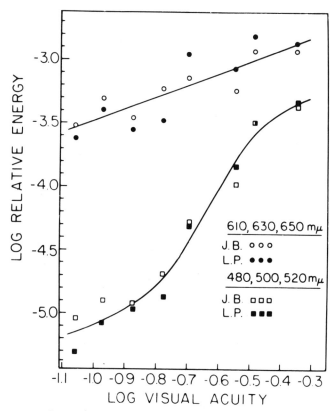

Figure 14. Log relative threshold energy as a function of log visual acuity required by the threshold criterion for each of two wavelength regions. Observers L.P., J.B.

length regions. It is clear that two quite different functions are required for the long and the short wavelength results.

With the aid of the data presented in Figure 14, it is possible to compute weighting constants for the equation given above for any visual acuity between 0.09 and 0.45. The possible application of such a function was investigated with the aid of data which have been presented above in Figure 9. Luminance thresholds for three of the spectral conditions portrayed in Figure 9 were calculated with the aid of spectral sensitivity curves derived for each of the eight visual acuity levels involved. The wavelength distributions of the illumination were obtained with the use of a Corning 3480 filter for orange light, Wratten neutral filters, and a Corning 5543 filter which provided a deep blue light. The computations are illustrated in Figure 15 by the dashed lines. The original data have

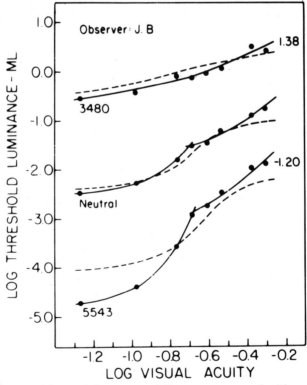

Figure 15. Three of the curves of Figure 9 compared with computed fits (dashed lines) based on the data in Figures 12 and 14

been fitted with solid curves. The results for the 3480 filter and the neutral filters are in reasonably good agreement with experimental data. In the case of the blue (5543) filter the computed curve is much too shallow. The prediction from the calculations outlined apparently is based on an excessively low estimate of sensitivity in the short-wavelength part of the spectrum. It must be concluded that additivity of various wavelength components of a broad band illuminant can only be predicted to a very rough approximation on the basis of the functions generated. Thus the results presented in Figure 10 and in Figure 14 are of only very limited value for the specification of luminous efficiency.

An experiment was undertaken to provide additional information as to the nature of the transition from photopic to scotopic vision during dark adaptation (Brown, Kirchenstein, and Shupp, 1960). The data presented in Figures 9 and 12 suggest that even when the eye is completely dark adapted, thresholds may depend upon the combined action of photopic and scotopic processes. With increase in the visual acuity required to reach the criterion of threshold, the relative contribution of the photopic processes is apparently increased.

The time course of dark adaptation was measured with each of two visual acuity criteria and with light detection. Two different spectral distributions of the test flash were employed, a yellow and a blue. The results are presented in Figure 16 for each of two observers for the three distributions of the test flash. The ordinate axis is expressed in scotopic luminosity. Computations were based on the C.I.E. scotopic luminosity function. Prior to performance of the experiment it was assumed that the late branches of the dark adaptation curves for light detection would be of identical form for each of the spectral distributions of the test flash. The assumption was also made that, with the introduction of a visual acuity criterion of threshold, there would be an increased contribution by a photopic process and that this might be more marked with the yellow test illuminant than with the blue. The late branches of dark adaptation curves were therefore expected to be of somewhat different form for the two acuity criteria of threshold.

It is evident that these predictions were not borne out. The results presented in Figure 16 indicate the late branches of dark adaptation curves are virtually identical for yellow and blue test illuminants when acuity criteria are employed, but they differ in form for light detection. The light detection curve is steeper with blue light than with yellow light. This implies that the curve obtained with blue light represents a different receptor population than that represented by the curve obtained with yellow light. With increased dark adaptation there is a relatively greater increase in the scotopic contribution when blue light is employed. The curve repre-

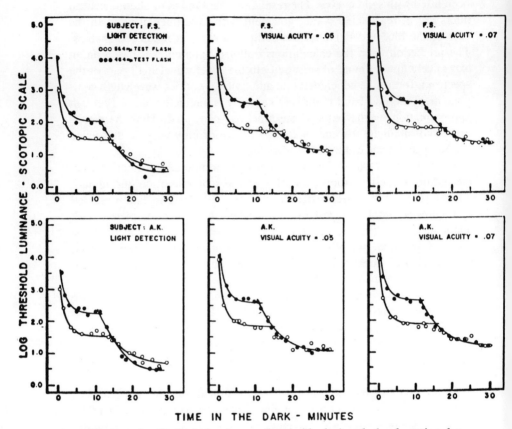

Figure 16. Scotopic luminance thresholds during dark adaptation for each of three threshold criteria, light detection and visual acuities of .05 and .07. Test flashes were monochromatic light at 564 or 464 nanometers

senting results with blue light thus falls somewhat more steeply than that for yellow light. These results are quite different than some results reported by Rushton (1961) several years ago. In Rushton's experiment a yellow and blue light were employed but threshold determinations for the yellow and blue were made alternately in a single experimental session. The results reported here were obtained with only a single wavelength distribution of the test light in any given experimental session. The explanation of the difference may rest with the criterion of threshold. Observers report that the appearance of the light at threshold is very different with short wavelength illumination as contrasted with longer wavelength illumination. The blue light was seen at threshold as a vaguely defined blur. On the other hand, the yellow light was always seen as a reasonably well defined circular test field, if seen at all. Observers had little difficulty in adopting a criterion of threshold which was appropriate to the conditions prevailing and which afforded a minimum threshold determination. On the other hand, when observers are required to detect both blue test flashes and yellow test flashes in a single session, there may be a tendency for them to adopt a criterion of appearance which is somewhat more uniform for both conditions. This would tend to elevate the blue threshold determination and might alter the nature of the receptor population involved. The use of an acuity criterion forces the observer to depend on spatial resolution in order to reach that criterion. The same criterion must be reached quite independent of differences in the wavelength distribution of test illumination. The conclusion is reached that a more nearly pure scotopic process is involved in the determination of dark adaptation curves which represent light detection thresholds with a blue test light than under other conditions. It may be impossible to obtain pure scotopic function unless the test illuminant is restricted to relatively short wavelengths and no spatial resolution requirement need be met. The results of this experiment place in question the validity of using even the standard scotopic luminous efficiency function under a variety of conditions which are presently considered to meet the criteria for scotopic function.

Rod-Cone Interference

A question that has concerned a number of investigators is the possible inhibitory effect of one receptor process on the function of the other. For example, the function of the rods is presumed to be inhibited, or at least obscured, by the function of cones when relatively high luminances are available, even though these may be made available in a brief flash to the dark adapted eye. On the other hand, it has been suggested that photopic, or cone, responses may be detrimentally affected by the function of the scotopic receptor process in the dark adapted eye.

An experiment was performed to test the possibility that scotopic or rod function might interfere with photopic or cone function (Brown, Metz, and Yohman, in preparation). Dark adaptation was measured with each of two procedures after light adaptation with either a yellow or blue adapting light. Photopic luminosity of the two adapting lights was a constant so that the adapting effect on the photopic process would be the same for each of the two adaptation conditions. Under these circumstances the scotopic process would receive a higher level of light adaptation with exposure to the blue adapting light than with exposure to the yellow adapting light. During dark adaptation the scotopic process would there-

fore be expected to recover more gradually and the late branch of the dark
adaptation curve should therefore appear somewhat later following blue
light adaptation than following yellow light adaptation. The two procedures
by which dark adaptation was measured included one in which subjects
were required simply to detect the presence of a test flash and another in
which they were required to identify the orientation of a grating test pattern.
The acuity level for the test pattern was such that only photopic processes
were believed to play any role in the achievement of threshold.

The results are presented in Figure 17 for each of two observers. Vir-

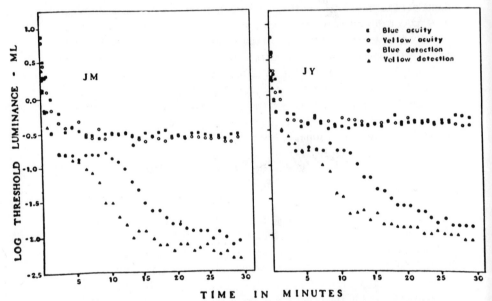

Figure 17.   Dark adaptation following light adaptation at 1500 millilamberts for 5 minutes with either yellow or blue light. Upper curves
represent luminance thresholds for identification of a 0.25 visual acuity
criterion and lower curves represent luminance thresholds for light
detection. Test flashes were white light. Observers J.M. and J.Y.

tually identical curves of dark adaptation are obtained following yellow
and blue light adaptation when the criterion of threshold is the identification of an acuity test object. The curves come down to a final level
smoothly and there is no late branch characteristic of scotopic function.
When light detection is employed a late branch is observed and it is later
in appearing when a blue light adaptation field has been employed. The
question of interest was whether or not there would be any evidence of

curve at the point in time where the scotopic process appears. The upper curve for an acuity criterion represents relatively pure photopic function and the lower curve shows each of two points in time where the scotopic function comes in, depending upon the prior condition of light adaptation. If the emergence of the scotopic process caused interference it should be identifiable in the photopic curve for an acuity criterion and the point of its appearance should depend upon the wavelength distribution of the adapting light. The results presented in Figure 17 provide no evidence for the conclusion that photopic processes are inhibited with the emergence of the scotopic process.

In summary, it has been shown that the specification of luminance efficiency of radiant energy is immensely complicated by the presence in the visual system of two spectral response processes which may interact in a variety of ways. Specification of luminance efficiency depends not only on the level of adaptation of the observer, but on a knowledge of the visual task which is to be performed, as well as the retinal region involved and the temporal characteristics of the stimulus. Under a variety of conditions there may be interaction between photopic and scotopic processes.

References

Admiralty Research Laboratory (Great Britain) (1942A). ARL/N4/0.360, October.

Admiralty Research Laboratory (Great Britain) (1942B). ARL/N6/0.360, November.

Brown, J. L. (1956). Review of the cone-to-rod ratio as a specification for lighting systems. *Illum. Eng. 51*, 577-584.

Brown, J. L., H. E. Alder, and M. P. Kuhns (Report in preparation). Experimental evaluation of the cone-to-rod ratio.

Brown, J. L., A. Kirchenstein, and F. Shupp (1960). Contributions of rods and cones to visual thresholds during dark adaptation. ONR Contract N-ONR 551 39 Report.

Brown, J. L., R. J. Metz, and W. J. Yohman (Report in preparation). Rod-cone interference.

Brown, J. L., L. Phares, and D. E. Fletcher (1960). Spectral energy thresholds for the resolution of acuity targets. *J. Opt. Soc. Amer. 50*, 950-960.

Brown, J. L., and L. K. Woodward (1957). Rod-cone interaction in the dark adapted eye. *Optica Acta 3*, 108-114.

Chapanis, A. (1949). How we see: A summary of basic principles. *Human Factors in Undersea Warfare*. National Research Council.

Cone, R. A. (1967). Early receptor potential: Photoreversible charge displacement in rhodopsin. *Science 155*, 1128-1131.

Crawford, J. M., P. W. Gage, and K. T. Brown (1967). Rapid light-evoked potentials at extremes of pH from the frog's retina and pigment epithelium and from a synthetic melanin. *Vision Research 7,* 539-551.

Hecht, S., and Y. Hsia (1945). Dark adaptation following light adaptation to red and white lights. *J. Opt. Soc. Amer. 35,* 261-267.

Hecht, S., and R. E. Williams (1922). The visibility of monochromatic radiation and the absorption spectrum of visual purple. *J. Gen. Physiol. 5,* 1-33.

Kuhns, M. P., J. L. Brown, and H. E. Adler (Unpublished report). Red light sensitization.

McLaughlin, S. (1952). A facilitative effect of red light on dark adaptation. Research report, Project No. NM001 059.28.01 U.S. Naval Air Station, Pensacola, Florida.

Rushton, W. S. (1961). Dark adaptation and the regeneration of rhodopsin. *J. Physiol. 156,* 166-178.

Adriana Fiorentini

Excitatory and Inhibitory
Interactions in the Human Eye

The knowledge of the functional properties of the neural network in the eye and in the brain is relevant to the study of the human visual performance, especially as regards the ability of the eye to distinguish detail. And, indeed, two main factors may be expected to limit visual resolution: the sharpness of the retinal image, and the ability of the retina-brain system to resolve the retinal images.

The light distribution in the retinal image can be measured by objective methods in the intact living human eye. The most recent measurements (Campbell and Gubisch, 1966) indicate that the optical quality of the human eye is substantially better than it was estimated before. For a precisely focused eye and with an optimal pupil diameter, the half width of the retinal image of a bright line is only about one minute of arc. Under these optimal conditions visual resolution is limited more severely by the performance of the neural visual system than by the poor optics of the eye (Campbell and Green, 1965).

Our knowledge of the mechanisms involved in the visual processes is based mainly on recordings of the response to light of single neurons in the eye or brain of animals. The fact that the response of a receptor to a light stimulus is affected by the simultaneous excitation of neighboring receptors seems to represent a general rule for the visual system, although these spatial interactions present different characteristics in animals of different species. In the invertebrates, the response of a single receptor unit is inhibited by simultaneous illumination of neighboring units (Hartline et al., 1956; Hartline and Ratliff, 1957, 1958; Ratliff and Hartline, 1959). In the much more complex eye of vertebrates spatial interactions are more complex. A retinal ganglion cell may respond to illumination of any part of a relatively large region of the receptor layer—the receptive field of

269

the ganglion cell—and separate stimuli to different parts of the receptive field may yield different response patterns. In particular, the central and the peripheral part of the receptive field may have antagonistic functions, so that a stimulus to the periphery inhibits the response to a central stimulus (Kuffler, 1953; Barlow et al., 1957; Hubel and Wiesel, 1960). The receptive fields of different ganglion cells overlap to some extent in the retina and further interaction may occur at a higher level in the visual system (Hubel and Wiesel, 1962; Baumgartner et al., 1965).

Recordings of the activity of single neurons cannot yet be obtained from human subjects. The results of psychophysical experiments, however, strongly suggest that lateral interactions occur in the human visual system, similar to those that are observed directly in animals (Beitel, 1936; Ratoosh and Graham, 1951; Battersby and Wagman, 1964; Ratliff, 1965).

Reasoning by analogy, it has been assumed that the human retina is made up of partially overlapping receptive fields, consisting of a central excitatory region and of a surrounding inhibitory region, and estimates of the sizes of the two regions of the "receptive field" have been derived from the results of various psychophysical experiments (Von Békésy, 1960; Glezer, 1965; Bryngdahl, 1966; Thomas, 1965). Moreover, several attempts have been made to find a quantitative expression for the spatial interaction in the human visual system and to investigate the characteristics of spatial summation of antagonistic responses in the human "receptive field."

Psychophysical Methods for the Study of Spatial Interaction

Most of the psychophysical research on spatial interaction has been carried out by taking as a variable a spatial parameter of the stimulus pattern and by finding the stimulus threshold for various values of that parameter.

One of the most useful methods for the study of spatial resolution is the determination of the contrast required for resolving sinusoidal gratings of variable spatial frequency. This method was introduced by Schade (1956) and extensively applied later on. The function relating the reciprocal of the contrast threshold to the spatial frequency is called the contrast sensitivity function or modulation transfer function (MTF) of the visual system, following the terminology of linear system analysis (Fourier analysis in the space domain).

From the known transfer function it is possible to compute the line (or point) weighting function of the visual system, that can be regarded as the spatial distribution of the subjective brightness in the perceived image of a line (or point) source, and can be used to compute the response to other stimulus patterns. As all other applications of Fourier analysis, these computations are meaningful only if the system to which they are applied is linear. This is certainly true for the optical component of the visual system, but may not be true for the neural component.

The MTF of the neural visual system has been obtained by measuring the contrast sensitivity for a sinusoidal interference pattern formed directly on the retina without the intervention of the optical system of the eye (Campbell and Green, 1965). The contrast sensitivity (reciprocal of the contrast threshold) has a peak at a frequency of a few cycles per degree and falls off exponentially for frequencies higher than ten cycles per degree.

The fall off in the psychophysical contrast sensitivity at low spatial frequencies is ascribed to lateral inhibition, and the fall off at higher frequencies to spatial summation. Recently, the methods of Fourier analysis have been applied to the study of contrast sensitivity of retinal ganglion cells of the cat (Enroth-Cugell and Robson, 1966). The contrast sensitivity function of one kind of ganglion cell has been found to resemble the psychophysical transfer function, although the two functions occupy different positions along the scale of spatial frequencies.

A quite different method for the study of spatial interaction is the measurement of the (threshold) response to a test patch of variable size. When this method was applied to the study of neural interaction in the receptive field of ganglion cells, the antagonistic activity of the central and peripheral regions of the receptive field was demonstrated by the existence of an optimal diameter for the test spot (Barlow, 1953; Barlow et al., 1957). Some difficulties have been encountered in the attempt to duplicate psychophysically these electrophysiological findings: determinations of the contrast threshold for a test spot of variable size show that there is no optimal diameter for the test spot (Barlow, 1958).

Recently, however, Westheimer (1965, 1967) has successfully modified this technique by measuring the contrast threshold for a small test spot of constant size that was flashed against a circular background of variable size. Starting with a background field having the same size as the test spot and then increasing the background diameter, the contrast threshold for the incremental spot is found to increase up to a maximum and then to decrease. These effects have been interpreted as showing spatial interaction of background (adapting) stimuli: excitatory interaction would occur in the retinal region that immediately surrounds the test spot, and inhibitory interaction in the region beyond the critical diameter of the background.

This interpretation is based on the assumption that a change in the threshold of the test spot reflects a similar change in the level of adaptation set up by the background stimulus: the higher the level of excitation in the background, the more energy must be in the test stimulus for it to be detected.

Still another method can be used to investigate spatial interaction: the (threshold) response to a small spot of light is measured in the presence of a second spot appearing at a variable distance from the first (Beitel, 1936; Ogle, 1962; Thomas, 1965). A similar method was used by Kuffler (1953) to demonstrate excitatory and inhibitory interaction in the receptive field of the cat's ganglion cells.

By a suitable choice of the stimulus conditions this method can offer a quantitative description of spatial interaction, as shown by the following experiment.

Interaction Between Two One-Dimensional Stimuli

A bright test line is superimposed to a uniformly illuminated background ($17cd/m^2$). A second line (inducing line) having the same length and width as the first is viewed against the same background. The inducing line is parallel to the test line and can be set at a variable separation from the test line (Figure 1-a).

The luminance of the inducing line is adjusted to a level just below its increment threshold (that was determined preliminarily in the absence of the test line) and is kept constant at that subliminal level. Then the incre-

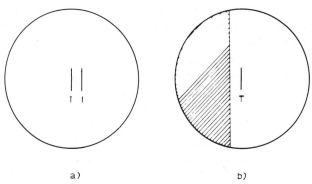

a)   b)

Figure 1. (*a*) Test line (T) and inducing line (I) viewed against a 3° circular background. The test line was brighter than the background; the inducing line was brighter in one experiment and darker in another. (*b*) Test line (T) viewed against a step illumination pattern, 3° diameter. The test line was centered against the background and the inducing line or the light-dark edge were set at a variable distance from the test line ment threshold for the test line is measured: (1) in the presence of the subliminal inducing line, for various separations of the two lines, and, (2) in the absence of the inducing line.

Threshold determinations are performed by the method of adjustment. Both lines are presented continuously and the subject fixates the test line while adjusting its luminance to threshold.

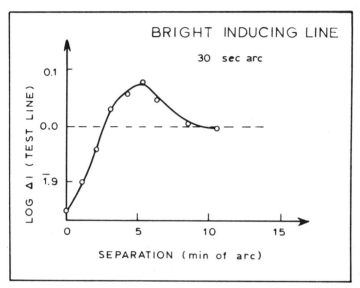

Figure 2. Log threshold for the test line as a function of separation from the incremental inducing line of 30 sec arc width. Zero on the axis of ordinates denotes the threshold obtained for the test line in the absence of inducing line

Figure 2 shows the results obtained with two lines of thirty second arc width and thirty minute arc length. Viewing was monocular, with a 2.4 millimeter artificial pupil and at a distance of one meter. The points in Figure 2 represent the test line threshold obtained for various separations of the two lines. Each point is the average of fifteen to twenty threshold determinations. The broken line indicates the threshold obtained in the absence of the inducing line.

As expected, the changes in the threshold that are obtained in the presence of the subliminal inducing line are very small. It is clear from the figure, however, that the subliminal line does not affect the threshold for the test line and that the effects are opposite according to whether the two lines are very close to each other or separated further apart.

The most obvious interpretation of these effects is that a summation occurs when the separation of the lines is less than about 2.5 minutes of arc and (mutual) inhibition for greater separations, up to about 8 minutes of arc. These figures compare fairly well with other estimates of the spatial extent of excitatory and inhibitory interaction in the light adapted human fovea (Glezer, 1965; Bryngdahl, 1966; Westheimer, 1967).

Almost identical results are obtained with an inducing line of two minutes of arc width (Fig. 3).

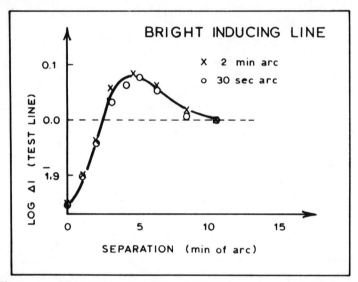

Figure 3. The same as for Figure 2, but for an incremental inducing line of 2 min arc width (crosses). The open circles are replotted from Figure 2 for comparison

If the incremental inducing line is replaced by a line just dimmer than the background, the changes in the threshold for the test line are approximately the negatives of those obtained in the previous case: the threshold is higher when the lines are close by, and lower when the lines are separated further apart. The results shown in Figure 4 have been obtained with a decremental inducing line of two minutes of arc, whose luminance was 0.02 log units below that of the background (Fiorentini and Mazzantini, 1966).

These result can again be interpreted as showing that excitatory interaction occurs within a small range of stimulus separations and inhibitory interactions for greater separations. And indeed, we may expect that, when the dim line is imaged within the excitatory region, the incremental luminance required to detect the test line has to be increased in order to

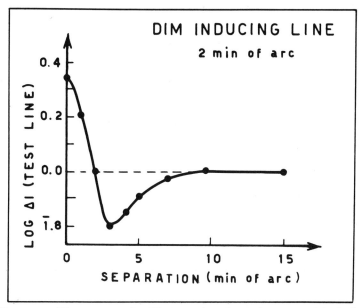

Figure 4. The same as for Figure 2, but for a decremental inducing line of 2 min arc width. From Fiorentini, A., and Mazzantini, L., *Atti Fond. G. Ronchi, 21:* 737–747 (1966)

compensate for the local decrement in background illumination. If instead the dim line is imaged in the inhibitory region, the response to the test line is partially released from inhibition by the background and a lower luminance is required for the test line to reach threshhold.

It has to be noted, however, that the threshold curve obtained with the decremental inducing line is not exactly the negative of the curve for the incremental inducing line. In particular, the peak thresholds in the inhibitory region occur at somewhat different locations. This is better seen in Figure 5 where the threshold data relative to the incremental and decremental lines are replotted on a linear scale. The curves represent the "amplitude" $A(x)$:

$$A(x) = \frac{\Delta I(\infty) - \Delta I(x)}{\Delta I(\infty) - \Delta I(o)}$$

where $\Delta I(x)$ is the threshold (in linear units) measured for a separation x of the lines, $\Delta I(o)$ is the threshold measured when the two lines are superimposed, and $\Delta I(\ )$ is the test line threshold measured in the absence of the inducing line.

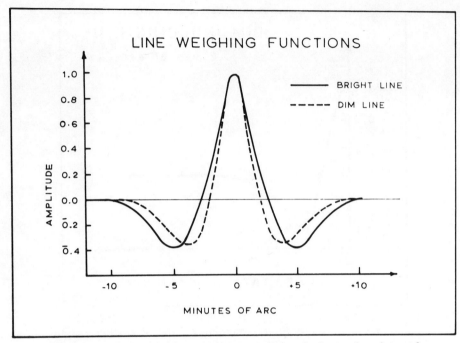

Figure 5. Line weighting functions computed from the data of Figure 2 (solid line) and of Figure 4 (broken line)

The two curves of Figure 5 can be regarded as the line weighting functions of the visual system, relative to the conditions of the above experiment (Ratliff, 1965). The differences in the two weighting functions indicate the existence of non-linearities.

As an independent check of linearity, the inducing line can be replaced by a step pattern of illumination. This consists of two uniform fields of different luminances with a common straight edge (Figure 1-b). The test line is parallel to the edge and the increment threshold is determined for the test line at various distances from the edge.

If spatial summation of excitatory and inhibitory effects were linear, the function that relates the threshold to the distance from the edge should be symmetrical. That it is not exactly so is shown in Figure 6, where the test line threshold is plotted against distance from a just perceptible contrast edge (Fiorentini and Zoli, 1967).

The threshold is higher on the dimmer side of the edge and lower on the brighter side, but the highest and lowest thresholds are obtained at somewhat different distances from the edge and the changes in threshold on opposite sides of the edge are not exactly symmetrical.

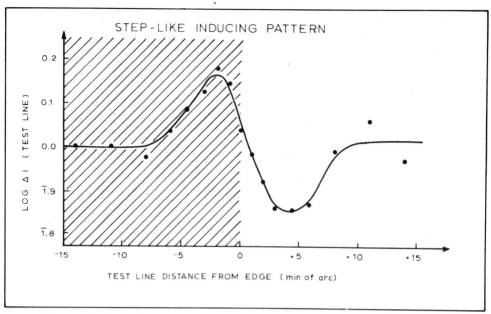

Figure 6. Log threshold for the test line as a function of distance from a light-dark straight edge of luminal contrast. Luminances: dimmer field (on the left in the figure) 17 cd/m²; brighter field 17.25 cd/m²

Small-Signal Non-Linearities in Spatial Summation

The existence of non-linearities in the processes of spatial summation is important especially in view of the application of Fourier analysis to the study of spatial resolution.

Electrophysiological experiments have shown that summation over the receptive fields of retinal ganglion cells is linear for some cells, but is very non-linear for others (Enroth-Cugell and Robson, 1966). Obviously non-linearities may exist at higher levels in the neural system, where individual neurons can be activated by a number of overlapping retinal units or be influenced by non-retinal inputs (Jung, 1961; Hubel, 1963).

Non-linearities are observed in psychophysical experiments involving high-contrast patterns (Fry and Alpern, 1953; Marimont, 1963; Fiorentini and Zoli, 1967) or changes in the mean level of adaptation (Kinkaid et al., 1960; Patel, 1966), while spatial integration is usually assumed to be linear near the contrast threshold.

Very recently, however, Thomas (1968) has offered evidence for non-linearities of spatial integration with stimuli of low contrast. Thomas measured the changes in perceived brightness of bright or dim lines that are obtained by flanking the lines with a variable number of rectangular inducing fields. The effects obtained by removing some of the inducing patterns are not exactly symmetrical to the effects of adding inducing patterns. Moreover the weighting functions relative to experiments with the bright line differ from those obtained with the dim line.

In the experiments described in the previous section, the contrast is even lower than in Thomas' experiments, and nevertheless there is evidence for non-linearities.

On the contrary, measurements of the liminal contrast required for resolution of various grating patterns seems to indicate that the results are predictable from those of sinusoidal gratings by application of simple linear analysis (Campbell, 1965).

These conflicting evidences could be explained if processes with different spatial characteristics were involved in different experiments, according to the spatial configuration of the various patterns used to test spatial summation.

Thomas offers an interpretation of his results in terms of the excitatory and inhibitory components of the "human receptive field" and suggests that spatial integration is linear for excitatory effects and non-linear for inhibitory effects.

The differences found in experiments with incremental and decremental stimuli could also be ascribed to different mechanisms of interaction at a brain level. For instance, two separate incremental stimuli could involve reciprocal inhibition of neurons of the same type (say, on-center cells), while a pair of incremental and decremental stimuli could involve reciprocal facilitation of neighboring neurons of antagonistic types (on-center and off-center cells).

A fact that was observed during one of the experiments reported above seems to be relevant to this point. The dim inducing line used in that experiment had subliminal contrast and could not be perceived when presented alone. When, however, the bright test line was set for a threshold determination at three or four minutes from the inducing line, the dim line was easily perceived. This fact of course had no counterpart in the experiment with the bright inducing line and it might be related to the difference found in the two line weighting functions.

Two-line Interaction and Grating Resolution

An attempt can be made to compare the results obtained from different experiments on spatial contrast sensitivity. In particular, it is interesting to compare data of contrast sensitivity for gratings with data of contrast threshold for lines.

From the line spread function of a system it is possible to compute its transfer function, provided that the system can be regarded as linear. The integrations are easier if the line spread function is an exponential function (Jones, 1958). The MTF can also be easily computed if the line weighting function can be expressed by the differences of two Gaussian functions (Enroth-Cugell and Robson, 1966; Bryngdahl, 1966). If we make the assumption that the line weighting functions of Figure 5 represent the response of two independent linear processes, we may compute the corresponding transfer functions. To do this, each of the weighting functions of Figure 5 has been fitted by the difference of two Gaussian functions with suitable values of the parameters. The corresponding MTF's are also the difference of two Gaussian functions.

The computed MTF's are shown in Figure 7. They are similar in shape, but occupy somewhat different positions along the spatial frequency scale and have their peaks at about six and eight cycles per second, respectively.

Compared with the experimental contrast sensitivity curves obtained with gratings, the MTF's computed from the data of line contrast threshold are narrower. Some data obtained by Campbell and Green (1965) for the contrast sensitivity of the visual system are replotted in Figure 7 for comparison. In the low frequency range the grating data are better fitted by the curve for the bright inducing line, while at higher frequencies there is better agreement with the curve for the dim inducing line.

Most of the contrast sensitivity curves obtained with gratings are broader than the curves of Figure 7 and some of them present a secondary peak (Patel, 1966). The psychophysical contrast sensitivity function also appears to be broader than the contrast sensitivity functions of single ganglion cells of the cat (Enroth-Cugell and Robson, 1966). These authors have offered a possible explanation for the relatively broad form of the human contrast sensitivity function.

They suggest that the human retina, like the cat's retina, contains ganglion cells with receptive field of different sizes. The contrast sensitivity functions of single cells would be relatively narrow and occupy different positions along the spatial frequencies scale, according to the different sizes of the receptive fields. The psychophysical contrast sensitivity at any one spatial frequency would be determined by those cells that are most sensitive at that frequency. Thus the human transfer function would be broader than the transfer function of single cells.

By extending this assumption to the whole retina-brain system, we may suppose that the processes with relatively narrow transfer functions that are

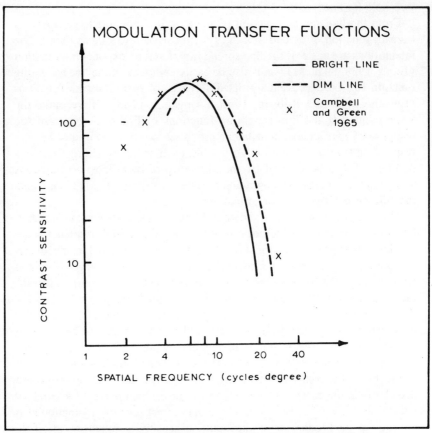

Figure 7. Modulation transfer functions computed from the line weighting functions of Figure 5. The curves have been arbitrarily displaced along the vertical axis, to facilitate comparison with experimental data of contrast sensitivity (crosses) replotted from Campbell and Green (1965)

separately involved in the experiments with incremental or decremental line stimuli, contribute to the resolution of gratings of lower and higher spatial frequencies, respectively.

Conclusive Remarks

The characteristics of spatial summation in the human visual system depend on many other stimulus parameters that have not been considered here, such as the mean illumination level (Glezer, 1965; Patel, 1966), the retinal location of the stimulus pattern (Westheimer, 1965, 1967; Bryngdahl, 1966), the orientation of one-dimensional patterns in the visual field (Campbell et al., 1966; Mitchell et al., 1967), the duration of exposure (Schober and Hilz, 1965; Nachmais, 1966), the temporal frequency for time-modulated patterns (Robson, 1966; van Nes et al., 1967), etc. Thus it may be difficult or even meaningless to compare data obtained with different experimental procedures.

Nevertheless, there is considerable agreement among the available data, as regards the general form of transfer functions or line weighting functions obtained under comparable conditions. Moreover, the similarity of the psychophysical results with the neurophysiological findings is suggestive and from new comparative studies on humans and animals it should soon be possible to obtain information to answer many questions about spatial interaction and its functional significance.

References

Barlow, H. B. (1953). Summation and inhibition in the frog retina. *J. Physiol. 119*, 69-88.

Barlow, H. B. (1958). Temporal and spatial summation in human vision at different background intensities. *J. Physiol. 141*, 337-350.

Barlow, H. B., R. Fitzhugh, and S. W. Kuffler (1957). Change of organization in the receptive fields of the cat's retina during dark adaptation. *J. Physiol. 137*, 338-354.

Battersby, W. S., and I. H. Wagman (1964). Light adaptation kinetics: The influence of spatial factors. *Science 143*, 1029-1031.

Baumgartner, G., J. L. Brown, and A. Schulz (1965). Responses of single units of the cat visual system to rectangular stimulus patterns. *J. Neurophysiol. 28*, 1-18.

Beitel, R. J. (1936). Inhibition of threshold excitation in the human eye. *J. Genl. Psychol. 14*, 31-61.

von Békésy, G. (1960). Neural inhibitory units of the eye and skin. *J. Opt. Soc. Am. 50*, 1060-1070.

Bryngdahl, O. (1966). Perceived contrast variations with eccentricity of spatial sine-wave stimuli. *Vision Res. 6*, 553-565.

Campbell, F. W. (1965). Visual acuity via linear analysis. *Proc. Symp. Information Processing in Sight Sensory Systems,* Calif. Inst. Tech., Pasadena, Calif.

Campbell, F. W., and D. G. Green (1965). Optical and retinal factors affecting visual resolution. *J. Physiol. 181*, 576-593.

Campbell, F. W., and R. W. Gubisch (1966). Optical quality of the human eye. *J. Physiol. 186*, 558-578.

Campbell, F. W., J. J. Kulikowski, and J. Levinson (1966). The effect of orientation on the visual resolution of gratings. *J. Physiol. 187*, 427-436.

Enroth-Cugell, C., and J. G. Robson (1966). The contrast sensitivity of retinal ganglion cells of the cat. *J. Physiol. 187*, 517-552.

Fiorentini, A., and L. Mazzantini (1966). Neural inhibition in the human fovea: A study of interactions between two line stimuli. *Atti Fond. G. Ronchi 21*, 738-747.

Fiorentini, A., and M. T. Zoli (1967). Detection of a target superimposed to a step pattern of illumination. *Atti Fond. G. Ronchi 22*, 207-217.

Fry, G. A., and M. Alpern (1953). The effect of a peripheral glare source upon the apparent brightness of an object. *J. Opt. Soc. Am. 43*, 189-195.

Glezer, V. D. (1965). The receptive fields of the retina. *Vision Res. 5*, 497-525.

Hartline, H. K., H. G. Wagner, and F. Ratliff (1956). Inhibition in the eye of limulus. *J. Genl. Physiol. 39*, 651-673.

Hartline, H. K., and F. Ratliff (1957). Inhibitory interaction of receptor units in the eye of limulus. *J. Genl. Physiol. 40*, 357-376.

Hubel, D. H. (1963). Integrative processes in central visual pathways of the cat. *J. Opt. Soc. Am. 53*, 58-66.

Hubel, D. H., and T. N. Wiesel (1960). Receptive fields of optic nerve fibers in the spider monkey. *J. Physiol. 154*, 572-580.

Hubel, D. H., and T. N. Wiesel (1962). Receptive fields, binocular interaction and functional architecture in the cat's visual cortex. *J. Physiol. 160*, 106-154.

Jones, R. C. (1958). On the minimum energy detectable by photographic materials. Part III. Energy incident on a microscopic area of the film. *Phot. Sci. Eng. 2*, 198-204.

Jung, R. (1961). Neuronal integration in the visual cortex and its significance for visual information. *Sensory Communication.* W. A. Rosenblith, ed. J. Wiley and Son, New York.

Kinkaid, W. M., H. R. Blackwell, and A. B. Kristofferson (1960). Neural formulation of the effects of target size and shape upon visual detection. *J. Opt. Soc. Am. 50*, 143-148.

Kuffler, S. W. (1953). Discharge patterns and functional organization of mammalian retina. *J. Neurophysiol. 16*, 37-68.

Marimont, R. B. (1963). Linearity and the Mach phenomenon. *J. Opt. Soc. Am. 53*, 400-401.

Mitchell, D. M., R. D. Freeman, and G. Westheimer (1967). Effect of orientation on the modulation sensitivity for interference fringes on the retina. *J. Opt. Soc. Am. 57*, 246-249.

Nachmais, J. (1967). Effect of exposure duration on visual contrast sensitivity with square-wave gratings. *J. Opt. Soc. Am. 57*, 421-427.

van Nes, F. L., J. J. Koenderink, H. Nas, and A. Bouman (1967). Spatio-temporal modulation transfer in the human eye. *J. Opt. Soc. Am. 57*, 1082-1088.

Ogle, K. N. (1962). Blurring of retinal image and foveal contrast thresholds of separated point light sources. *J. Opt. Soc. Am. 52*, 1035-1039.

Patel, A. S. (1966). Spatial resolution by the human visual system. The effect of mean retinal illuminance. *J. Opt. Soc. Am. 56*, 689-694.

Ratliff, F. (1965). *Mach Bands: Quantitative Studies on Neural Networks in the Retina*. Holden-Day, San Francisco.

Ratliff, F., and H. K. Hartline (1959). The response of limulus optic nerve fibers to patterns of illumination of the receptor mosaic. *J. Genl. Physiol. 42*, 1241-1255.

Ratoosh, P., and C. H. Graham (1951). Areal effects in foveal brightness discrimination. *J. Exp. Psychol. 42*, 367-375.

Robson, J. G. (1966). Spatial and temporal contrast-sensitivity functions of the visual system. *J. Opt. Soc. Am. 56*, 1141-1142.

Schade, O. H. (1956). Optical and photoelectric analog of the eye. *J. Opt. Soc. Am. 46*, 721-739.

Schober, H. A. W., and R. Hilz (1965). Contrast sensitivity of the human eye for square-wave gratings. *J. Opt. Soc. Am. 55*, 1086-1093.

Thomas, J. P. (1965). Brightness-contrast effects among several points of light. *J. Opt. Soc. Am. 55*, 323-327.

Thomas, J. P. (1968). Linearity of spatial integrations involving inhibitory interactions. *Vision Res. 8*, 49-60.

Westheimer, G. (1965). Spatial interaction in the human retina during scotopic vision. *J. Physiol. 181*, 881-894.

Westheimer, G. (1967). Spatial interaction in human cone vision. *J. Physiol. 190*, 139-154.

Glenn A. Fry

The Visibility
of Square Wavė Gratings

Square wave gratings have been used in various ways to study how the eye perceives fine detail. Shlaer (1937–38) showed that at high levels of luminance the visual acuity reaches a maximum level which appeared to be related to the coarseness of the retinal mosaic. Fry (1946) recommended the clinical use of this procedure for assessing the coarseness of the retinal mosaic. In this respect it supplements the double slit interference pattern used by Byram (1944) to show the relation of the coarseness of the retinal mosaic to resolving power. The double slit interference pattern has also been used by O'Brien and Miller (1952) for measuring visual acuity.

For spacings of the gratiṅg lines greater than the spacing of the photoreceptors, one can use a square wave grating as a test pattern and vary the contrast between the bright and dark bars to determine the threshold of visibility.

The contrast can be expressed in terms of the modulation (M).

$$M = \frac{L_1 - L_2}{L_1 + L_2},\tag{1}$$

where L_1 is the luminance of the bright bars and L_2 the luminance of the dark bars.

Figure 1 shows the modulation threshold data of DePalama and Lowry plotted as a function of the center-to-center distance ($\bar{s}$) between bright bars.

284

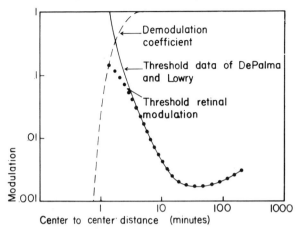

Figure 1. Threshold data (solid curve) of DePalma and Lowry for square wave gratings replotted in terms of the threshold modulation as a function of the center-to-center spacing of the bright bars. The dotted curve shows the computed values for the threshold retinal modulation based on the assumption that the line spread function of the optical image on the retina is a Gaussian distribution having a standard deviation of 0.212 minute of arc. The grating used by DePalma and Lowry had a total luminance of 20 fL and subtended a visual angle of 6°. It was viewed at a distance of 14 inches with both eyes unrestricted by artificial pupils.

One must differentiate between the modulation of the test object presented to the eye and the modulation ($\overline{M}$) of the retinal image, which is defined as follows:

$$\overline{M} = \frac{E_1 - E_2}{E_1 + E_2},\qquad(2)$$

where E_1 is the retinal illuminance at the centers of the retinal images of the bright bars and E_2 is the retinal illuminance at the centers of the retinal images of the dark bars.

The image-forming mechanism of the eye degrades the modulation of the retinal image by reducing it to a level lower than that of the test object. This involves convoluting the square wave grating used for the test object with the line spread function for the image formed by the eye.

When the eye is in focus for the target, one can assume that the line spread function for the eye is Gaussian (Fry and Cobb, 1935; Fry, 1965A), as shown in Figure 2.

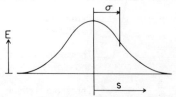

Figure 2.   The line spread function of the
human eye which is assumed to be Gaussian.

$$E \sim \exp\left[-\frac{1}{2}\,(s/\sigma)^2\right], \tag{3}$$

where E is the retinal illuminance, s the distance from the center of the
image, and σ the standard deviation of the distribution. The value for σ
depends on the size of the pupil, the wavelength composition of the light,
and the aberrations of the eye. Fry and Cobb (1935) have proposed a
method of determining an index of blur ϕ from measurements of the
thresholds of bars of different widths. For the Gaussian distribution

$$\phi = 2.5066\sigma. \tag{4}$$

This assessment of blur includes a certain amount of physiological irradiation (Fry, 1965B). One can circumvent this problem by using monochromatic, coherent light from a point source and determining the standard
deviation of the Gaussian distribution which has the same ϕ value as the
Fraunhofer image of the point source (Fry, 1955, pp. 72-77). For the
Fraunhofer image

$$\phi = 2.03\,\frac{\lambda}{\bar{g}}\,, \tag{5}$$

where λ is the wave length in microns, $\bar{g}$ is the radius of the entrance pupil
in millimeters, and ϕ is the index of blur in minutes of arc measured at the
second nodal point.

Fry (ibid., pp. 84-85) has developed a method for computing the index
of blur for heterochromatic light. It should also be noted that if the point
spread function is Gaussian, the line spread function is also Gaussian
(ibid., pp. 14, 22), and the value for σ is the same for both.

The process of convoluting a square wave grating with a Gaussian
spread function is illustrated in Figure 3, where the standard deviation of
the spread function is constant from grating to grating, but the spacing
changes. In the case of wide spacing, the blur affects only the borders; but

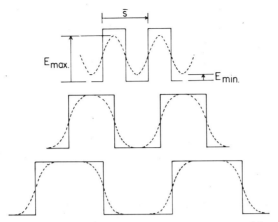

Figure 3. The process of convoluting a square wave with a Gaussian spread function. The standard deviation (σ) of the spread function is constant from grating to grating and is equal to 1/18 of the center-to-center distance in the coarsest of the three square wave gratings.

as the spacing gets finer, the grating becomes transformed to a sinusoidal grating in which the amplitude varies with the spacing.

The solid curve in Figure 4 shows how the modulation varies with the

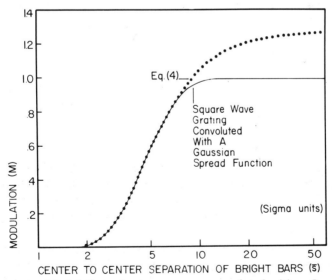

Figure 4. Modulation of a square wave convoluted with a Gaussian spread function.

spacing. As indicated in both Figure 3 and Figure 4, the modulation levels off at a value of unity as the spacing increases. The fact that a square wave convoluted with a Gaussian spread function does become transformed into a sine wave at high spatial frequencies may be shown as follows.

Engel (19   ) has shown that a grating of bright lines uniformly spaced and convoluted with a Gaussian spread function gives rise to a distribution of illuminance E expressed by the following formula:

$$E \sim \sigma v \, [1 + 2 \exp [-2 \, (\pi \sigma v)^2] \cos (2\pi vs)$$
$$+ 2 \exp [-8 \, (\pi \sigma v)^2] \cos (4\pi vs)$$
$$+ 2 \exp [-18 \, (\pi \sigma v)^2] \cos (6\pi vs) \ldots] \qquad (6)$$

where $v =$ the spatial frequency,
  $\sigma =$ the standard deviation of the Gaussian spread function,
  $s =$ the distance across the grating from an arbitrary zero point located at one of the bright bars, and
  $E =$ the flux per unit area.

As the spacing gets finer and finer, all the cosine terms except the first drop out, and thus the distribution is a pure sine wave.

It follows that any periodic sequence of line elements such as a square wave grating must give rise to a series of sine waves out of phase which sum to give a pure sine wave (Fry, 1968).

The formula for a high frequency square wave of unit modulation convoluted with a Gaussian spread function is as follows:

$$E \sim \tfrac{1}{2} + (2/\pi) \exp [-2 \, (\pi \sigma v)^2] \sin (2\pi vt). \qquad (7)$$

From Equation (7) we can derive a simple formula for the modulation ($\overline{M}$) of the retinal image of a high frequency square wave of modulation (M).

$$\overline{M} = [(4/\pi) \exp [-2 \, (\pi \sigma v)^2]] \, M. \qquad (8)$$

The term $[(4/\pi) \exp [-2 \, (\pi \sigma v)^2]]$ represents the demodulation coefficient by which the modulation of a square wave is reduced by the optical system of the eye. This equation accurately describes the data for high frequency gratings as shown in Figure 4, but it does not hold for low frequency gratings.

According to Shlaer, the upper limit of visual acuity was found to occur when the center-to-center spacing of the bright bars in a square wave grating is about 1.07 minutes of arc. Since the minimum value for De-

Palma and Lowry is 1.38 minutes, it may be assumed that the visual acuity in this case is not limited by the coarseness of the retinal mosaic.

The dashed curve in Figure 1 represents the demodulation coefficient for the image-forming mechanism of the eye and is the same as the solid curve in Figure 4. The standard deviation of the Gaussian spread function has been arbitrarily set at 0.212 minute, which gives a 1.9 percent demodulation coefficient at a center-to-center spacing of one minute of arc. The retinal modulation (dotted curve in Figure 1) is obtained by multiplying the ordinates of the dashed curve by the threshold modulations for the square wave grating. It is obvious that the threshold retinal modulation is a function of the spatial frequency. Up to a center-to-center spacing of 40 minutes, there is a gradual decrease in the threshold retinal modulation. Whether this involves simple physiological irradiation of excitation (Fry, 1965B) and inhibition (Fry, 1948, 1963) or a breakdown of the mechanism for maintaining borders remains to be seen. The role of micronystagmus must also be considered.

The number 0.212 was selected as the value for σ because it is the largest value of σ which will give a retinal modulation curve that gradually increases as the spacing decreases toward the limit of resolution. The effect of using a larger value of σ is illustrated in Figure 5, where the value

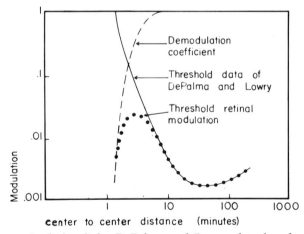

Figure 5. Analysis of the DePalma and Lowry data based on a σ value of 0.75.

for σ is 0.75 minute. After reaching a maximum value at $\bar{s} = 3$ minutes, the retinal modulation decreases as the spacing gets finer. The value of

0.212 is still somewhat larger than the theoretical value 0.160 for a three-millimeter pupil and a wavelength of 0.589 micron.

Let us turn now to consider what happens as the center-to-center distance increases from 40 minutes upward. It is obvious from Figure 3 and Figure 4 that, as the spacing increases, the demodulation coefficient reaches its maximal value of unity and remains constant. Furthermore, the gradient at each border is constant and independent of the spacing. On this account one might expect that the contrast threshold would be constant and equal to the threshold for a single border. Since all of the borders are at the threshold, one would expect that the threshold for any one of the borders would not be affected by the presence of the other borders. However, it may be noted in Figure 1 that the threshold goes through a minimum at a bar width of about 20 minutes, and as the width increases, the threshold increases until it reaches a maximum which may be assumed to be the same as for a single border. This upturn of the curve has also been found by Schober and Hilz (1965). They showed that it disappears or becomes much less pronounced with short exposures and low luminance levels. The upturn of the curve has also been reported by Fry (1959).

If a single bar is used instead of a grating and if its width is gradually increased, the threshold decreases until it reaches a minimum and thereafter, further increase in width has no effect (Fry and Cobb, 1935; Fry, 1965B). The threshold is the same as for a single border. In the case of a square wave grating one can say that the threshold for a coarse grating is the same as for a single border. Then we have to say that up to a point the bars become more visible as the spacing becomes finer. One can note introspectively that with wider spacing the borders are sharper at the threshold of visibility. With narrower spacing the bars appear as fuzzy striations at the threshold.

It cannot be claimed, therefore, in the case of a square wave grating, that the visibility of the border between a given pair of bars is not affected by the presence of the borders on the right and left. In terms of Figure 6 the threshold for the border at B is lower because of the presence of the borders at C and D.

This means that the Dittmer and Blachowski effects (Fry and Bartley, 1935) will have to be reinvestigated because it has been assumed that, when a pattern like that shown in Figure 7 is used to study the effect of the border between B and C on the contrast threshold of the border between A and B, the contrast threshold is at a minimum when the contrast between B and C is zero. From the data obtained with gratings, it must be concluded that the minimum does not occur when the contrast between B and C is zero, because the threshold must be lower when the contrast

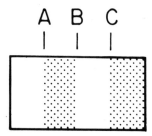

Figure 6. Dependence of the threshold of a border on the presence of adjacent borders.

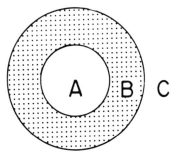

Figure 7. The Dittmer and Blachowski effects. The Dittmer effect is the effect of the contrast of the border between C and B on the contrast threshold of the border between B and A. The Blachowski effect is the effect of the distance between the two borders.

between B and C is equal to that between B and A. It would be helpful to study this phenomenon with stabilized images.

In trying to visualize how the eye processes information about borders, one may postulate a frequency-equalizing mechanism that works toward maintaining equal frequencies in adjacent ganglion cells. This tends to wash out a frequency difference on the two sides of a border, and there must be some unique feature of the mechanism that maintains this frequency difference at a superthreshold border.

The mechanisms involved in irradiation of excitation and inhibition can be thought of as occurring at a lower level in the system.

The same procedure can be used in analyzing Shlaer's data at high luminance levels. Shlaer manipulated the retinal modulation by varying the width of the opaque bars in his gratings. He used gratings in which the ratios of the width of the opaque bars to the width of the clear bars were 1 to 1, 1 to 3, and 1 to 7. Such gratings, when demodulated with a Gaussian spread function have a σ value of 0.212, produce retinal modulations as indicated in Figure 8. At low frequencies the demodulation coefficient curves have been worked out using the technique illustrated in Figure 3. However, at high frequencies the following formulas apply:

For an opaque to clear ratio of 1 to 3,
$$\overline{M} = [0.600 \exp [-2 \, (\pi \sigma v)^2]]M. \tag{9}$$
For an opaque to clear ratio of 1 to 7,
$$\overline{M} = [0.279 \exp [-2 \, (\pi \sigma v)^2]]M. \tag{10}$$
Equation (8) covers the case of the 1 to 1 ratio.

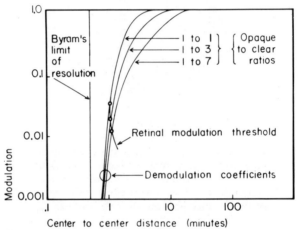

Figure 8. Analysis of Shlaer's data at high luminance levels. The grating pattern was disk-shaped and had a diameter of 4°. It was viewed with one eye through an artificial pupil 2 millimeters in diameter. The light diverging from each point of each bright bar and entering the eye was not coherent and hence Equation (5) does not apply. The luminance of the bright bars was constant (not specified, but at least higher than 100 fL), and the surround was equal to the average luminance of the grating.

Shlaer found the threshold center-to-center distances for his three gratings to be 1.07, 1.09, and 1.15 minutes of arc, and these thresholds are represented by the open circles in Figure 8. The curve through these thresholds represents the threshold modulation curve for high levels of luminance. The sharp rise of this curve at its fine end indicates that the visual acuity is limited by the coarseness of the retinal mosaic, but it must be noted that when the diameter of the artificial pupil was increased from 2.00 to 2.35 millimeters, the just resolvable center-to-center distance decreased from 1.07 to 0.94 minutes of arc.

It is interesting to note that Byram (1944) found in the case of a double slit interference pattern which can be made finer than the retinal mosaic that the lines break up into curved segments which appear to shimmer or flutter. The absence of this effect with a grating viewed through a circular aperture would indicate that blur produces fusion of the lines before this effect can occur. Complete fusion of the lines of the interference pattern did not occur until the center-to-center spacing was reduced to 0.5 minute of arc.

Conclusions

When a square wave grating is viewed with coherent light at a high level of luminance and the eye is in focus, the resolving power approximates the coarseness of the retinal mosaic but is still limited by the blur of the retinal image. A double slit interference pattern must be used to measure the coarseness of the retinal mosaic.

When the eye is in focus, blur of the retinal image helps reduce the resolving power of the eye in the range of center-to-center distances from zero up to five minutes, but it has little or no effect on the visibility of coarse gratings.

In the range of center-to-center distances from the threshold of resolution up to about 40 minutes of arc the resolving power is degraded by the irradiation of excitation and inhibition, which demodulates the impression initiated in the retina. A sine wave grating can be used more effectively than a square wave grating in assessing these effects.

Above 40 minutes of center-to-center distance the major factor affecting resolution is the interaction of borders. This interaction is related to the Dittmer effect and the Blachowski effect. Also in this range inhibition improves the sharpness of the gradients at the borders.

However, this effect at the borders eventually becomes independent of the distance between borders. Thus one must look for an explanation of the variation of the modulation threshold at low spatial frequencies in terms of the frequency-equalizing mechanisms which create and maintain contrast borders.

References

Byram, G. M. (1944). Physical and photochemical basis of visual resolving power, Part II. *J. Opt. Soc. Am. 34*, 718.

Engel, R. E. (19   ). The spread function of line gratings with Gaussian blur.

Fry, G. A. (1946). Monocular measurement of visual acuity corrected. *Optometric Weekly 37*, 1795-1799.

Fry, G. A. (1948). Mechanisms subserving simultaneous brightness contrast. *Am. J. Opt. 25*, 162-178.

Fry, G. A. (1955). *Blur of the Retinal Image.* Ohio State University Press, Columbus, Ohio, pp. 72-77.

Fry, G. A. (1959). The relation of blur and grain to the upper limit of useful magnification. Report (RADA-TN-59-267) from the Ohio State University Research Foundation to the Rome Air Development Center under Contract No. AF30(602)-1580.

Fry, G. A. (1963). Retinal image formation: Review, summary, and discussion. *J. Opt. Soc. Am. 53*, 94-97.

Fry, G. A. (1965A). Distribution of focused and stray light on the retina produced by a point source. *J. Opt. Soc. Am. 55*, 333.

Fry, G. A. (1965B). Physiological irradiation across the retina. *J. Opt. Soc. Am. 55*, 108-111.

Fry, G. A. (1968). Square wave gratings convoluted with a Gaussian spread function. *J. Opt. Soc. Am. 58*, 1415-1416.

Fry, G. A., and S. H. Bartley (1935). The effect of one border in the visual field upon the threshold of another. *Am. J. Physiol. 112*, 414-421.

Fry, G. A., and P. W. Cobb (1935). A new method of determining the blurredness of the retinal image. *Trans. Am. Acad. Ophth. and Oto.* pp. 423-428.

O'Brien, B., and N. D. Miller (1952). Resolving power of the retina for chromaticity contrast and for luminance contrast. *J. Opt. Soc. Am. 42*, 289.

Schober, H. A. W., and R. Hilz (1965). Contrast sensitivity of the human eye for square wave gratings. *J. Opt. Soc. Am. 55*, 1086-1091.

Shlaer, S. (1937–38). The relation between visual acuity and illumination. *J. Am. Physiol. 21*, 165.

Daniel Kahneman

Changes in Pupil Size and Visual Discriminations During Mental Activity

A large number of interrelated physiological changes accompany any substantial involvement in mental activity. Figure 1 presents what is

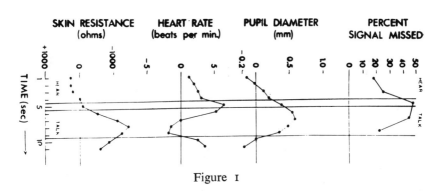

Figure I

certainly a small sample of the changes that occur during the performance of a particular task that I have often used: the subject hears four digits from a tape recorder, each digit spoken on a beat of a metronome at one per second; he pauses for one second, then responds with a new string of digits, adding one to each digit that he has heard (e.g., 3480-4591). His speech is paced by the metronome. The figure presents four indices, of which two fall within the province of visual science. The four are skin conductance, heart rate, size of pupil, and an index of visual deficit. All four indices show the same trend: a sharp rise while listening to the string, a peak at the pause or fairly early in the report, a rapid relaxation that anticipates the termination of the task. The correspondence between these trends is unlikely to be accidental. Had we added a theoretical curve of the changing level of effort demanded by the task at any time, that fifth curve would have been rather similar to the rest. Autonomic changes in the sympathetic direction and a reduction of visual discrimination both follow second-by-second variations in the intensity of mental effort. The remainder of my remarks are devoted to the blinding effect and to the pupil response.

First, to the blinding effect. The top graph of Figure 1 is redrawn from an experiment in which large illuminated letters were successively exposed at a 5 per second rate while the subject was engaged in the digit transformation task. He was to report whether the letter K was one of the letters presented. It is obvious in Figure 1 that the letter K is most likely to be missed when shown at the time of maximal involvement in the task. Several additional points should be made:

1. Blinding is very substantial during the pause between listening and report. No significant stimuli are presented at that time, and there is no overt activity on the subject's part. At that time, the subject is exclusively involved in silent thinking activity.

2. Photographs of the eye indicate that blinding occurs even though the eye is open and on target. A similar conclusion was reached by Baker some years ago in his study of the well-known vigilance decrement.

3. We are all familiar with what fiction writers describe as a "blind, unseeing stare" which appears to consist of several components: reduced blinking rate, ocular divergence, enlarged pupils, and, we suspect, relaxed accommodation. However, the blinding effect is not caused by either the enlarged pupil or an accommodation change; our subjects see no better when they stare blindly through an artificial pupil.

The occurrence of the blinding effect need not surprise us if we assume that perceiving demands attention and that attention cannot be everywhere at once. Some conceptions of attention make such a derivation particularly compelling. Hernandez-Peon, among others, has used the searchlight analogy to describe the functioning of attention. With a single searchlight, naturally, most of the space must be dim and only one area can be illuminated at any one time. The conception of the organism as a single channel device expresses the same idea in different terms. From yet a different angle, we have the suggestion most recently made explicit by Lacey, that intake and processing of information are incompatible physiological states which are controlled, at least in part, by the autonomic nervous system. Thus, an increased heart rate accompanies, and perhaps makes possible, any intense mental involvement, whereas the heart beats more slowly during alert attention to outside stimuli.

I shall present some experimental data to suggest that a conception of blinding as an automatic consequence of mental activity is overly simple. Specifically, I should like to suggest that we are not blind because we think; we are blind in order to think, and we are only as blind as we absolutely need to be in order to deal with the task at hand. The distinction between the two views is the following: according to any version of the searchlight theory, what one is blind *to* should not make any particular difference to the intensity of the blinding effect; we are blind when otherwise engaged. On the functional theory, what we are blind to makes a critical difference. If vision conveys stimuli that are irrelevant and distracting, then vision will be impaired. In a second experiment, our subjects again perform the transformation task. Sometime during the task a single letter is presented for 120 msec, on a computer-controlled display scope. The letter is invariably preceded and followed by 50 msec of visual noise. The subject is to recall what the letter was, at the termination of the task. The critical variable is the nature of the visual material which precedes the sequence masking field—target letter—masking field. There are two conditions; in one, a steady fixation cross is shown, and in the other, digits are flashed at a 5/second rate, a display that one finds quite compelling and demanding of attention.

Now to the predictions. The main point of the functional theory is that the subject should be blinder to the digits than to the fixation cross because the digits, if he were to see them, would be more detrimental to the task. No such prediction can be derived from any variant of the searchlight theory. The role of the target letter is that of a measuring probe; we assume that the letter is unlikely to be seen if the subject is virtually blind at the instant of exposure. The functional theory therefore predicts that more letters will be missed when they are preceded by a distracting stimulus. Finally, the prediction may be made that the differential effect of digits and cross should be greatest at the time of maximal involvement in the task. The distracting digits are most likely to interfere when the task is most demanding, and the need to suppress them is therefore most urgent at that time. When the task is easy, the channel may remain open to the digits as well as to the cross.

The results are shown in Figure 2, and they are just what the functional view of the blinding effect would lead us to expect. The flashing digits indeed cause more blinding than the fixation cross, but the effect is essentially restricted to the pause and to the early phase of the report, the period of maximal task-load.

A few additional remarks conclude the discussion of this experiment: (1) on some occasions, our subjects certainly see the letter and simply forget it. This of course is not a blinding effect. However, forgetting cannot account for the particular shape of the function in Figure 2. Our sub-

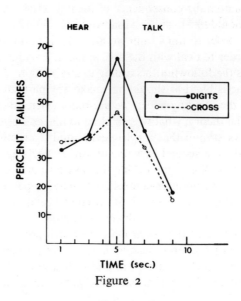

Figure 2

jects, by the way, clearly distinguished blinding from forgetting. (2) We interpreted this experiment in functional terms. The subject is blinder to the digits than to the cross because the digits would interfere more severely with the mental task, if he were to see them. In terms of mechanism, this view implies a feedback loop which the searchlight theory does not require: the organism periodically samples various sensory channels and evaluates whether the information arriving in them is relevant to the task. If the information is irrelevant, its distractiveness is evaluated; if it is highly distracting, blinding occurs, as required to protect the performance of the task. The experimental results lend support to the notion that such a feedback loop exists.

There is no need here to emphasize the adaptive value of such a device.

I now turn to my other topic, the changes of pupil size that accompany mental activity. The existence of these changes has long been known. Descriptions of the effect were given in the 1880's, and mental effort was occasionally mentioned in Lowenstein's writings on the pupil. However, the current surge of interest in psychological determinants of pupil dilations can certainly be traced to the work of Hess and Polt at Chicago. I shall not discuss their controversial studies of dilations and constrictions that accompany pleasant and unpleasant affect. The starting point of my own work was their report of precise correlations between the difficulty of arithmetic problems and the size of the dilations during the solution of these problems.

Precision is indeed the operative word in the discussion of pupillary changes. This may be cause for surprise, since sophisticated descriptions of hippus have appeared that tend to present the pupil as an undercontrolled, noisy system. The facts of the matter are not in doubt. Anyone who has observed the pupil of a resting subject must have noticed the rhythmic appearance of waves of dilation-constriction. However, an equally important fact has attracted less attention than it deserves. As soon as a problem is given, such as one of Hess' mental arithmetic questions, the noise of the pupil virtually disappears. There is an immediate dilation and the pupil remains steady with its size enlarged. Any major constriction at that time can be ascribed with near-perfect confidence to the subject taking a break from the assigned task. This disappearance of pupillary noise under task involvement is obviously crucial to a comprehensive theory of pupillary control. I do not know, at present, how to fit it into current models.

Some additional characteristics of the pupil response may be observed in Figure 3, which is drawn from an experiment on pitch discrimination that I published last year with Jackson Beatty. The subject hears a ready signal; after a variable interval, a standard tone of 850 Hz at a comfortable loudness; four seconds later, a comparison tone that may have any of eleven frequencies from 820 Hz to 880 Hz; four seconds later, a final signal that instructs the subject to say "high" or "low" depending on whether the second tone is higher or lower than the first. Infrared pictures of the eye are taken at 1/second through the trial. Figure 3 includes average curves for two frequencies of the comparison tone: 850 Hz, an insoluble problem, and 880 Hz, the most discriminable tone. The average curves include data from ten subjects, each of whom had five trials on each frequency. Several conclusions may be drawn: (1) the pupil response is fast. Following the presentation of the comparison tone, the pupil dilates and constricts within two seconds. From other experiments we have learned that this fast response is characteristic of intermediate levels of illumination, when resting pupil diameter is approximately four mm. Dilation and constriction are both sluggish when illumination is weak. With very strong illumination, the dilation may fail to occur. (2) The dilation to the tones is not a reflex response to sensory stimulation. It corresponds to the complexity of what the subject does with the sensory information received. Note that the curve for the more difficult discrimination is much more elevated than its neighbor, and that the separation is complete one second after the presentation of the comparison tone that sets the problem. Note also that the presentation of the 850 Hz tone as standard hardly elicits a dilation; its constant repetition in that role has robbed it of any informational value. On the other hand, the presentation

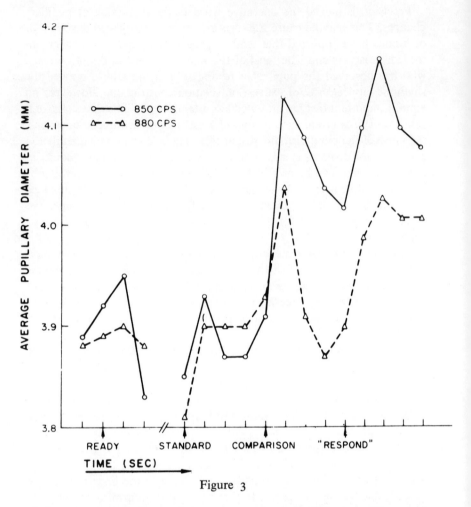

Figure 3

of the same tone as comparison sets a particularly difficult problem, and elicits a correspondingly large dilation. Figure 4 illustrates the precision with which the size of the dilation reflects problem difficulty. (3) The habituation of the various components of the response over five blocks of trials confirms the present interpretation of the pupil response. Habituation is not general. The standard tone and the ready signal that precede it both become less significant in the course of the experiment and the pupillary responses to them tend to vanish. On the other hand, the comparison tone sets a problem that does not become much easier with practice and the response to it remains intact. Grey Walter has made a very similar point in his study of the negative contingent variation in the

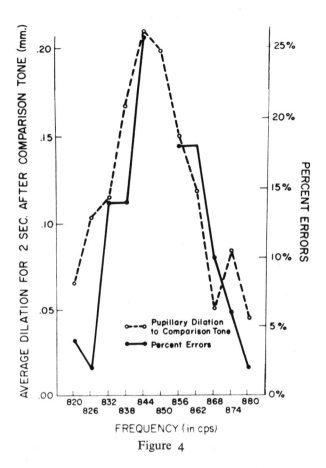

Figure 4

EEG. (4) Dilations occur when there is work to be done. In general, the mere anticipation of a problem occasions no substantial response. That point was made most clearly in a recent attempt by Scott Peavler and myself to repeat and modify the pitch-discrimination experiment. There was no standard tone in the new version and both tones were informative. We told the subject at the ready signal how difficult the next problem would be, expecting that announcement to affect the size of the pupil during the waiting period. In fact, anticipated difficulty had no effect during the waiting period, and the only effect of difficulty occurred after the presentation of the comparison tone, when the subject was actually engaged in solving the discrimination.

The pitch-discrimination and the digit-transformation tasks are but two of several experimental situations in which we have studied the pupil response during the last two years. Other experiments tell the same story: pupillary changes provide an exceedingly sensitive indication of second-by-second variations in mental effort. They are far more sensitive in that respect than other indicators of sympathetic arousal, and much easier and more economical to measure than the blinding effect which I discussed earlier.

One additional result will sharpen the interpretation of the pupillary response. We have seen that there is nothing unique or magical to this response. It is merely the best currently available indicator of the sympathetic arousal that accompanies thought. However, the question of whether it indeed accompanies thought may still be raised, in view of the long tradition that associates sympathetic arousal with fight-flight emotional states. Specifically, I have often been asked the question of whether the pupil dilates because our subject is straining to solve a mental task or because he is anxious about his ability to do so.

The pupil unquestionably dilates under any kind of emotional stress. It is possible to show, however, that pupil size is almost a pure measure of effort under task conditions that are relatively neutral. One experiment that makes this point is the following: subjects are successively presented by tape-recorder with eight digit-noun pairs, e.g., 3-barn. They are to learn the noun that goes with each digit. No normal subject can retain eight such pairs on a single presentation. In order to guide the learning effort, half of our subjects are paid one cent for each noun correctly paired with an odd digit, and five cents for each noun correctly paired to an even digit. The payoffs are reversed for the other subjects. Figure 5 presents the average pupil responses of ten subjects. On second one, they hear the digit which immediately identifies the item as high-reward or low-reward; on second four they hear the noun that is to be associated with this digit.

If pupil size is a measure of emotional arousal, we might expect differential dilations to follow the digits which identify each trial as important or unimportant. Figure 5 shows that this is not the case. The only consistent difference between the responses in the two reward conditions occurs after the subject hears the noun. There is work to be done at that time, and subjects evidently work harder for five cents than for one cent. Significantly, those subjects who show the largest difference between the pupillary responses to high-reward and low-reward items also show the largest difference between the number of items of the two types that they correctly recall. The correlation between the two variables is .92, for ten subjects. There is no obvious way of deriving this result from the idea that pupil size is a measure of emotional arousal, whereas it follows strictly from the idea that pupil size is a measure of effort, under the conditions of the present experiment.

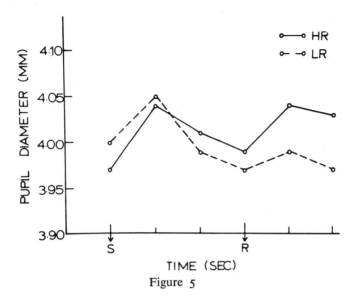

Figure 5

In summary, the second part of this paper has suggested that a pupil size is a surprisingly sensitive index of short-term variations of involvement in mental tasks. The first part of the paper had described variations of visual discriminations that also accompany mental effort. It was shown that this blinding effect is probably mediated by a sensitive feedback loop, and that incoming sensory information is evaluated in terms of both relevance and distractiveness before the decision is taken to shut down or attenuate a sensory channel.

In closing, I would like to raise the question of the biological function of having a dilated pupil at a time when one is functionally blind. Why does the pupil dilate when we think? Why, in fact, does the pupil dilate during one phase of the orienting reflex? Obviously, the story cannot end with the old cliché that a pupillary dilation lets more light in, and thereby increases visual sensitivity. For a diurnal animal in daylight, central vision is generally impaired by an enlarged pupil.

A former colleague, Colwyn Trevarthen, suggested a possible resolution for this apparent paradox. If an enlarged pupil enhances peripheral sensitivity, for example to moving objects, even as it reduces central acuity, then everything else might fall into place. It would be highly adaptive for an animal to post an increased guard at the periphery of the visual field while preparing to go more or less blind to what goes on in the center. A similar increase of peripheral sensitivity might be a useful part of the "what-is-what" reflex pattern. However, having no direct evidence that pupillary dilations actually function in this manner, we must look forward to the hope of getting some enlightenment in this complex matter in the future.

References

Baker, C. H. Further towards a theory of vigilance. Buckner, D. N. and McGrath, J. J. (eds.), *Vigilance: A Symposium*, McGraw-Hill, 1963, pp. 127-53.

Hernandez-Peon, R. Physiological mechanisms in attention. Russel, R. W. (ed.), *Frontiers in Physiological Psychology*. New York and London: Academic Press, 1966, pp. 121-147.

Hess, R. H., and J. M. Polt. Pupil size in relation to mental activity during simple problem-solving. *Science*, 1964, *143*, pp. 1190-1192.

Kahneman, D., and J. Beatty. Pupillary responses in a pitch-discrimination task. *Perception & Psychophysics*, 1967, *2*, pp. 101-105.

Kahneman, D., J. Beatty, and I. Pollack. Perceptual deficit during a mental task. *Science*, 1967, *157*, pp. 218-219.

Kahneman, D., and W. S. Peavler. Incentive effects and pupillary changes in association learning. *J. of Experimental Psychology*, 1969, *79*, pp. 312-318.

Kahneman, D., B. Tursky, D. Shapiro, and A. Crider. Pupillary, heart rate and skin resistance changes during a mental task. *J. of Experimental Psychology*, 1969, *79*, pp. 164-167.

Lacey, J. I. Somatic response patterning and stress: Some revisions of activation theory. M. H. Appley and R. Trumbull (eds.), *Psychological Stress*. New York: Appleton-Century-Crofts, 1967.

Lowenstein, O. Experimentelle beitrage zür lehre von den katatonischen pupillenveranderungen. *Monatschrift fur Psychiatrie und Neurologie*, 1920, *47*, pp. 194-215.

Yves Le Grand

Unsolved Problems in Vision

Strictly speaking, the title of this paper is without significance: as Henri Poincaré wrote seventy years ago, a scientific problem is neither solved, nor unsolved, but more or less solved. Curiously enough, the more a specialist is working hard on a particular problem, the less he thinks that this problem is solved, because as his work is progressing, he sees with more precision the difficulties, mysteries, and contradictions of the question. The only solved problems are those of which we know nothing, because in their case there is no difficulty. For example, I have myself never worked on visual pigments, and consequently I have the impression that rod pigments are fairly well known. I should be surprised if my friend George Wald had the same opinion, because he spent many years on the rhodopsin problem and I guess that for him it is still full of mysteries. He is right and I am wrong, and this is good for young workers because the more science progresses, the longer is the way in front of us.

The unsolved problems in vision I shall speak about are actually problems in physiological optics now in progress in my laboratory at the Museum of Natural History, or in the laboratories in the Institute of Optics in Paris. The Museum is a research place and the Institute of Optics is a school where I give each year a course on physiological optics. Some of you have visited these laboratories, and as to the others, I hope to see them soon in Paris, in spite of political and financial difficulties which, happily, do not alter the friendship between scientists everywhere in the world.

Corneal Topography

This problem has been studied for about three centuries, and we have still a lot of work to do before reaching a statistically good understanding of it. Dr. Roger Bonnet of my laboratory is the French specialist of corneal topography and he has devised several methods of measurement. In a point M of a plane section of any surface, in rectangular coordinates x and y, let R be the radius of curvature and d the angle between the normal and the x axis (Fig. 1). Gullstrand's classical equations may be written:

$$x = - R \sin d \qquad\qquad y = R \cos d \qquad\qquad (1)$$

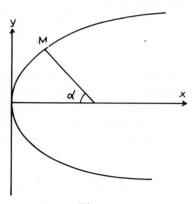

Figure I.

Starting from a preliminary stereophotogrammetric study of several corneas, Bonnet proposed the following formula:

$$R - R = A \cdot 10 \qquad\qquad (2)$$

where R is radius of curvature at the ophthalmometric pole S. The coefficient A characterizes the extent of the optical zone, and k is a flattening coefficient. With a special topographic ophthalmometer, Dr. Bonnet has verified formula (2) on a large number of subjects: it is a good approximation for 70% of subjects, and for 50% the k coefficient is in the range 2.8–3.3 rad^{-1}. For k = 3 formula (2) agrees with Berg's expression. An indirect verification of formula (2) is the success of a practical method of fitting contact lenses, derived from this theory and now largely used in France.

If a narrow beam of parallel rays is sent upon the cornea in M, these rays having an angle 2 with the x axis and being reflected parallel to this x axis, it is possible to measure y directly and to calculate

$$x = \text{tg } dy$$

This is the basis of a new photokeratoscopic method in progress in our laboratory; it will be applied first to animal eyes, in order to investigate the hormonal process which stabilizes the corneal form. It seems that in man some antibiotics cause corneal deformations, in particular a diminution of R (and not a flattening as is sometimes said). Photokeratoscopy will be a useful tool in such researches.

Aphakic Vision

Professor Arnulf and his team in the Institute of Optics in Paris have shown that, when an optical system is affected by large aberrations, it is still possible to choose the focusing so that the diffraction figure is not larger than the perfect Airy's disc. The only difference is that the maximum of illumination at the center of the diffraction figure is less than for the stigmatic system.

Dr. Bonnet has applied this theory to the aphakic eye when corrected by a contact glass. In this case, there exists an important spherical aberration, about 2.5 diopters for a pupil of 5 mm diameter. For the best focusing, the diffraction figure is quite the same as an Airy's disc, so acuity is not spoiled. However, this is true only at high luminances, because the loss of contrast due to the fact that the diffraction figure is less sharp will produce a decline of acuity at low luminances.

At high luminances, it is even possible to maintain a good acuity (pattern of parallel bars with 1 minute width) when focusing is not the best: this explains the appearance of pseudo-accommodation in a range which sometimes is as high as 2 diopters. Direct measurements by Dr. Bonnet on several sphakic subjects have confirmed both the spherical aberration and the apparent depth of focus (Figs. 2 and 3).

Transfer Function of the Eye

It is an old idea in acoustics to evaluate the quality of any receiving system by its response to sine waves of equal energy and various frequencies, but in optics the same principle applied to spatial frequencies is quite young. The quality of an optical instrument, and in particular its resolving power, is now expressed by its *modulation transfer function*.

The application of this transfer function to the visual apparatus has been in recent years the subject of a large number of papers, and the eye response to spatial sine waves has been measured by many authors. (You will perhaps be surprised to know that as early as 1935 I myself did acuity measurements by interference fringes directly produced on the retina by

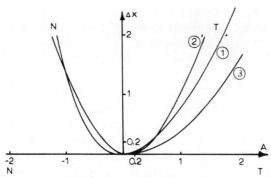

Figure 2. Spherical aberration of the aphakic eye wearing a contact glass; abscissae : distance from the pupil center in mm; ordinates : aberration in diopters. Experimental curves (1 and 2) and theoretical curve (3) (after Bonnet).

Young's two holes method.) I will not try to review these researches, but only to summarize two of them performed in the Institute of Optics in Paris in 1967, one by Mrs. Françoise Berny and the other by Miss Odette Dupuy.

The aim of Mrs. Beny's work was to measure the effect of the spherical aberration of the eye upon the quality of the retinal image. But what exactly is the meaning of "retinal image"? In my opinion, one must make a sharp distinction between the *optical image* given by the media of the eye (cornea, aqueous humour, lens, vitreous), irrespective of the retina itself, and the true *retinal image*. Physically the optical image of the eye should be the continuous distribution of illumination on a fictitious surface placed in contact with the retinal epithelium. The true retinal image, on the contrary, is affected by diffusion in retina itself, and also by the discrete structure of cones (Stiles-Crawford effect).

The method used by Mrs. Berny is Foucault's shadow technique, with a broad slit. In this case it is the optical image only that is used (with a narrow slit, diffusion in the retina might play its role). The apparatus is shown in Figure 4. The slit is illuminated by an incandescent lamp, with filters in order to cut IR and project on the subject's retina by a lens L_1 and a half-silvered plate L. In the plane of the image of the retina given by the lens L_2, the knife E moves so that its edge is parallel to the image of the slit. A photographic objective projects in S_1 on a photographic film an image of the subject's pupil S. When everything is ready, an electronic flash, replacing the incandescent lamp, gives the image of the pupil. A

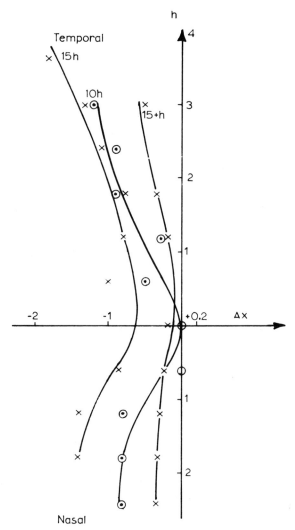

Figure 3. Spherical aberration of the aphakic eye wearing an ordinary correction; abscissae : aberration in diopters; ordinates, distance from the pupil center in mm. The three curves were made on the same day at 10.00 and 15.00 and the following day at 15.00 (after Bonnet).

densitometric study of the photographs corresponding to various positions of the knife edge allows the calculation of the wave surface and hence the spherical aberration.

Although the accommodation is fixed by a point of fixation P for a given

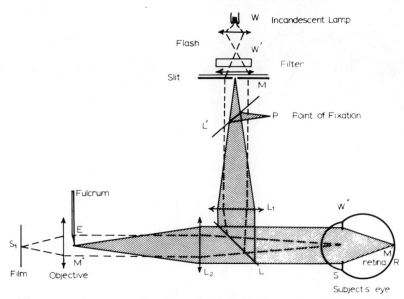

Figure 4. Apparatus for the evaluation of the quality of the retinal image by Foucault's shadow method (after Mrs. Berny)

subject and a given accommodation, the results are different from one photograph to another because of the well-known fluctuations of accommodation. For the calculation of wave surfaces, it is necessary to admit that diffusion by choroid suppresses the phase distribution due to the first passage through the eye. This hypothesis has been verified by projecting two coherent images of a slit on the retina; no interference is visible in the pupil image after diffusion. It is possible that this hypothesis would be wrong if human eyes had a "tapetum."

It is also necessary to adopt a criterion for the "best" image. Mrs. Berny supposed it is the one having the highest illumination in the center of the diffraction figure. Then it is possible to calculate the spherical aberration. Measurements were performed on 4 young subjects; the results differ from one to another. Sometimes (Fig. 5) the eye is under-corrected for small value of accommodation, but the curve twists at higher values; in other cases it is twisted for all values of accommodation. Generally speaking the aberration diminishes when accommodation increases.

The transfer function for the optical image is easy to calculate from these results. For 2 subjects in 4, the eye is excellent for a pupillary diameter of 6.5 mm and all values of accommodation, the eye fulfills Lord Rayleigh's criterion, it works like a perfect instrument. For the other 2 it

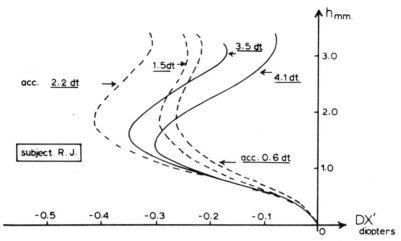

Figure 5. Spherical aberration for various accommodations; in abscissae, the aberration in diopters; in ordinates the distance from the pupil center in mm. Subject R.J.

is not true. For Mrs. Berny herself, the eye is quite good for the accommodation 3.4 diopter, but rather bad for small accommodations.

For subjects who are older, it seems that heterogeneity between the lens sectors spoils the image. Measurements with Mrs. Berny's photographic method are precise but long. She is now modifying her apparatus in order to obtain fast measures by a photoelectric method. It will then be possible to obtain statistical results of transfer functions for the optical image in the human eye, in which all causes of degradation will be maintained (microfluctuation, heterogeneities, irregular aberrations, and so on).

To what extent is it possible to apply to the true retinal image these transfer functions, determined for the fictitious optical one? It is difficult to say. The only point demonstrated by Mrs. Berny is that the calculated change of focusing with pupil diameter is identical with this change measured subjectively by the best acuity. The apparatus is shown in Figure 6: the subject looks simultaneously at optotypes seen through constant artificial pupil of 2 mm diameter in one part of the field, and through a variable one (from 2 to 8 mm) in the other. But it is evident that diffusion in the retina itself will produce a degradation of the transfer function at high spatial frequencies; some work has already been done on this point, but much more is necessary before the question is settled. The same remark applies also to the Stiles-Crawford effect.

A third "transfer function" is frequently mentioned. It applies to the

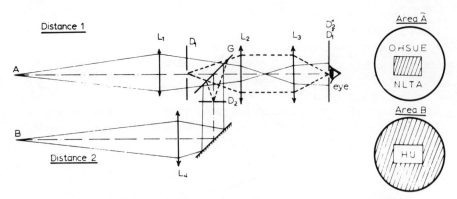

Figure 6. Apparatus for measuring the change of acuity with the pupil diameter: tests A and B are seen simultaneously by transmission and reflection on the half-silvered glass G; different artificial pupils D_1 and D_2 are projected by the lens system L_2 L_3 on the pupil of the observer's eye (after Mrs. Berny).

total visual receptor (optical system of the eye, retina as a physical medium, retina as a receptor of radiation, transmission of visual messages to the lateral geniculate nucleus and to the cortex). This function is analyzed in a paper of Miss Dupuy, a paper which has been published in *Vision Research*. Miss Dupuy has studied more than 65 publications on this subject, and her conclusion is the following: there is no transfer function, strictly speaking, for the total visual system, because this concept implies a mathematical linearity that is lacking in vision. A better term would be *modulation perception function*. This function has been already studied in particular conditions. For example, Lowry and De Palma (1961) used Mach bands as test, and showed that the optical system of the eye is a low-pass filter and the receptor a high-pass filter, so that there is a preferential medium zone for spatial frequencies (about 12 min of arc). Bryngdahl (1966) has worked on a sinusoidal pattern and measured the subjective contrast of the test. But the distribution of illumination in the perceived image is perhaps not sinusoidal (work on this point is beginning in the Institute of Optics). An interesting conclusion given by Bryngdahl is that, in opposition to optical instruments, the modulation perception function for a given spatial frequency changes with the object modulation.

As Miss Dupuy points out, we are far from complete knowledge of the total modulation perception function. We are beginning to understand the first step, that is, the transfer function of the optical system of the eye, and the progress of Mrs. Berny's work will soon give us a fair statistical idea of this function. Knowledge of the second step, including specially the diffusing properties of the retina, is less advanced, but we have some indications, for example, the paper of Campbell and Gubisch (1966) in which they compare the transfer function obtained from the ophthalmoscopic examination of the retinal image of a narrow slit (Miss Flamant's method) with a direct comparison of the contrasts of two sinusoidal patterns, one seen in incoherent light (total pupil) and the other in coherent light (Maxwellian view). These last results were obtained by Arnulf and Dupuy (1960) and by Campbell and Green (1965) with a fair agreement. The "attenuation ratio" (Fig. 7) shows the degradation at high spatial

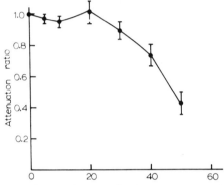

Figure 7. Attenuation ratio. In abscissae the spatial frequency in cycles per deg.

frequencies due to retinal diffusion. But the third step, that is, the knowledge of the total modulation perception function, is scarcely tackled. For the enormous work which remains to be done, it would be very useful if an international agreement were to give standard values for the parameters (wavelengths, pupillary diameters, accommodation, luminances) and for the way of expressing spatial frequencies (cycles/deg, cycles/min, $grad^{-1}$, $radians^{-1}$, mm^{-1} on the retina for a given focal length of the eye).

Spatial Localization

A very different kind of research is being done now by Miss Clotilde Bourdy in my laboratory. Miss Bourdy is a specialist in space vision and stereopsis. Until now her main work has been on aniseikonia and some problems related to Luneburg theory. Two years ago she spent a year in Columbus, Ohio, to work in collaboration with Dr. Fry and Dr. Blackwell.

A very interesting and new idea of Miss Bourdy's was to try to find some correlation between the directional sensitivity of retina (measured by the Stiles-Crawford effect) and the problem of visual direction in space, and especially the monocular and binocular fixation. It is well known that there exists no evidence of an anatomical point of fixation in the fovea. However, as Javal pointed out long ago, as soon as two points are seen separated, the subject knows what point he is fixating, and it is not a psychological attention effect, for when he looks at the second point, there is a slight movement of the eye. It seems that the fixation point is defined at about one min of angle. You may object that fixation disparity is much larger, but it is related to other phenomena. Accordingly, it is generally agreed that the fixation point is not an anatomical entity, a sort of hereditary king among foveal cones, but a physiological one, in relation with the maximum of visual acuity, something like the best cone, a good president in a democratic state.

It is, then, very tempting to think that the directional efficiency of cones, probably linked to the direction of their axis more or less perpendicular to the retina, which changes the "trapping" effect upon photons, is linked with the notion of spatial direction through the maximum of visual acuity. On the other hand, it is well known that binocular spatial direction is in relation with movements of the eyes and muscular balance in the eyes. Accordingly Miss Bourdy is performing new measurements on various subjects chosen so that they represent a large span of heterophories. These measures are:

1. Precise measurement of the maximum of efficiency inside the pupil, in relation with the ophthalmometric pole of each eye. The Stiles-Crawford effect is determined by the classical method of direct equalization of the halves of a bipartite field, each part being seen by Maxwellian view through different points of the pupil; luminance of white light: 13.6 cd/m^2. Measurements were done every half millimeter on a horizontal diameter of the pupil, the origin being the ophthalmometric pole determined by the corneal Purkinje image. Each value is the mean of five measures. Until now, Miss Bourdy has used five esophores (4Δ and 5Δ). It is found that for orthophoric subjects the maximum of efficiency is centered in the pupil (Fig. 8) whereas for heterophoric subjects there is a large excentration (Fig. 9).

2. Determination of stereoscopic acuity, using two rods. One is continuously seen, the other appears during a fraction of a second, and the subject must say if it is in front or behind the other; the stereoscopic threshold is measured per the constant stimuli method with 150 presentations. The lateral separation of the rod varies between 0 and 1 degree. In this foveal domain, there appear important variations of the stereoscopic threshold, and dissymetries variable from one subject to another. For

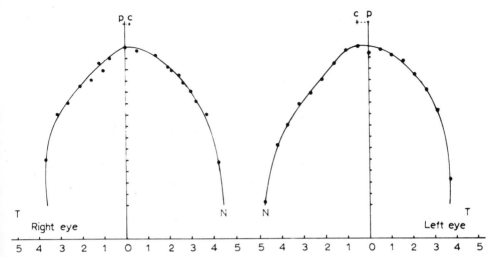

Figure 8. Stiles-Crawford effect for an orthophoric subject. In abscissae, distance from the ophthalmometric pole P in mm. In ordinates, relative efficiency (after Miss Bourdy).

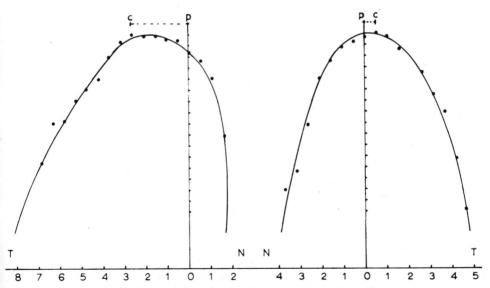

Figure 9. As in Figure 8 but for an heterophoric subject; for one eye there is a large difference between the ophthalmometric pole P and the point C of maximum efficiency.

orthophoric observers, the central position gives the best results, for heterophoric subjects the acuity is not so good and the best results are found for 10 to 20 minutes of arc of separation.

3. Determination of the systematic error in vertical alignment. With the same apparatus, vernier alignment is measured, one half of a rod being seen by each eye. The results seem difficult to interpret; the systematic error does not present any correlation either with heterophory or with pupil efficiency.

4. Monocular disparity of fixation. A haploscope is used, with each arm being able to rotate around an axis passing through the center of rotation of an eye. The subject sees binocularly a fixation point in front of him, and through half silvered mirrors two point sources, each on one arm of the haploscope. Accommodation and convergence may vary independently. Disparity of fixation is measured on each eye, in functions of accommodation and convergence. This last part of Miss Bourdy's work is just beginning.

Color Metrics

A last "unsolved" problem which is now being studied in my laboratory concerns the metrics of color space, or, more simply, the metrics in the surface of constant luminance. Von Helmholtz who mentions it in the second edition of his classical *Handbook* (it disappeared in the third, the only one to be translated in English), stated it is generally agreed that the differential element is a Riemannian one (quadratic form), but not a Euclidean one (it is impossible by any linear transformation to equalize to one the coefficients of dx_i^2 and to 0 those of $dx_i \cdot dx_j$). Like Einstein's universe, the color universe should be curved. In fact, there is no proof that color metrics is Riemannian, although Stiles gave arguments for a physiological basis of a quadratic form, and I argue for a statistical basis. Experiment must decide. MacAdam's well-known ellipses seemed to confirm the quadratic hypothesis, but actually he drew ellipses at best through a small number of experimental points, and it is not a proof that an ellipse is the best curve to fit the data.

Dr. François Parra tried a fresh approach to this difficulty question. The first step was to build a differential colorimeter, in which a beam of white light gives, by rotation through three primary filters (at a speed sufficient to avoid flicker) a light of any chromatic composition in one half of the field, and of a slightly different composition in the second half. One knob allows this slight change at constant luminance, the chromaticity point being moved on a fixed straight line of the diagram. A Wollaston prism rotating at half the speed of the light beam (Fig. 10) allows the separation

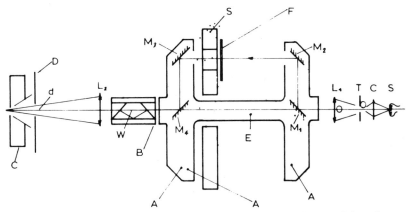

Figure 10. Apparatus for the measurement of differential color threshold (after Parra).

line to be at rest. The field (near 4 degrees diameter) is surrounded by an adaptation field of 52 degrees diameter. Both are seen through an artificial pupil of 2.5 mm diameter. The luminous level varies between 820 and 1230 trolands for the test field, and up to 320 trolands for the adaptation one, so that all measurements are surely photopic.

The differential threshold was measured by the method of limits at 6 points in the chromatic diagram. For each point, 26 different thresholds were determined in directions at 15 degrees from each other. More than 20,000 measures were done by one observer (Dr. Parra), and the general trends of the results were confirmed by two other observers (all normal in color vision) with less detail.

The results obtained give thresholds which are in good agreement with MacAdam, but the technique employed and the large number of points around each center allows the following conclusion: the curves differ systematically from ellipses and show some humps that indicate preferential directions in the diagram; it seems that these directions point toward the confusion centers for protanopes and tritanopes. It is the first time, so far as I know, that measurements of differential thresholds for normal subjects give directly some indications about the fundamentals of color vision (Fig. 11).

Dr. Parra is now extending his work to anomalous color vision. He hopes also to discriminate among normal observers the two tendencies which end respectively as protanomalous and deuteranomalous vision. As a conclusion to the question raised about color metrics, it seems now certain that color universe is not Riemannian. It is not very surprising. Why

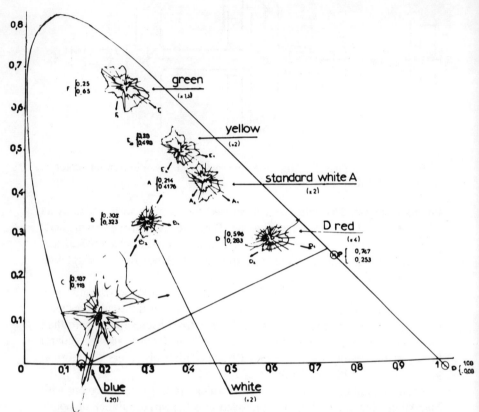

Figure 11. Differential thresholds for 6 points in the chromaticity C.I.E. diagram (after Parra).

should Nature obey our simplistic rules? Going back to the universe of physics, it was shown also by Lichnerowicz that Einstein's world was not Riemannian if electromagnetics were included.

Vision is a fascinating science, and, although old, it is always new, for more and more questions are put to scientists. Vision of man is, in principle, the easest to study, because subjective and objective approaches are both possible. For animals, difficulties are greater. And when astronauts will meet living things on another planet, a new chapter will open. So, fortunately, future workers will never be short of unsolved problems.

References

Arnulf, A., and O. Dupuy (1960). *C. R. Acad. Sci. Paris 250,* 2757.

Berg, F. (1929). *Acta Ophthal. 4,* 386.

Berny, F. (1969). *Vision Research 9,* 977.

Bonnet, R. (1964). *La Topographie Corneenne.* Desroches ed., Paris.

Bourdy, C. (1968). *J. Amer. Opt. Assoc. 39,* 1085.

Bryngdahl, O. (1966). *J. Opt. Soc. Amer. 56,* 811.

Campbell, F. W., and D. G. Green (1965). *J. Physiol.* (London) *181,* 576.

Campbell, F. W., and R. W. Gubisch (1966). *J. Physiol.* (London) *186,* 558.

Dupuy, O. (1968). *Vision Research 8,* 1507.

Flamant, F. (1955). *Rev. d'Opt. 34,* 433.

Le Grand, Y. (1935). *C. R. Acad. Sci. Paris 200,* 490.

Le Grand, Y. (1949). *Rev. d'Opt. 28,* 261.

Lowry, E. M., and J. J. De Palma (1961). *J. Opt. Soc. Amer. 51,* 740.

Parra, F. (1966). Thesis. Paris.

Stiles, W. S. (1946). *Proc. Phys. Soc. 58,* 41.

Part IV

Clinical and Applied Aspects
of Visual Science

Irvin M. Borish

The Relation of Visual Science
to Refractive Procedures

The clinical refractive procedures bear a similar relation to the investigative visual sciences, to use a trite and oft employed parallel, to that engineering bears to research in pure science. The visual scientist is concerned obviously with the disclosure or determination of those factors which explain the visual processes as we know them: the physical structure of light itself and the means whereby the course, expansion or constriction, organization, or alignment of light can be accomplished; the physiological and anatomical propensities of the visual apparatus and the relations, responses, and mechanisms involved in this apparatus, and the physiological and psychological attributes of higher perceptual processes, including not only the determination of the foregoing upon these phenomena but the reaction and association of the qualities of intellect, experience, and the other attributes composing the whole personality of the receptor upon the ultimate imagery. Whether the knowledge discovered has any application to the practical procedures for the determination of the individual so-called refractive error would be merely coincidental, and surely the least of the urges motivating any respective investigation. It is the aim of this paper to comment upon some areas of the field of clinical refraction and to emphasize the need or gap between the operations routinely employed in the ordinary clinical tests of visual performances and the knowledge in the field of purer visual science, available or yet to be determined, but thus far unapplied to these procedures.

Perhaps the most rudimentary clinical practice, and one upon which is placed an import and an explication that far exceed the interpretation merited by the premises underlying the procedures used, is the measurement and recording of "visual acuity."

"Visual acuity," however we may modify, elaborate, or compound our clinical rationale, is among the chief motives influencing our judgment of both the need for refractive or other visual correction and the measure of that correction we should apply. It is only in extraordinary circumstances, or for specific and mutually understood objectives, that a refractionist today would prescribe a spectacle correction that actually reduced the measured visual acuity of his patient. It is likewise "subnormal" visual acuity, at either far or near, that in turn often motivates most patients to seek relief, although other influences may be of equal import. Until very recently, this mere measure of visual acuity comprised the major modus operandi for both school and other special services screening and selection, and, even with modification, it is still the basic element in selection of the child presumed to require more detailed attention. The anomalies of such applications are discussed in comprehensive detail in the Orinda study (Peters et al., 1959).

Yet the average refractionist employs a method of measuring visual acuity that ignores, except in the rudimentary essentials, much of the information about vision and visual cognizance that has been uncovered by workers in pure visual science.

The Snellen test chart, based upon Snellen's progression, was first offered in 1862. Shortly thereafter Green introduced his progression, based upon a different relative size of line to line, and finally modified it to a geometrical progression ratio of 1.26, in which each larger line is that fraction of the preceding one. Ogle recently confirmed this ratio as most desirable. However, uniformity of progression is not necessarily a premise of various charts, particularly in charts used in different countries or of different manufacture. Three different forms of progression are possible: one involving a geometrical progression based upon the relation of the height of the letter to the test distance, a second using arithmetical progression based on a decimal system of Monoyer's for notating vision, and a third based upon the same ratios as the first but with an arithmetic rather than a geometric progression.

The differences between charts using these different forms of progression are illustrated when one considers that the first system would require six steps to move from 6/6 to 6/20, while the same six steps would move only from 6/6 to 6/15 in the second system; the second would require, however, only one step from 6/30 to 6/60 while the third would require six steps for the same transition. Thus one form of progression would seem to provide far more accurate, or at least finer and more detailed, delineation for acuities of a lowered quality. Similarly, the second system requires ten steps to go from 6/6 to 6/3 while the last goes from 6/6 to 6/0 in one step, an obvious impossibility.

Although some agreement exists in accepting the progression of 1.26, the $\sqrt[3]{2}$, for line to line sequence, the best system of progression (arithmetical or geometric, based on fraction or decimal notation) has not actually been determined. Ogle, for example, favored the geometric fractional method while Sloan has favored an arithmetical fractional system. Further difference of opinion exists as to whether a chart designed for a given test distance results in truly representative values at a different test distance, although such use of charts is more common clinically than otherwise.

In addition to the matter of progression, the entire question of the optotypes used for each line is far from definite. The one minute angle of minimum resolution, harking back over one hundred years to an elementary alternate cone averment, is still the basis for the construction of optotypes, although its premise lies in a concept of retinal distribution and geometry and of neural distribution and cerebral anatomy long replaced by more modern physiological and anatomical investigation. In addition, some disagreement of opinion exists as to the value of serifed versus non-serifed letters, although Roberts in 1965 found little difference between them.

Most charts ignore the fact that the one minute basis, itself, is ignored from one specific letter of the alphabet to another, or in the same letter on serifed as compared to non-serifed form. Some 20 deviations of letters, serifed versus non-serifed, of varying geometric proportions and design, have been offered from Snellen in 1862 to Prince in 1956.

The Landolt C was accepted as a standard in 1909, sixty years ago. Theoretically, all other letters are presumed to compare to it in degree of relative legibility. Studies of the difficulty of recognition of various letters have been made by Sheard (1921), Sloan (1952), Coates and Woodruff (1947), Lebensohn (1935), among others. Full agreement is not found, probably due to different forms of optotypes and other variations of test procedures. Standardization of either optotype forms or progression is still lacking.

Even the desirable methods of recording show variation from one authority to another, one time to another, or one country to another. The use of the Snellen fraction in either feet or meters is common, but the basic test distance, 20 feet or 6 M, is not, 5M being the standard for continental testing. As noted before, the accuracy of results using a chart designed for one test distance at another is subject to question. In addition, visual acuity has been expressed in minutes of arc, in a decimal system, in the Armagniac system (decimal x 10); as oxyopter (Snellen x 60), and in reversed Snellen fraction as "visual badness" (Swaine). The objectives of the various representations are to give designations of vision independent of test distances (decimal system), or in forms of the threshold visual angle (minutes of arc), or to provide a uniform ranking of acuity levels (Armagniac), or to avoid the inference that 20/40 is 50% of the acuity of 20/20 (oxyopter). As with most other factors involved in acuity, standardization is absent although the Snellen fraction predominates. Cowan has pointed out that the visual angle, however, is geometric, not linear, hence misrepresented by all designations.

When nearpoint charts are considered, the categorization of vision becomes even more highly arbitrary. The Jaeger chart, most commonly employed, is not standard in wording, size of letters, in optotype selection, spacing, or in relation to angles of resolution from one size to another, or from one card to another. The selection is based more upon availability from the printer's font than upon visual science criteria. Lebensohn has introduced a chart based upon 10 vertical and 10 horizontal rows of the same 10 letters, no two rows revealing the same order of letters; the chart was photographically reduced to accurate angular representation. The ten letters represent letters of different degrees of difficulty of visibility, and the chart is presented at varying distances as required.

A myriad of objective techniques has thus far been limited to essentially laboratory or experimental use. Further development for accuracy of technique, ease of use within the office, and more precise correlation with accepted subjective evaluation would be highly desirable.

Most of this discussion has ignored many of the known physiological and perceptual elements involved in visual acuity and revealed in detail by research in pure visual science. Most of these are ignored just as completely in the clinical test procedures concerned with acuity.

For example, the influence of the size of the pupil or of intensity of illumination upon relative visual acuity is recognized and it has been investigated by many. Almost no cognizance of either is revealed in the ordinary clinical test procedures. Some foot-candle norms have been published, but with the use of projection charts, the question of luminosity becomes one of the surround more than of the chart itself. Since the luminance in the room affects pupil size within limits, the two become mutually involved. Ogle has reported a pupil of 2-2.5 mm as the limiting size for improving acuity by reducing aberration, while Jenkins reports a 3 to 4 mm pupil diameter as optimum for visual testing. Fifty lumens per square foot are considered the upper limit of illumination for improvement of vision (Duke Elder); Gilbert and Hopkins (Ogle) report an improvement of one line of acuity for each increase of illumination ratio of 10:1. Over 1000 feet lux results in a drop of acuity (Fixel). The optimum is 500 lux on the chart and 50 for the background (Oshima). Pease and Allen found that a drop in room illumination required an increase in luminance of background. A relation between visual acuity, luminance, and age has been shown, with the threshold falling 4% from age 22 to 43 (Walsh). Similarly, the spherical and chromatic aberrations of the eye are ignored. Although the focus of red (650μm) has been reported .036 mm behind the retina and violet (436μm) 0.48 before the retina (Nutting) and a range between the colors of 1.22 to 1.66D at far (Ames and Proctor), and .58 to 1.57 D at near (averaging 1.05 D) (Jenkins) has been noted, most tests of acuity ignore totally the spectral characteristics of their chart or slide or the fact that as lamp life increases, the transmission properties may alter. Ryan has emphasized that both the intensity of light and the wave length vary with the smallest section of the caustic. The theoretical emmetropic eye actually has the smallest part of the caustic in the eye located 0.60 D. in front of the paraxial focus of the 55μm waves upon the retina. An eye in focus for approximately 600μm at far is also in focus for 540μm at 33 cm.

The influence of age upon light threshold has been mentioned. Weymouth has shown varying age discrepancies for ranges of acuity, but the progression of most charts is too gross to denote accurately such alterations in usual clinical procedures.

The effect of stray light as a reduction of contrast to create blur has been noted by Weymouth. Sloan finds it not significant if the contrast is not less than 84%, in which case a variation of $\pm 25\%$ exerts a negligible effect upon the test. Little control of either stray light or of measurement of uniformity of contrast is available to the clinician.

Similarly, the influence of time is evident daily to any clinician. Numerous patients exhibit the phenomena of improvement of acuity with increased study time of the chart, but no means of clinical evaluation of this influence is offered. Other factors, essentially psychological, such as the guessing bias disclosed by Fry and Prince, the influence of training upon interpretation of blurred images, or of fatigue or boredom (so evident in aged or very young patients), while known, still lack means for metric qualification of the acuity.

Finally, of greatest significance to the refractionist is a means of closely correlating refractive error and acuity. Sloan has established that 2.8 x error will equal the resolvable angle in minutes. Singh and Jain created the formula Y (refractive error) $= 1.6322 - 1.6355$ x (decimal V.A.). Since acuity is influenced by so many other variables, it is evident that such precise mathematical approximations must ignore the ranges of distribution per error which would exist due to them even if ordinary physiological variations were not also likely. Peters has published a series of tables of the relation of error to acuity which considers astigmatic as well as spherical error and also the effects of age, and represents the most likely predictive data in this regard.

Flom has offered a system of measuring visual acuity that more closely considers the variables inherent in any psychological or perceptive measurement than heretofore offered. This has been appropriately titled "Psychosometric Analysis." A number of charts of Landolt C's and illiterate E's in varying position, each chart bearing letters of uniform and designated angular size, are presented. Acuity is recorded in terms of the percentage or number of images per chart correctly reported. The number of correct responses is plotted against the size of letters presented. A threshold can be selected, such as 5 correct out of 8, or 7 correct out of 8, etc. This offers a different and potentially more accurate evaluation, but charts for its utilization in regular office routine have not been promulgated.

In summary, the measurement of visual acuity is, in the main, still performed upon charts based on progressions and optotypes, presented early in the development of visual science and not uniformly accepted or validated, with the entire body of recent disclosures which might affect the findings totally ignored, uncontrolled, and unapplied to the judgments made. Time, contrast, luminosity, pupil size, color, surround are all ignored, or non-quantitively considered.

One important aspect, peripheral visual acuity, although significantly investigated by Weymouth, Low, Fenberg, and others, remains almost totally ignored outside of the laboratory. The closest approximation to the monitions of these researchers found in clinical practice lie in the techniques applicable to the charting of visual fields, an area not contained in the constant routine of examination. Only recently have these principles been even vaguely applied in this area.

The evaluation of the focus of the target as a measure of peripheral acuity as compared to the recognition of the target as a representation of presence alone involves consideration of the many factors affecting peripheral visual acuity compared to the elements involved in the mere apprehension of a target by means of contrast. The measurement of peripheral field isopters by quantitative perimetric methods approximates a concept of estimating by means of peripheral visual acuity, although Hotchkiss, Washer, and Raspbery (Fankhauser and Enoch, 1964) emphasize that contrast appears best when the retinal image is at minimum diameter, while acuity is best when the image of a point has the most light at the center, irrespective of its diameter, indicating different and non-coincident planes for the two types of responses. The evaluation of peripheral acuity by isopter size calls for development of clinical technique, based upon investigation of the relationship which might exist. Time, adaptation of the retina, exposure time, luminance, rate of movement, etc., are involved in both peripheral acuity and field chart measurements. Since most tests measure a differential light sense, the independence or mutual dependence of the two need definition. The influence of the refractive error is usually ignored, as is the simple optical magnification, minification, or distortion of the retinal images in relation to the size of the projected fields—an incredulous oversight in a technique dependent upon size of the isopters or blind spot for diagnostic import.

Some clinical improvement is evident in the introduction of qualitative and quantitative techniques, although instrumentation is still in the main clumsy, tedious, and time consuming, and the techniques subject to both willful and involuntary error. Despite the differences in relative acuities at different angles of presentation, the technique of perimetry offers a uniformly sized target to all areas of varying retinal sensitivity, while the tangent screen techniques inversely distort or reduce target size at the more peripheral areas where lowered acuity is prevalent. Targets sufficient to elicit response in low acuity areas are far above threshold of higher acuity areas.

Similarly, the relation of color sense and color deficiency to color field charting is only superficially considered, and even the evaluation of the import of color isopters is questionable. Whether the color merely provides a different degree of intensity or contrast to the test, as claimed by some, or actually measures a different sensory aspect is still being debated.

Other elements of retinal sensitivity and its response to varying stimuli are potentially significant in judging the health and function of the retina. Fankhauser and Enoch (1962) and Harrington (1963) employed ultra-violet illumination and sources to measure peripheral isopters. Teissler and Vyskoech (1958) employed X-ray techniques. Francois and Verriest (1954) introduced dim light perimetry leading to the concept of scotopic and mesopic measurement. Binocular methods such as red-green filters (Baisinger, 1955) or polaroid (Sachsenweger, 1956) grew out of the use of the Stereo-campimeter but proved applicable to perimetric and campimetric techniques. The measurement of Flicker Fusion Fields (Miles, 1950, and others) introduced an entirely different approach to determining peripheral retinal function and response as has the application of Haidinger brushes (Sloan and Naquin, 1955) to the same purpose. These latter methods, although of intense interest in experimental visual physiology, have not developed into readily usable clinical apparatus or techniques. The recent introductions of "screeners," like visual acuity screeners, exhibit sufficient deficiencies to create ultimate self-limitations of use.

Some fields of clinical application derived directly from initial perceptual researchers. Such a field is that of aniseikonometry, which introduced an added element to the question of anisometropic and unequal retinal and cerebral imagery. Simpler, more precisely measuring, and perhaps more critical clinical methods are required to establish this phase in the usual visual examination routine. The space-eikonometer, the basic device of clinical applicability, is still a highly subjective device and its readings are not totally without conjectural elaboration. It indicates deviations from image parity by interpretations of spatial localization, which may or may not involve errors adjunctive to the patient's complaint. Since equal ocular images need not result from equal retinal images, the distinction of inherent aniseikonia, created by intrinsic causes from those produced extrinsically by lenses needs disclosure. As Halass notes (1966), the eikonometer is concerned with primitively organizing retinal correspondence and harmonious coordination of various perceptual factors could exist despite their lack. If spatial organization is well organized or developed with unaided vision, the problem is to avoid disrupting them by unwanted size changes caused by corrective lenses. The clinical area is certainly closely related to the continuing laboratory study of perception and cognition, although simple office techniques of evaluation are totally absent.

Other routine examination procedures also exhibit dependence upon purer forms of visual investigation. The need for defining the cornea construction and topography more accurately involves both keratometry and the fitting of contact lenses. Descriptive and investigative applications of Drysdale's principle (Bennett, 1964) and the applications of Moire's Fringes (Mandell, 1966) have been well described and even applied, but no standard clinical application to the problem was presented that was superior in principle (if more facile in mechanical details) to the ophthalmometer which harks back to 1854 (Mandell, 1962 and 1964). The photokeratoscope, introduced in 1896, has only recently been modified for reasonable accuracy (Knoll, 1961). While much knowledge of corneal physiology and respiration has been gleaned (Hill et al.) in recent years, and its influence upon contact lens wearability delineated, no clinical device or readily relevant method for valuation of this effect is yet provided that is as important as such quantitative measures would appear to be for prognosing the suitability of a given contact lens to a given cornea. Recently, the question of even the simple relation of keratometry as we know it to the corneal curves apparently exhibited have been questioned (Ludlam and Wittenberg, 1966).

In the highly psycho-physiological compass involving ocular motility, accommodation, convergence, and the relationship between them, a definite need to more closely parallel results attained by clinical techniques to those attained by laboratory haploscope methods exists. This is particularly true since the results of clinical tests are interpreted and analyzed explicitly toward diagnosis of the applicable functional facility of the patient submitting to the clinical examination. Certainly methods should be developed to measure "responses" of the functions rather than interpolations of response based upon "stimuli" to them. As Morgan (1968) has pointed out, it is also essential to develop "open-loop" rather than "closed-loop" test methods of the associations between the functions, so that feedback does not become an influence included as part of the resultant. The knowledge of variation of individual muscle action in various directions of gaze is totally ignored in most standard test procedures in which visual performance is assayed while in the primary plane of fixation. The relation of ocular dominancy to perceptual skills or as a factor in selectivity where anisometropia or amblyopia exist, as well as in the corrective determination in conditions of monocular aphakia, needs more precise clinical application. The entire question of "dominant eye" versus "dominant field" is still moot (Flax, 1966). Similarly, the conjectural use of objective means of measuring heterophorias, such as photoelectric observance of eye position during the cover test, has scarcely been approached.

In the most directly applicable clinical procedure, the subjective test of refractive status, all that has been said about the neglect of already prevalent knowledge as concerned with visual acuity measurements applies equally here, since the appraisements assessed in this test involve judgments performed upon the identical acuity test chart. The use of the identical chart for acuity and for testing is in itself a debatable point. The astigmatic dial of Green, a century or so old, is still the prototype of newer forms of astigmatic dials except for the comparatively recent departure exemplified in the Raubitchek arrow. Although used in Europe since 1927, binocular fixation during subjective test procedures are barely appreciated in this country with the development of polaroid techniques.

Above all, subjective appraisal—the gauging of improvement by survey of two slightly altered situations—is still performed in sequential overtures: first one image is presented, then another to be compared with it. This involves not only memory, but all the other elements, such as fatigue or inattention, which may and do mislead selection between the lenses creating the images. Only recently, in regard to the cross cylinder, have Matsuura and Biessels developed devices for presenting two images simultaneously, from which one is to be selected. Certainly the greater validity of a method offering similar applications to all of the subjective procedures, such as the selection between two spherical powers, would appear obvious. Such an improvement would increase patient reliability markedly.

What has been said about these limited phases of refraction applies equally for phase after phase of clinical practice.

This paper appears to reflect a most critical and even supercilious judgment of current clinical techniques, but this is not my intent. Our methods have stood the test of time, and for that matter daily, even today, demonstrate their utility and value. The question being raised is whether they are the best we have or are capable of having in the light of present day knowledge.

References

Bennett, A. G. (1964). A new keratometer and its application to corneal topography. *Brit. J. Phys. Opt. 21.*

Bennett, A. G. (1965). Ophthalmic test types. *Brit. J. Phys. Opt. 22.*

Berk, M. M. (1962). A critical evaluation of color perimetry. *Int. Ophth. Clinics 2.*

Berner, G., and D. Berner (1953). Relation of ocular dominance, handedness and the controlling eye in binocular vision. *Arch. Ophth. 50.*

Best, W., and K. Bohnen (1955). Visual acuity and central visual fields. *Ber. Dtsch. Ophthal. Ges. 59.*

Biessels, W. J. (1967). The cross cylinder simultans test. *J. Am. Opt. Assoc. 38, 6.*

Cowan, A. (1954). A suggestion for a method of evaluating the central visual acuity. *Proc. XVII Int. Congress Ophth.*

Enoch, J. M. (1965). *The Current State of Receptor Amblyopia.* Docum. Ophthal. Liege.

Enoch, J. M., and L. E. Glasman (1966). Physical and optical changes in excised retinal tissue. *Invest. Ophthal. 5.*

Fankhauser, F., and J. M. Enoch (1962). The effect of blur upon perimetric thresholds. *Arch. Ophth. 68.*

Fixel, C. A. P., and W. R. Stevens (1955). Measurement of visual acuity. *Brit. J. Phys. Opt. 12.*

Flax, N. (1966). The clinical significance of dominance. *Am. J. Opt. and Arch. Am. Acad. Opt. 43.*

Fletcher, R. J. (1965). Instruments used for objective examination of the eye. *Principles and Practice of Refraction.* G. Giles, ed. Second edition.

Flom, M. C. (1966). New concepts in visual acuity. *Opt. Weekly 57,* July.

Francois, J. and G. Verriest (1954). Campimetry in dim light and neuro.-ophthalmology. *Bull. Soc. Belg. Ophth. 107.*

Halass, S. (1966). Aniseikonia: A survey of the literature. *Am. J. Opt. and Arch. Am. Acad. Opt. 43.*

Harrington, D. O. (1953). Perimetry with ultra violet (black) radiation and luminescent test objects. *Arch. Ophth. 49.*

Harrington, D. O., and W. F. Hoyt (1955). Ultra violet radiation perimetry with monochromatic blur stimulus. *Arch. Ophth. 58.*

Havener, W. H., and J. W. Henderson (1954). Comparison of flicker perimetry with standard perimetric methods. *Arch. Ophth. 52.*

Jenkins, T. C. A. (1962). Aberrations of the eye and their effect on vision. *Brit. J. Physiol. Opt. 20.*

Knoll, H. A. (1961). Corneal contours in the general population by the photo-keratoscope. *Am. J. Opt. and Arch. Am. Acad. Opt. 35.*

Lebensohn, J. E. (1965). Visual charts. *Int. Ophth. Clinics. Refraction 5.*

Ludlam, W. M., and S. Wittenberg (1966). The effect of measuring corneal toroidicity with reference to the line of sight. *Brit. J. Phys. Opt. 23,* 178-185.

Lynn, J. R. (1962). Current trends in quantitative perimetry. *Int. Ophth. Clinics: The Retina 2.*

Mandell, R. B. (1964). Corneal areas utilized in ophthalmometry. *Am. J. Opt. and Arch. Am. Acad. Opt. 41.*

Mandell, R. B. (1966). Corneal curvature measurements by aid of moire fringes. *J. Am. Opt. Assoc. 37.*

Mandell, R. B. (1962). Refraction point ophthalmometry. *Am. J. Opt. and Arch. Am. Acad. Opt. 39.*

Matsuura, T. T. (1961). The Matsuura auto cross. *Opt. Weekly 52,* 2153.

Meyer-Schwekeroth, G., and K. Bohnen (1957). Projection perimetry with polarized light. *Klein. Nibl. Augenheilk. 131.*

Miles, P. (1950). Flicker fusion fields: II. Findings in early glaucoma. *Arch. Ophth. 43.*

Mitchell, D. W. A. (1953). Investigating binocular difficulties. *Brit. J. Phys. Opt. 10.*

Morgan, M. W. (1967). *Accommodation and Convergence, 4th Prentice Lecture.* Am. Acad., Chicago.

Ogle, K. N. (1953). On the problem of an international nomenclature for designated visual acuity. *Am. J. Ophth. 36.*

Oshima, S., T. Enimoto, S. Shinodu, and M. Takayaki (1962). Standardization of visual acuity charts. *Arch. Soc. Ophth. Jap. 66.*

Pease, P. L., and M. J. Allen (1967). Low contrast visual acuity and the effect of ambient illumination, filters, and scatter. *Am. J. Opt. and Arch. Am. Acad. Opt. 44.*

Peters, H. B. (1961). The relationship between refractive error and visual acuity at three age levels. *Am. J. Opt. and Arch. Am. Acad. Opt. 38.*

Peters, H. B., H. L. Blum, J. W. Bettman, F. Johnson, and V. Fellows, Jr. (1959). The Oneida vision study. *Am. J. Opt. and Arch. Am. Acad. Opt. 36.*

Prince, J. H., and G. A. Fry (1956). The effects of errors of refraction on visual acuity. *Am. J. Opt. and Arch. Am. Acad. Opt. 33.*

Sloan, L. L. (1951). Measurement of visual acuity. *Arch. Ophth. 45.*

Sloan, L. L. (1959). New test chart for the measurement of visual acuity at far and near distance. *Am. J. Ophth. 48.*

Sloan, L. L (1961). Area and luminance of test object variables in examination of the visual field in projection perimetry. *Vision Research 1.*

Sloan, L. L., and A. Altman (1954). Factors involved in several tests of binocular depth perception. *Arch. Ophth. 52.*

Sloan, L. L., and H. A. Naquin (1955). A quantitative test for determining the visibility of the Haidinger brushes: Clinical applications. *Am. J. Ophth. 40.*

Weymouth, F. M. (1955). Visual acuity, an analysis of the stimulus situations. *Am. J. Opt. and Arch. Am. Acad. Opt. 32.*

John H. Carter*

Age Trends in Certain Refractive Findings

Introduction

For many years it has been recognized that parameters beyond the ocular static refraction can influence both visual efficiency and comfort. Of these, perhaps the most significant concern the relationships between accommodation, convergence, and physical object location, and Donders (1864), Percival (1892), and Sheard (1928) discussed the clinical importance of these variables during the early days of modern refraction. Indeed, Percival and Sheard proposed rules of thumb to be used to predict comfort when a tentative prescription is contemplated for use in a nearpoint application.

Unfortunately, the relationship between clinically derived findings and the physiological mechanisms upon which their magnitudes depend has not always been apparent. However, in 1943, Fry (1943) suggested that five physiological variables are significantly involved in the relationship between accommodation and convergence of the eyes. The present study concerns the magnitudes of these variables among 506 subjects within the age limits of ten and twenty and among 490 subjects within the age limits of thirty-seven and forty-seven years.

The data presented herein reflect the variability of the physiological parameters being measured as well as the variability introduced by data-taking by thirty-four different (though well experienced) examiners. All examiners presumably used similar testing methods. All were members of a postgraduate group and received their O.D. degrees from the Pennsyl-

* The author wishes to express his appreciation to Dr. Nathan Brod of the Psychology Department of the Pennsylvania College of Optometry for his suggestions concerning the statistical analysis.

vania College of Optometry in February, 1968. Members of the group had received their earlier professional training variously from Columbia University, Indiana University, the Ohio State University, the University of Havana (Cuba), and the Massachusetts College of Optometry. Clinical experience among examiners varied from a minimum of seven through a maximum of thirty-one years.

Each subject used in this investigation was free from recognized ocular pathology, exhibited a corrected acuity in his poorer eye of not less than 20/25, and presented substantially equal accommodative amplitudes in his two eyes. All findings were obtained with the individual subject wearing his distance refractive correction.

Theoretical Considerations

The five fundamental variables postulated by Fry (1943) are as follows:
1. the phoria at distance
2. the ACA ratio
3. the amplitude of accommodation
4. the amplitude of positive fusional vergence
5. the amplitude of negative fusional vergence.

Functional interactions of these various mechanisms are generally presumed to occur in a manner somewhat like that illustrated by Figure 1, a block diagram acknowledged to be somewhat speculative from the point of view of the anatomist.

An accommodative center is assumed to emit impulses, directed both to the ciliary muscle and to a multiplication operator which defines the ACA ratio. The resulting accommodative-vergence signal passes to a center where it sums with a second input. This second input signal represents fusional vergence, and can be either positive or negative in sign, depending upon the required direction of eye movement to yield foveal bifixation of the object of regard. To produce an overt motor act, the summed accommodative vergence and fusional vergence signals sum with tonic impulses to the medial recti. At the same time, their phase-inverted complement sums with tonic impulses to the lateral rectus muscles.

Thus, accommodation appears to give rise to an open-loop convergence adjustment which, in most instances, reduces to a small value the amount of supplemental vergence change required for perfect bifixation. Fine convergence adjustment, in whatever amount may be required, is then accomplished by a closed-loop fusional vergence mechanism which acts to prevent diplopia.

Frequently, clinical findings are plotted on Cartesian coordinates. It

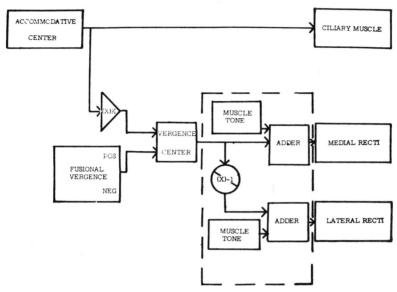

Figure 1.   A model demonstrating the functional interaction between accommodative and fusional vergence

has become traditional to plot accommodation on the ordinate against convergence on the abscissa, even though it is now generally agreed that accommodation is the independent and convergence the dependent variable.

Figure 2 shows, in idealized form, the zone of clear, single binocular vision (after Fry). Its shape is approximately that of a parallelogram whose lateral position is governed by the tonic position of rest of the eyes and by the relative magnitudes of the positive and negative fusional vergence amplitudes. Its base (the x-axis) corresponds to the accommodative zero level. The parallelogram slopes up and to the right since there is a direct relationship between accommodation and convergence. The right and left borders of the parallelogram represent the limits of positive and negative fusional vergence, respectively. These limits are defined *approximately* by the clinical base-out-to-blur and base-in-to-blur lines. The slope of either clinical line can be used to estimate the ACA ratio, though the base-out-to-blur line is usually more conveniently used for this purpose. The graph altitude reflects the accommodative amplitude.

The line that is parallel to both the base-in-to-blur and base-out-to-blur lines and that also passes through the distance phoria point corresponds to the locus of hypothetical phoria points for which the accommodative response of the eye equals the stimulus to accommodation. The clinical

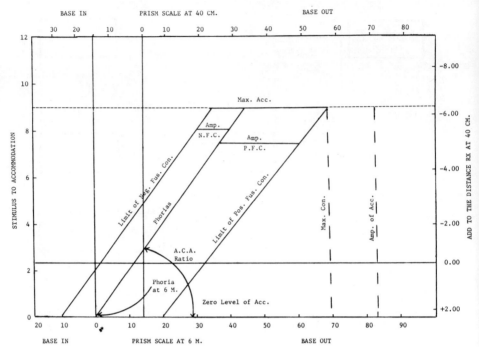

Figure 2. The zone of clear, single, binocular vision (after Fry)

phoria line, on the other hand, is the straight line that best fits the group of phoria points determined when convergence is measured and specified as a function of accommodative *stimulus*.

Several significant deviations of the clinical graph from the parallelogram model exist:

1. At near-zero levels of accommodation.
   a. The *apparent* negative fusional vergence amplitude diminishes significantly and base-in-break points are encountered without prior blurs. This is due to a limitation upon relaxable accommodation and makes the base-in-to-blur line inconvenient to use for determining the ACA ratio.
   b. The blur-point indication of the positive fusional vergence amplitude diminishes slightly due to a small plus bias inherent in typical subjective refraction data.
2. At increasing levels of accommodation, the apparent boundaries of the zone of clear, single binocular vision diverge due to an increased depth of focus associated with reduced pupil size. As a result, the base-out-to-blur line tends to yield a moderate overestimate of the ACA.

3. As the near limit of accommodation is approached, the base-in-to-blur, base-out-to-blur, and phoria lines all tend to inflect to the right. The mechanical accommodative components here operate at reduced efficiency and this gives rise to an increased amount of accommodative convergence associated with each successive accommodative increment.

Methods, Results, and Discussion

Each of the thirty-four optometrists whose data are included in the present study submitted mean and standard deviation values for his own sample. These data from individual optometrists were then combined by the determination of weighted means for the group and by the calculation of averaged standard deviation values by the method of McNemar (1965).

Amplitude of accommodation was measured uniocularly by the method of Donders and referred to the primary nodal plane. While Donder's method produces an estimate significantly biased by depth of field, it is nevertheless uncomplicated and represents perhaps the best known *clinical* method for amplitude measurement. Table 1 shows the mean, the standard error of the mean, the standard deviation of the distribution, and Donder's tabular data for amplitude of accommodation. While Donder's data and the present findings agree for older subjects, the mean value for amplitude of accommodation is significantly lower than Donder's table suggests it should be for the younger group. While the distribution of ages within a given age group is not known, even the unlikely assumption that all members of the 10 through 20 year age group were of age 20 yields an amplitude estimate which is lower than Donder's tabular value for the same age to a degree significant at the 1% level.

Table 1. Amplitude of Accommodation

Age Interval	Donder's Tabular Value at Midpoint of Age Interval	Weighted Mean (X)	Standard Deviation (σ)	Standard Error of the Mean (σX)
10 through 20	12 D.	9.5 D.	1.8 D.	0.08 D.
37 through 47	4.1 D.	4.1 D.	1.6 D.	0.07 D.

Heterophoria at distance was measured by the Von Graefe technique. The mean phoria value for the 10 through 20 year age group was orthophoria and for the 37 through 47 year group, it was 0.2 prism diopters of exophoria. The ratio of this difference between means to its standard error was found to be only 0.8, so the apparent change in phoria value with age seems not to have been statistically significant.

Blur points, base-in and base-out, were obtained for each subject at each of two fixation distances. Use of any fixation distance within a half diopter of either the zero level or the amplitude limit of accommodation was appropriately avoided.

ACA ratio estimates were based upon the slope of the base-out-to-blur line. The mean ACA value was found to be 6.3/1 for the younger and 6.6/1 for the older age group. These values are generally higher than those which have been found in other studies of ACA (1957, 1950). This is not unexpected, however, since most clinical methods of ACA determination are strongly influenced by the accommodative response/stimulus ratio, and, since this is often substantially less than unity, such methods generally lead to underestimate of the magnitude of the ACA. On the other hand, *use of the base-out-to-blur line as an index does tend to result in an ACA overestimate* due to the increasing depth of focus of the eye as the pupil constricts in response to a decreasing fixation distance.

Perhaps more important than the numerical value of the ACA ratio is whether or not the ACA tends to increase directly with age. This question has relevance to fundamental accommodative theory. Specifically, it relates to the question of whether maximum ciliary contraction is required to yield the full accommodative amplitude regardless of the age of the patient.

Davis and Jobe (1957) failed to find a statistically significant age-increase in ACA, while Alpern (1950) (and Hirsch) found the ACA to diminish somewhat with advancing age. Fry (1959), on the other hand, used long-term longitudinal data to demonstrate that his own ACA ratio increased with age.

While the present data may seem to suggest that a moderate increase in ACA ratio occurs with age, the "z" score for the difference between the means in this study was only 1.7, a value which suggests (two-tailed hypothesis) a lack of significance at the 5% level.

A literature review provided little data concerning the fusional vergence amplitudes, although it seems to be commonly believed that the age-decline in accommodative vergence is at least partially compensated by an increase in fusional vergence amplitude. The present data, however, fail to support this assumption. Table 2 shows that the positive fusional vergence amplitude exceeds the negative for each age group but indicates that the amplitudes of positive and of negative fusional vergence both *decline* with age. The reduction of positive fusional vergence amplitude with age is significant at the 1% level while the decline in negative amplitude is significant at the 5% level (two-tailed hypothesis).

Table 2. Amplitudes of Positive and Negative Fusional Vergence

Parameter	X_1	Parameter	X_2	Difference $(\overline{X}_1 - \overline{X}_2)$	Z	P
PFC, 10-20 age	18.1$\triangle$	NFC, 10-20 age	13.6$\triangle$	4.5$\triangle$	8.89	$<<.001$
PFC, 37-47 age	16.6$\triangle$	NFC, 37-47 age	12.6$\triangle$	4.0$\triangle$	8.20	$<<.001$
PFC, 10-20 age	18.1$\triangle$	PFC, 37-47 age	16.6$\triangle$	1.5$\triangle$	2.65	$<.01$
NFC, 10-20 age	13.6$\triangle$	NFC, 37-47 age	12.6$\triangle$	1.0$\triangle$	2.18	$<.05$

Summary

Fry's five fundamental variables were studied for a group of 996 subjects. These were distributed among two age groups, 10 through 20, and 37 through 47.

Amplitude of accommodation declined with age in anticipated fashion although young subjects manifested somewhat lower amplitudes than would have been predicted from Donder's table. No age-change, significant at the 5% level, was found for either distance phoria or ACA ratio. The amplitude of positive fusional vergence exceeded the corresponding negative fusional amplitude for both groups of subjects and the amplitudes of positive and negative fusional vergence both declined with age.

This study appears to raise at least two significant questions: (1) why do cross-sectional data on ACA age-change generally appear to conflict with available longitudinal data, and (2) why do so few early presbyopes seem to experience significant asthenopic symptoms, since their partial loss of accommodative convergence appears to be compounded by a reduction of positive fusional vergence amplitude?

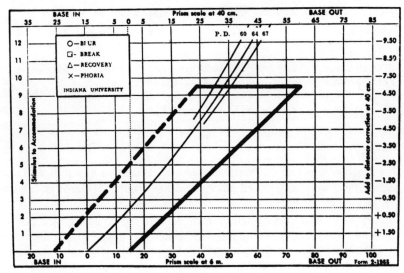

Figure 3A

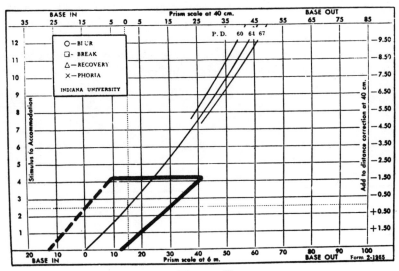

Figure 3B

References

Alpern, M. (1950). The zone of clear single vision at the upper levels of accommodation and convergence. *Am. J. Optom. and Arch. Am. Acad. Optom. 27,* 491-513.

Davis, C. J., and F. W. Jobe (1957). Further studies on ACA ratio. *Am. J. Optom. and Arch. Am. Acad. Optom. 34,* 16-25.

Donders, R. C. (1864). *On the Anomalies of Accommodation and Refraction of the Eye.* The New Sydenham Society, London, England. English translation by W. E. Moore.

Downie, N. M., and R. W. Heath (1965). *Basic Statistical Methods, second edition.* Harper and Row, New York.

Fry, G. A. (1943). Fundamental variables in the relationship between accommodation and convergence. *Optometric Weekly 34,* 153-155, 183-185.

Fry, G. A. (1959). The effect of age on ACA. *Am. J. Optom. and Arch. Am. Acad. Optom. 36,* 299.

Percival, A. S. (1892). The relation of convergence to accommodation and its practical bearing. *Ophthalmic Review 2,* 313-328.

Sheard, C. (1928). Zones of ocular comfort. *Trans. of the Amer. Acad. of Optometry 3,* 113-129.

William M. Lyle

The Inheritance
of Corneal Astigmatism*

Refraction by the front surface of the cornea provides about two thirds of the total refractive power of the eye; hence a toroidal front surface of the cornea produces characteristic alterations in the refractive properties of the eye.

Most eyes show some corneal astigmatism (Jackson, 1933; Biro, 1948; Duke-Elder, 1949; Borish, 1954). Because some corneal astigmatism is almost universal, the pedigrees examined in the present study do not depend on being ascertained just because of the discovery of a single index case.

It is frequently stated that astigmatism is, in part at least, a genetically transmitted trait (MacElree and Morrow, 1934; Biro, 1948; Hamilton, 1951; Francois, 1961; Southall, 1961). Howe, in 1911, when discussing the heredity of corneal astigmatism and of ocular muscle anomalies said, "Indeed in the whole range of medicine probably no better examples of heredity can be found than those which show themselves in the eye." Francois (1961) said "the only certain data about the inheritance of optic refraction concern corneal refraction and astigmatism." An editorial

* The study reported here is based upon an extensive investigation of certain pedigrees by H. W. Hofstetter and the present author. Some of these data formed a part of a thesis submitted by W. M. Lyle to the Graduate School of Indiana University in June, 1965.

The final two years of this investigation were supported by a Research Grant (NB 04863-01) from the Division of Neurological Diseases and Blindness of the United States Department of Health, Education, and Welfare.

in *The Lancet* (July 28, 1962, p. 184) commented concerning Heredity in Refraction of the Eye as follows:

> Probably the most convincing evidence of a genetic basis comes from studies of astigmatism. High degrees of astigmatism placed at unusual axes have been observed repeatedly as a dominant affection; but this represents a special form of refractive error, as the anomaly is almost entirely corneal. All refractive states, emmetropia and ametropia (whether determined by correlation deficiency or anomalous components) are genetically determined.

Although these observations have been supported by many investigators, there is not universal agreement concerning the inheritance of astigmatism nor of refractive errors. Typical comments concerning the inheritance of refractive errors are the following paraphrased remarks:

Stenstrom (1946), Refractive power of the eye is determined in such a complex way that its behavior in heredity is complicated and difficult to estimate.

Sorsby (1951), The inheritance of retinal size determines refraction, but little is known of the inheritance of refractive error.

Manchester (1955) and *Francois* (1961), Consistently accurate predictions concerning the inheritance of refractive error cannot yet be made.

Duke-Elder (1949) and *Manchester* (1955), Small refractive errors showed dominant transmission although irregular dominance or even recessive transmission have been seen. Large refractive errors followed a recessive mode of inheritance.

Wixson (1946), Refractive error had more of the characteristics of a recessive than a dominant hereditary trait. He found the highest correlation between mother and daughter.

Mann (1957), Pointed out the difficulty of reducing the inheritance of errors of refraction to a genetic order because of their ubiquity.

F. A. Young, From his study of monkeys found a very low correlation between refractive error and heredity.

van Alphen (1961), Proposed that the essential cause of refractive error was an aberration of the "stretch" of the eye in response to the intraocular pressure, thus the corneal curvature and the final size of the eye were modified from the values which would have been determined by genetic factors. This analysis does not seem to allow for the fact that the intraocular pressure might be genetically determined too. Nevertheless van Alphen believes that both corneal curvature and corneal diameter are subject to a strong hereditary influence.

Wibaut (1932), Found so high a correlation between the corneal power of parents and their children that he concluded that this optical component was wholly determined by heredity.

Stern and Rosenberg (1940), Based their opinion on the analysis of a large volume of clinical data and concluded that corneal astigmatism did not show a clear pattern of genetic transmission.

Hofstetter and Rife (1953), From their study of twins suggested that astigmatism was almost entirely an environmentally determined trait.

Ida Mann (1957), When discussing high astigmatism and corneal errors of refraction said: "Corneal astigmatism exists in various hereditary forms. The common type is dominant, which probably accounts for the high incidence of hypermetropic astigmatism with-the-rule. The axis type and amount of the astigmatism is often transmissible."

Waardenburg (1961), Considered that at least four genes were involved in the determination of corneal refraction and that uncomplicated regular corneal astigmatism was an autosomally dominant character with incomplete penetrance and variable expressivity. He noted that other investigators thought that its mode of inheritance was polygenic and still others described a recessive pattern.

Duke-Elder (1962), Said corneal astigmatism was inherited as an irregular autosomal dominant. In 1963 he and Cook reaffirmed the view that in the great majority of cases the pattern was autosomal dominant and only rarely autosomal recessive.

Modes of Inheritance

The transmission of most physiological traits is determined by multiple genes. This is called polygenic or multifactorial inheritance, but there is a slight difference between the meaning of the two terms. Polygenic means that two or more gene pairs collaborate to produce the given trait, but no individual gene has a predominant influence. A reasonably Gaussian distribution is not proof of polygenic inheritance, but as Roberts (1963) noted, it is a necessary concomitant and an argument in its favor. McKusick (1964) uses the term graded characters for those characters whose genetic component is presumed to be polygenic. Polygenic inheritance is indicated for many of the eye's characteristics (Duke-Elder, 1949; Spooner, 1957). The iris color is controlled by at least four genes (Duke-Elder, 1962). Francois (1961) states that refractive error is probably dependent on polygenic inheritance. In Waardenburg's opinion both corneal power and corneal radius show signs of polymeric inheritance. Waardenburg considers that at least four independent, but isomeric (i.e., contributing equal values) genes with intermediate results in heterozygotes are involved in the determination of corneal refraction.

In multifactorial inheritance more than one gene acts but one is a major gene and the others serve mostly as modifiers. The term multifactorial includes the concept that several nongenetic factors appear to be involved as well as a number of genes. Multifactorial traits may be referred to as quantitative traits or cumulative factors; examples are intelligence or stature. Sorsby believes that corneal power is inherited in a multifactorial manner and that refraction is determined by multiple genes. The phenotype distribution reflects the influence of two kinds of agents:

1. *Genetic,* there are a number of genes involved, and although each is inherited in a Mendelian manner they have effects which are similar and they tend to supplement each other.

2. *Environmental,* environmental factors modify the phenotype and add to or subtract from the summated genetic effects.

Thus a continuous variation of the phenotype can result in spite of the discontinuous, quantal variation of the genotype. Similarly a discontinuous distribution can occur and simulate a typical Mendelian inheritance pattern. Grüneburg called this kind of discontinuous distribution quasicontinuous. For these traits the simple numerical laws of inheritance are not so apparent, the constituent factors are not easily isolated, and the inheritance pattern is not as readily identified. To analyze this type of inheritance requires a fusion of the Mendelian (genetical, discontinuous) and the Galtonian (biometrical, quantitative) methods. Although the distribution curves obtained resemble normal curves they are, according to Waardenburg, really population curves of mixed biotypes which are the result of a number of modification curves, each produced by one environmental factor acting upon the phenotype. When the environmental effects are not equally distributed in the positive and negative direction the curve will exhibit skewness.

If the inheritance of corneal astigmatism follows the polygenic, quantitative, or multifactorial mode of transmission, we can look for the following:

1. The probability that a child will inherit the trait in question increases in proportion to the number of his relatives who already manifest the trait. Use of a large sample is particularly desirable when examining a polygenic trait.

2. When a parent and one offspring are affected, then it is probable that the other parent is affected too. The offspring may show a larger

range of values than do the parents, but the children tend to show values which are intermediate between those of their parents. If the genes are strictly additive, then the average score of the offspring should equal the mid-parental value.

3. When one child is affected, the probability that his sib will be affected $= \sqrt{P}$, where $P =$ the population frequency. The derivation of this formula may be found in Edwards (1960). He shows that in the case of first degree relatives the correlation of the genotypes will be approximately half and in the more general case for n^{th} degree relations the correlation becomes $(\frac{1}{2})^n$.

R. A. Fisher discovered in 1918 that in cases of polygenic inheritance subject to certain assumptions, the degree of resemblance between relatives could be represented by a simple mathematical expression. Mc-Kusick (1964A) and Roberts (1963) list these assumptions as:

1. Inheritance alone is involved. An environment which modifies members of a family group all in the same direction will increase their phenotypic resemblance and this might be assumed to have a genotypic basis.

2. There is intermediate inheritance. The heterozygote is intermediate between the two homozygotes and the gene pairs do not show dominance or recessiveness.

3. There is random mating. Assortative mating increases all measures of family likeness.

To which might be added the further assumption that the genes are perfectly additive and are not modified by other genes.

Relatives have genes in common which they have received from a common ancestor. The regressions of child on parent, or parent on child, or sib on sib or that of a more distant relative on the proband are equal to the number of genes they have in common.

Let r' equal the regression coefficient. It is a measure of likeness which, when the distribution is symmetrical, becomes equal to r, the Pearson product moment coefficient. Generally, the regression coefficient has to be calculated by finding the covariance and variances. When studying the regression: (1) of a child on the mean value of his parents and (2) of the mid-parental value on the child, these two regression coefficients differ from each other because of the different variances of the two classes. The regression of the child on mid-parent should be unity (linear regression) since all the child's genes come from his parents. Accordingly Penrose says linear regression of child on mid-parent suggests perfectly additive genes. The regression of mid-parent on child has an expected value of 0.5 because only half of the genes of the two parents are identical with those of the child. The expected correlation coefficient is the geometric mean between the two regression coefficients (Penrose, 1963; McKusick, 1964A).

$$r = \sqrt{(r'_1)(r'_2)}$$

In the example of child and mid-parent, this becomes:

$$r = \sqrt{(1)(\tfrac{1}{2})} = \sqrt{\tfrac{1}{2}} = +0.71$$

Table 1 shows the proportion of genes in common for a number of relationships and also shows the theoretical correlation coefficients.

Table 1. Proportion of Genes in Common and Theoretical Correlation Coefficients

Relationship to proband	Proportion of genes in common	Theoretical correlation coefficient
Monozygotic twin	1	$\sqrt{(1)(1)} = +1.00$
Dizygotic twin	$\tfrac{1}{2}$	$\sqrt{(\tfrac{1}{2})(\tfrac{1}{2})} = +0.50$
Mid-parent, i.e., mean of two	1	$\sqrt{1 \times \tfrac{1}{2}} = +0.71$
Mid-grandparent, i.e., mean of four grandparents	1	$\sqrt{1 \times \tfrac{1}{4}} = +0.50$
Parent, child, sib	$\tfrac{1}{2}$	$\sqrt{(\tfrac{1}{2})(\tfrac{1}{2})} = +0.50$
Grandparent, grandchild, uncle, aunt, nephew, niece, half-sib	$\tfrac{1}{4}$	$\sqrt{(\tfrac{1}{4})(\tfrac{1}{4})} = +0.25$
First cousin	$\tfrac{1}{8}$	$\sqrt{(\tfrac{1}{8})(\tfrac{1}{8})} = +0.125$
Spouse	0	$\sqrt{(0)(0)} = 0.00$

In cases involving dominant or recessive genes these correlations are reduced, the amount of reduction depends on the gene frequency. Let dominant gene frequency equal p and recessive gene frequency equal q, then according to Penrose we may expect to find correlations which can be derived from the two following formulae:

for parent-child relationship $\quad r = \dfrac{q}{1+q}$

for sib-sib relationship $\quad r = \dfrac{1+3q}{4(1+q)}$

Dominance effects can be tested by finding the regression of child on mid-parent. Dominance and recessiveness not only lower the regression of child on mid-parent from the theoretical value, but also make the regression non-linear (Roberts, 1963). Vogel and Kruger (1967) describe three criteria for separating multifactorial inheritance from simple dominance with incomplete penetrance. Unfortunately none of their criteria are applicable in the model assumed here because the trait is one with a high population frequency and no significant sex difference.

Dominance and recessiveness may be thought of as the two extremes of a continuum of variable types of hereditary transmission, all of which are basically similar. Whether the phenotype is described as dominant or recessive depends, to a certain extent, on the acuteness of the methods used for detecting the results of gene action. A term which best describes a certain type of inheritance is intermediate inheritance which may also be referred to as incomplete recessiveness or incomplete dominance. In cases of intermediate inheritance the heterozygous individual differs phenotypically from either the homozygous recessive or the homozygous dominant. Normally a recessive gene reveals its presence only in an individual homozygous for that trait or when it is located on the x-chromosome of a male. Schmidt's sign appears to be an example of the partial expression of a recessive gene in a heterozygous female (Hirsch, 1963, p. 303). Probably most individuals are heterozygotic for the genes which determine refraction.

If there is a recessive component then the sib-sib resemblance should be greater than the parent-offspring resemblance. If sib-sib correlation is less than expected, then the environment is probably increasing the variability. Environmental effect can be modified by: (1) interaction as a result of differences in effects of the same genes acting in different environments, (2) maternal effects as a result of a common intra-uterine environment, (3) effects of the common environment upon the members of a family.

Most geneticists agree that the major part of the variability of the human phenotype is due to the variability of the genetic make-up of the individual and the minor part is due to the variability of the environment, but all phenotypes are the product of the genotype and the environment. Two of the established techniques for the genetic study of human material are: (1) twin method and (2) pedigree analysis.

Studies of monozygotic and dizygotic twins can often provide definitive information regarding heredity. Twin studies in general show a high concordance between both members of a pair of monozygotic (uniovular) twins in total refraction and in its components (e.g., Holste, 1940; Hofmann, 1942; Sorsby, 1962). Francois (1961, p. 193) says "the identity of ametropia in identical twins lasts throughout life, which shows that the evolution of refraction is hardly influenced by external factors. When refraction is identical, particularly if the ametropia is accompanied by astigmatism, there is a strong possibility that the twins are monozygotic; non-identity suggests dizygotic twins." Hofstetter and Rife (1953) from a study of twins concluded that, "in view of the generally recognized reliability and validity of the astigmatism findings, these low correlations may well represent the true picture. They indeed suggest that astigmatism is almost entirely an environmentally determined trait."

A method of pedigree analysis was utilized in the study reported here. The validity of the statistical procedures employed depends upon the assumption of a suitable model for analysis.

Method

Corneal astigmatism measurements were made on 604 subjects who were members of fifty-one families*. Each family extended over at least three generations. Most of the data were abstracted from the files of the author and his predecessor. Clinical measurements provide two kinds of information concerning corneal astigmatism. These are the magnitude of the corneal astigmatism and the orientation of its axis.

For the purposes of this study, the amount of corneal astigmatism was determined by the difference in refractive power between the two principal meridians of the cornea. These powers were measured by the ophthalmometer (keratometer) and expressed in diopters although the instrument really measures the radius.

The axis meridian of the corneal astigmatism was taken to be that principal meridian which had the least refractive power. The usual clinical notation for axis position begins with 0° at the subject's left side of his eye and proceeds counterclockwise so that 90° is up and continues to 180° at the subject's right. The so-called standard notation for axis position is the one suggested in England by an Optical Society Committee in 1904 and adopted in 1917 by a German committee (the Technischer Ausschuss fur Brillenoptik) and hence called the TABO notation (Council of British Ophthalmologists, 1921). The 0°-180° base line is considered to be parallel to a line joining the centers of rotation of the two eyes.

If the meridian of least power (flattest meridian) was located at axis position 180° or within 30° of this orientation, the astigmatism was considered to be of the with-the-rule type and was arbitrarily assigned a

---

* These pedigrees were diagrammed and form part of the thesis. Copies may be obtained from University Microfilms, Ann Arbor, Michigan.

positive value. If the flattest meridian was located at axis position 90° or within 30° of this orientation, the astigmatism was considered to be of the against-the-rule variety and was assigned a negative value. All other axis positions were considered to be oblique and the astigmatism was assigned a zero value.

To arrive at a mean value for the amount of corneal astigmatism of an individual, the corneal astigmatism of each eye as measured with the ophthalmometer was assigned a positive, negative, or zero value in accordance with the rules given above, then the value so obtained from the right eye was summed algebraically with that similarly obtained from the left eye and this total divided by two.

To arrive at a mean value for the axis orientation the following arbitrary system was used. For each eye the angular separation in degrees between the corneal meridian of least power, as determined by the ophthalmometer, and the horizontal meridian (180° line) was found. This amount was assigned a positive value regardless of the direction of rotation of the axis meridian away from the 180° line. The angular values thus found for the right and left eyes were summed and the total divided by two.

Galton, in 1889, when studying the inheritance of exceptional ability, used mid-parental values based on the average measurement of the two parents. Sorsby (1957) computed the correlation between mid-parent and offspring for corneal power and found this correlation to be $+0.61$ with a standard error of ±0.24. The mid-parental values used by Sorsby were found by calculating the mean value of the four eyes of the two parents. A comparable study by Holt (1961) of dermal ridge patterns makes use of a similar table of expected correlations. For her study she utilized for each individual a single numerical value which was obtained by totalling the ridge count on all the fingers. Similarly in the present report a mean value for the corneal astigmatism was obtained by adding the astigmatism of the right and left eyes of each subject and dividing by two. Francis (1963) and others (e.g., Duke-Elder, 1949; Phillips, 1952) report a high correlation between the astigmatism of the right and left eyes of individual subjects. The correlations found between the right and left eyes of the individuals in the pedigrees examined here were as follows: for amount of astigmatism, $r = +0.84$; for axis position, $r = +0.66$.

Pearson product moment correlations between various members of the 51 families whose pedigrees were available were calculated by means of program BMD05D which was converted to 3600 Fortran by Dr. Walter Chase. Then the correlations were calculated on the Control Data Corporation's 3600 electronic digital computer located at the Indiana University Research Computing Center.

Results

For the magnitude of the corneal astigmatism the correlations are shown in Table 2. In each case, the 95 percent confidence intervals were calculated by means of Fisher's r to z transformation (McNemar, 1963).

Table 2. Correlations Between Members of Families in Regard to
Amount of Corneal Astigmatism

Man vs.				Woman vs.		
	$<r<$				$<r<$	
−0.13	−0.05	+0.03	wife			
−0.05	+0.05	+0.15	son	−0.05	+0.05	+0.15
+0.02	+0.10	+0.19	daughter	+0.06	+0.14	+0.21
+0.21	+0.28	+0.35	sib	+0.10	+0.19	+0.28
−0.18	−0.05	+0.08	uncle	−0.11	+0.02	+0.14
−0.26	−0.17	−0.06	aunt	−0.06	+0.04	+0.13
−0.03	+0.09	−0.21	cousin	−0.25	−0.10	+0.05
−0.56	−0.31	0.00	paternal grandfather	+0.06	+0.29	+0.49
−0.33	−0.11	+0.13	paternal grandmother	−0.21	0.00	+0.21
−0.45	−0.21	+0.08	maternal grandfather	−0.27	−0.03	+0.22
−0.08	+0.07	+0.27	maternal grandmother	+0.02	+0.19	+0.35

$$
r
$$
$$
\downarrow
$$
$$
\frac{(z-1)}{\sqrt{N-3}} \qquad z \qquad \frac{(z+1)}{\sqrt{N-3}}
$$
$$
\downarrow \qquad \downarrow \qquad \downarrow
$$
$$
(\quad) < r < (\quad)
$$

For the mid-parental values versus the offspring the correlations were: with sons, −0.03 < +0.11 < +0.25; with daughters, −0.01 < +0.11 < +0.23.

Figure 1 is a scatterplot of the mean amount of corneal astigmatism of the parents (mid-parental value) against the mean amount of corneal astigmatism possessed by all their offspring. It is apparent that there is little correlation between these amounts. In cases of traits for which polygenic genes are the principal determiners we expect to find a close similarity between the mid-parental value and the mean value of all their children.

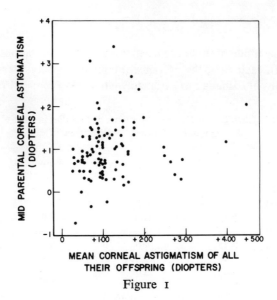

Figure I

Similarly, with regard to axis position, the correlations shown in Table 3 were found.

Table 3. Correlations Between Members of Families in Regard to
Axis of Corneal Astigmatism

Man vs.				Woman vs.		
	$<r<$				$<r<$	
+0.08	+0.17	+0.25	wife			
0.00	+0.11	+0.21	son	+0.12	+0.22	+0.31
0.00	+0.09	+0.18	daughter	+0.13	+0.21	+0.28
+0.11	+0.18	+0.25	sib	+0.05	+0.15	+0.24
−0.11	+0.02	+0.16	uncle	0.00	+0.13	+0.25
+0.07	+0.17	+0.27	aunt	−0.12	−0.02	+0.07
+0.22	+0.32	+0.43	cousin	+0.18	+0.33	+0.45
−0.44	−0.17	+0.13	paternal grandfather	−0.39	−0.16	+0.09
−0.49	−0.29	−0.05	paternal grandmother	−0.12	+0.09	+0.29
−0.07	+0.22	+0.47	maternal grandfather	−0.46	−0.24	+0.02
+0.18	+0.38	+0.52	maternal grandmother	−0.23	−0.06	+0.11

For the mid-parental values versus their offspring the following correlations were obtained for axis position: with sons, $-0.07 < +0.07 < +0.21$; with daughters, $-0.01 < +0.10 < +0.21$.

In Figure 2 the mid-parental corneal astigmatism is compared with the

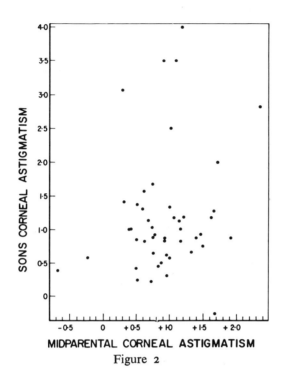

Figure 2

mean corneal astigmatism possessed by all their sons. The absence of any significant correlation is indicated.

Supplementary studies, forming part of the original investigation but not reported in detail in the current paper, confirmed that the distribution of the amount of corneal astigmatism approximates a normal curve. The distribution of the axis position does not fit a normal curve; it is bimodal and most frequently the axis is located near the 180° line. For these reasons scatter plots were prepared for this variable but again almost no correlation was evident.

In spite of the low correlations found, it is of some academic interest to consider the possibility of specific genetic patterns.

As noted earlier, in those patterns of inheritance in which the phenotype is determined, for the most part, by a number of genes each of which adds its influence, the average measurement of the offspring should approximate the mid-parental measurement. Sorsby and Fraser (1966) calculated correlation coefficients for the components of ocular refraction based upon their earlier measurements of twins. The coefficients they found are consistent, in their opinion, with an hypothesis of quantitative inheritance due to a number of genes with additive effects and without dominance.

From the scatterplot in Figure 1 we see that there is little correlation between the mean amount of corneal astigmatism of all the offspring as compared to the mid-parental amount of corneal astigmatism. Table 4 provides a comparison of observed versus expected correlations for a trait determined by a number of genes. In the group of subjects examined in this study and within the limits of the genetic model assumed, corneal astigmatism does not appear to be determined to any significant extent by such a group of genes. Some coefficients (e.g., $r = +0.29$ for the correlation between a woman and her paternal grandfather for the amount of corneal astigmatism) may be influenced by the small number of subjects falling within a particular subgroup.

Table 4. Observed vs. Expected Correlations

Relationship	Expected correlation if inheritance is polygenic	Observed Correlations Re amount of corneal astigmatism	Re axis of corneal astigmatism
Mid-parent vs. sons	+0.71	+0.11	+0.07
Mid-parent vs. daughters	+0.71	+0.11	+0.10
Man vs. son	+0.50	+0.05	+0.11
Man vs. daughter	+0.50	+0.10	+0.09
Woman vs. daughter	+0.50	+0.14	+0.21
Man vs. sibs	+0.50	+0.28	+0.18
Woman vs. sibs	+0.50	+0.19	+0.15
Man vs. uncle	+0.25	−0.05	+0.02
Woman vs. uncle	+0.25	+0.02	+0.13
Man vs. aunt	+0.25	−0.17	+0.17
Woman vs. aunt	−0.25	+0.04	−0.02
Man vs. paternal grandfather	−0.25	−0.31	−0.17
Woman vs. paternal grandfather	+0.25	+0.29	−0.16
Man vs. paternal grandmother	+0.25	−0.11	−0.29
Woman vs. paternal grandmother	+0.25	0.00	+0.09
Man vs. maternal grandfather	+0.25	−0.20	+0.22
Woman vs. maternal grandfather	+0.25	−0.03	−0.24
Man vs. maternal grandmother	+0.25	+0.07	+0.38
Woman vs. maternal grandmother	+0.25	+0.19	−0.06
Man vs. first cousin	+0.125	+0.09	+0.32
Woman vs. first cousin	+0.125	−0.10	+0.33
Man vs. wife	0.00	−0.05	+0.17

To evaluate the possibility that the trait corneal astigmatism is influenced by autosomal dominant or recessive genes the correlations of Table 5 were assembled.

Table 5. Observed vs. Expected Correlations when
Dominance or Recessiveness Occurs

	Expected Correlations			Observed Correlations	
Relationship	For common genes where p=q=½	For rare dominant genes (p rare)	For rare recessive genes (q rare)	Re amount of corneal astigmatism	Re axis of corneal astigmatism
Man vs. son	+0.33	+0.50	approaches zero	+0.05	+0.11
Man vs. daughter	+0.33	+0.50	″　″	+0.10	+0.19
Woman vs. son	+0.33	+0.50	″　″	+0.05	+0.11
Woman vs. daughter	+0.33	+0.50	″　″	+0.14	+0.21
Man vs. sib	+0.42	+0.50	+0.25	+0.28	+0.18
Woman vs. sib	+0.42	+0.50	+0.25	+0.19	+0.15

Because the sib-sib resemblance is greater than the parent-offspring resemblance (with regard to amount of corneal astigmatism) and because the sib-sib resemblance is of the order expected, one might infer that corneal astigmatism was inherited as a recessive trait. However, the high frequency of the occurrence of corneal astigmatism indicates that this condition is not likely to be the result of the expression of a rare recessive gene. In the eyes measured here over 96 percent had 0.12 D or more of corneal astigmatism. The observed correlations should therefore be compared with the expected correlations for a common gene. If corneal astigmatism is an example of intermediate inheritance and if most people are heterozygotic for the genes controlling refraction then it would be theoretically possible for a recessive gene to influence a large proportion of the population.

To investigate the possibility that corneal astigmatism is inherited as a sex-linked trait the correlations shown in Table 6 were tabulated. Expected values are from a table prepared by Penrose (1963). If the trait is a recessive one determined by genes on the x chromosome then the expected values would be slightly less than those shown, when dealing with pairs of relatives of which one or both members are female.

Table 6. Observed vs. Expected Correlations for a Sex-Linked Trait

		Observed Correlations	
Relationship	Expected correlations if sex-linked	re amount of corneal astigmatism	re axis of corneal astigmatism
Father-son	0	+0.05	+0.11
Father-daughter	+0.71	+0.10	+0.09
Mother-son	+0.71	+0.05	+0.22
Mother-daughter	+0.5	+0.14	+0.21

The data provide no evidence to support the view that corneal astigmatism is inherited as a sex-linked trait.

That there is some tendency for corneal astigmatism of 2.00D or more to occur in families is shown by the fact that while only about 14 percent of all people in these pedigrees have this much astigmatism, yet 33 percent of the sibs of those with 2.00D or more of corneal astigmatism have this amount or more. Similarly, 25 percent of the offspring of those persons who have 2.00D or more of corneal astigmatism have this relatively large amount of corneal astigmatism, too. The implication is that corneal astigmatism of 2.00D or more may be a segregating trait.

Because the correlations were generally low an effort was made to evaluate the influence of the coding system utilized in case these procedures had brought about an apparent reduction in an actual correlation between relatives. However, as will be evident from Table 7 the correlations remained low even when minimal coding procedures were utilized as they were for the last two groups in Table 7. Scatter plots were prepared and they, too, indicated little or no correlation.

Table 7. Observed Correlations for Specific Groups

| | | Observed Correlations | |
| | | re amount of corneal astigmatism | re axis of corneal astigmatism |
N	Relationship		
141 pairs	Husbands vs. wives	−0.05	+0.17
64 pairs	Males vs. females of about the same age, unrelated and not married to each other	+0.09	−0.18
65 pairs	Father vs. first son (right eyes only)	+0.07	+0.06
53 pairs	Mother vs. first daughter (left eyes only)	+0.09	−0.02

In spite of the clinical impression that corneal curvatures change little with age (e.g., Brungardt, 1968) cross-sectional data indicate that nominal changes do occur (Jackson, 1933; Ryer and Hotaling, 1945; Tait, 1954; Morgan, 1958; Sorsby, 1961). Landolt said, in 1886, corneal astigmatism generally remains stationary throughout the whole life. Stern and Rosenberg (1940) observed a statistical decrease in the percentage of cases showing with-the-rule astigmatism as age increased. The small size and infrequent incidence of this change as illustrated in Figure 3 (from our own data) suggests a negligible effect on the genetic interpretation of the present study.

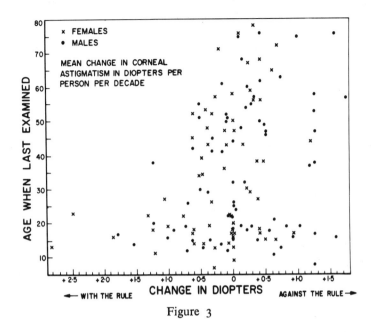

Figure 3

One can also be legitimately concerned with the extent to which the examiner's own preconceptions as to genetic influence could bias his recorded clinical findings. Therefore we must consider the possibility that the examiner of two or more familially related subjects might be influenced by the belief that refractive errors are inherited, or even by the conscientious desire to establish the degree of inheritance. This bias would certainly be most effective in instances where two or more members of a family are examined in immediate succession. Many an optometrist, for example, has examined a pair of twins, or even merely sibs, and upon completion of the second found himself going back to "take another look" at the first. There are ample evidences that this type of bias produces significantly greater apparent similarity between recorded values than actually exists. The tendency toward making measurements in conformity with one's prejudice would, of course, be inoperative if the examiner were, by some tactic, made unaware of the paired relationship between two successively measured eyes.

In the present study, all of the data were obtained by persons who were completely unaware of the fact that their data would be used in a genetic study. Further, it can be conservatively estimated that the great majority of the subjects in this study, perhaps as many as 80 percent, were examined singly, that is, in no successive or serially scheduled sequence that could have prompted comparisons either from direct reference to the records of familially related persons or from memory.

The fact that the present data show so very little heredity influence on astigmatism suggests that the above described sources of bias must have been negligible. If this influence was in fact operative, then the hereditary influence is even less than the statistical correlations indicate.

In view of the reports by virtually all other investigators that they found an apparent genetic influence, one must question whether or not they merely measured their own bias toward theory. In every instance, the other investigators would seem to have collected their data in awareness of the possible evidence for heredity. In the twin study in which the heredi- tary influence appeared to be almost completely absent the individual members of each pair of twins included were usually examined by different examiners (Hofstetter and Rife). The conclusion of Stern and Rosenberg (1940) from an analysis of a very large volume of clinical material was that the amount of corneal astigmatism did not show a clear pattern of genetic transmission.

The influence of heredity on corneal astigmatism, contrary to the ex- pressed opinion of almost all clinical authorities as reviewed in the earlier pages of this report, appears to be very small. It is doubtful that the heredi- tary factor is of clinical importance in the routine determination of the correcting opthalmic lens. Indeed the correlations between relatives of completely separated generations, i.e., not contemporaries, as represented in the lower groups of Table 8 indicate no consistently inherited com- ponent at all! The less completely separated or more nearly contemporary generation, represented by the middle group of Table 8 averages out to a correlation which is not significantly different from zero. Only the con- temporaries, the upper group in Table 8, show a consistent but very low indication of a genetic influence. One could argue that this latter evidence is merely an environmental effect shared by the contemporaries.

Table 8. Group Correlations with Regard to Amount
and Axis of Corneal Astigmatism

| | Observed Correlations | |
Correlations between	re amount of corneal astigmatism	re axis of corneal astigmatism
Male vs. sibs	+0.28	+0.18
Female vs. sibs	+0.19	+0.15
Male vs. cousin	+0.09	+0.32
Female vs. cousin	−0.10	+0.33

Table 8 (continued)

Male vs. son	+0.05	+0.11
Female vs. son	+0.05	+0.22
Male vs. daughter	+0.10	+0.09
Female vs. daughter	+0.14	+0.21
Male vs. uncle	−0.05	+0.02
Female vs. uncle	+0.02	+0.13
Male vs. aunt	−0.17	+0.17
Female vs. aunt	+0.04	−0.02
Male vs. paternal grandfather	−0.31	−0.17
Female vs. paternal grandfather	+0.29	−0.16
Male vs. paternal grandmother	−0.11	−0.29
Female vs. paternal grandmother	0.00	+0.09
Male vs. maternal grandfather	−0.20	+0.22
Female vs. maternal grandfather	−0.03	−0.24
Male vs. maternal grandmother	+0.07	+0.38
Female vs. maternal grandmother	+0.19	−0.06

The last statement cannot be dismissed lightly for it indicates a need for the study of every conceivable environmental factor which could affect the development of corneal astigmatism. Waardenburg (1961) discusses the influence of environmental factors upon the phenotype and the resulting modification of the distribution curves of corneal refraction. At what period in life the environmental influences are most effective is not discernible from the present study because measurable changes in corneal astigmatism appear to occur throughout life. Nevertheless, the pattern of corneal astigmatism for each individual seems reasonably stable. The fact that differences between generations are greater than differences within generations suggests subtle environmental influences which are dependent upon the complex variables associated with the different ways of life experienced by the peoples of successive generations.

Genetic pattern identification is more difficult for a trait which is common and which shows no sex difference. In this study over 96 percent of all eyes had 0.12 diopters or more of corneal astigmatism. The modal value of the amount of corneal astigmatism was 0.62 diopters. The influence of this high frequency of astigmatism may blunt the efficiency of the statistical methods used. The almost universal presence of some corneal astigmatism acts like noise in a system and conceals relationships which might otherwise be apparent. Therefore a further analysis of the data is being carried out in a way designed to minimize the effect of the high incidence of low amounts of astigmatism.

Another possibility is that the inheritance of corneal astigmatism follows a specific but different genetic pattern in different families. A well known example of a trait which exhibits this peculiarity is the condition called pigmentary retinosis which has been found to follow autosomal dominant, autosomal recessive and sex-linked pathways of inheritance. Pooling of data tends to obscure the various inheritance patterns if such exist within the group surveyed.

Ludlam and Wittenberg (1966) have shown that ophthalmometric measurements of the corneal contour which are centered on the line of sight will produce results quite different from similar measurements centered on a corneal apical normal. They define the corneal apical normal as the extension of the shortest instantaneous radius of the cornea. This line will correspond to the geometrical axis if the corneal contour is a surface of revolution. If ophthalmometrically made measurements of the cornea are not fundamental indices of corneal shape then it is not at all surprising that clinically determined corneal astigmatism fails to show hereditary transmission.

Astigmatism is a difference rather than a conventional dimension and in this sense does not exist as a structural entity, per se. If corneal astigmatism is accepted as a trait for analysis of its inheritance, then there is no philosophical limitation to the number of traits which could be considered, because any number of "differences," "ratios," or more complex mathematical representations of anatomical interrelations is possible. The presently obtained results with their lack of evidence for the inheritance of corneal astigmatism should not be considered as denying the usefulness of mathematical relationships in the descriptions of bodily structures, but mitigate against the use of certain mathematically determined values as traits for genetic study.

Historically clinically measured astigmatism has been almost universally regarded as a trait subject to substantial genetic control. This study indicates that corneal astigmatism is virtually free from the influence of heredity. Because the chief component of the total astigmatism of the eye is the corneal astigmatism the implication is that heredity plays little part in establishing the total astigmatism of the eye.

References

von Alphen, G. W. H. M. (1961). On emmetropia and ametropia. *Supplementum ad Ophthalmologica 142*. S. Karger, Basle, Switzerland.

Biro, I. (1948). Data concerning the heredity of astigmatism. *Ophthalmologica 115*, 156-166.

Borish, I. M. (1954). *Clinical Refraction*. The Professional Press, Inc., Chicago, pp. 45-60. Second edition.

Brungardt, T. F. (1968). Contact lens clinic. *Optometric Weekly*, February 15, pp. 47-48.

Council of British Ophthalmologists (1921). Report on the standardization of the notation of the axes of cylinders. *Brit. J. of Ophthal. 5*, 317-321.

Duke-Elder, Sir S. (1937). *Textbook of Ophthalmology 1*, 33, 742, 764, 781-782. C. V. Mosby Co.

Duke-Elder, Sir S. (1949). *Textbook of Ophthalmology 4*, C. V. Mosby Co., pp. 4243-4413.

Duke-Elder, Sir S. (1962). *System of Ophthalmology 7*, C. V. Mosby Co., pp. 3-113.

Duke-Elder, Sir S., and C. Cook (1963). *System of Ophthalmology 3*, Henry Kimpton, 306-307, 497-539, 554.

Edwards, J. H. (1960). The simulation of Mendelism. *Acta Genetica et Statistica Medica 10*, 63-70, S. Karger, Basle, Switzerland.

Francis, J. L. (1963). Myopia. *The Optician 3* (11), 595-596.

Francois, J. (1961). *Heredity in Ophthalmology*. C. V. Mosby Co., pp. 104-193, 290-297.

Hamilton, J. B. (1951). *The Significance of Heredity in Ophthalmology—A Tasmanian Survey*. A. H. Massina and Company, Melbourne, Australia.

Hirsch, M. J., and R. W. Wick (1963). *Vision of Children*. Chilton Books, pp. 5-68, 79-98, 145-172.

Hofmann, W. P., and E. T. Carey (1942). Congenital myopic astigmatism in identical twins. *Am. J. of Ophthal. 23* (12), 1495-1496.

Hofstetter, H. W., and D. C. Rife (1953). Miscellaneous optometric data on twins. *Amer. J. of Opt. and Arch. Am. Acad. Opt. 30* (3), 139-150.

Holste, A. (1940). Comparison of refraction in homozygous and heterozygous twins. *von Graefe Arch. Ophthal. 143*, 467-474.

Holt, S. B. (1961). Quantitative genetics of fingerprint patterns. *Brit. Med. Bull. 17* (3), 237-250.

Jackson, E. (1933). Changes in astigmatism. *Amer. J. of Ophthal. 16* (3), 967-974.

Ludlam, W. M., and S. Wittenberg (1966). Measurements of the ocular dioptric elements utilizing photographic methods. *Am. J. Opt. and Arch. Am. Acad. of Opt. 43* (3), 249-267.

MacElree, G. A., and J. R. Morrow (1934). Causative and interpretational factors in corneal astigma. *J. of Amer. Acad. of Opt. 8*, 176-195.

Manchester, P. T. (1955). Advising patients with hereditary eye diseases. *Am. J. of Ophthal. 40* (1), 412-417.

Mann, I. (1957). *Developmental Abnormalities of the Eye*. Brit. Med. Assoc., p. 355. Second edition.

McKusick, V. A. (1964A). *Human Genetics*. Prentice-Hall, Inc., pp. 108-110.

McKusick, V. A. (1964B). *On the X Chromosome of Man*. American Institute of Biological Sciences.

McNemar, Q. (1963). *Psychological Statistics*. John Wiley and Sons. Pp. 149, 384-385, second edition. Pp. 140, 426-437, third edition.

Morgan, M. W. (1958). Changes in refraction over a period of twenty years in a non-visually selected sample. *Am. J. of Opt. and Arch. Am. Acad. Opt. 35* (6), 281-299.

Penrose, L. S. (1963). *The Biology of Mental Defect*. Sidgwick and Jackson Ltd., pp. 98-119. Third edition.

Phillips, R. A. (1952). Changes in corneal astigmatism. *Am. J. of Opt. and Arch. Am. Acad. Opt. 29* (6), 379-380.

Roberts, J. A. F. (1961). Multifactorial inheritance in relation to normal and abnormal human traits. *Brit. Med. Bull. 17* (3), 241-246.

Ryer, E. L., and E. E. Hotaling (1945). *Ophthalmometry*. The American Optical Co.

Sorsby, A. (1951). *Genetics in Ophthalmology*. Butterworth and Co., London.

Sorsby, A., et al., (1957). *Emmetropia and Its Aberrations: A Study in the Correlation of the Optical Components of the Eye*. Medical Research Council Special Report, Series No. 293. H. M. Stationery Office, London.

Sorsby, A., and G. R. Fraser (1966). Statistical note on the components of ocular refraction in twins. *J. of Pediatric Ophthal. 3* (4), 33-36.

Sorsby, A., M. Sheridan, and G. A. Leary (1962). *Refraction and Its Components in Twins*. Medical Research Council Special Report, Series No. 303. H. M. Stationery Office, London.

Southall, J. P. C. (1933). *Mirrors, Prisms and Lenses*. Macmillan Company, New York, pp. 696-706.

Southall, J. P. C. (1961). *Introduction to Physiological Optics*. Dover Publications, pp. 18, 149.

Spooner, J. D. (1957). *Anatomy of the Eye*. The Hatton Press Ltd., pp. 11-16, 172.

Stenstrom, S. (1946). Untersuchungen über die Variation und Kovariation der optischen Elemente des menschlichen Auges. *Acta. Ophthalmologica*.

Stenstrom, S. (1948). Untersuchungen über die Variation und Kovariation der optischen Elemente des menschlichen Auges. *Am. J. of Opt. and Arch. Am. Acad. of Opt. 25* (5), 218-232; 6, 286-299; 7, 340-350; 8, 388-397; 9, 438-447; 10, 496-504. Translation by D. Woolf.

Stern, Clifford N. (in collaboration with H. H. Rosenberg) (1940). *The Relative Contributions of Cornea and Lens in the Changes of Astigmatism with Age*. Master's thesis. New York University.

Waardenburg, P. J., A. Franceschetti, and D. Klein (1961). *Genetics and Ophthalmology*. Blackwell Scientific Publications Ltd., Oxford, Vol. 1, pp. 97-98, 447-543.

Wixson, R. J. (1956). Statistical analysis of hereditary factors in ametropia. *Am. J. of Opt. and Arch. Am. Acad. Opt. 33* (7), 374-379.

Nathan Flax

Problems in Relating Visual Function to Reading Disability

On casual inspection it would seem that there should be a positive relationship between vision and reading since, with but rare exceptions, reading is accomplished by visual inspection of printed or written symbols. Yet the literature does not show very conclusive relationships (Bond, 1957; Spache, 1964; Gates, 1947; Harris, 1961; Vernon, 1957). The most obvious visual factors do not seem to correlate with reading. Visual acuity measurements do not serve to differentiate those who are good readers from those who have difficulty reading. Refractive error shows, at best, a minimal relationship to reading (Eames, 1959; Flom, 1963). The role of binocular function in reading is far from clearly established. Eye health would seem to bear little or no relationship to reading facility. Accommodative inefficiency is recognized as a detriment to reading in the mature population but its contribution to reading deficiency in school age children is equivocal. Strabismus does not seem to serve as a major deterrent to reading success. Color deficiency may interfere slightly in the first and second grades but does not seem to limit reading ability. Other functions, such as dark adaptation, peripheral vision, pupillary responses, flicker fusion frequency, and a host of other physiologic factors which comprise vision have never been subjected to any serious investigation for their contribution to the reading act simply because this never seemed to be a fruitful area for investigation. Why is there the paradox of reading being accomplished via vision while, on the basis of existing research, the most obvious clinical visual measurements do not demonstrate any clear relationship with reading?

There are a number of reasons that might explain this. The most obvious, poor research, is certainly true in this area of investigation but this is not the prime reason for the failure of any clear cut relationships to emerge. There is no doubt that the vast majority of existing research studies are guilty of inherent deficiencies in research design or execution which would preclude any meaningful conclusions (Huelsman, 1958). The problem cannot, however, be so easily explained away. Even in studies where there has been more rigorous research design and control there has likewise been a failure for stronger relationships to appear. It would seem that either there are no relationships to be determined or that the fundamental structuring of the problem masks whatever relationships might be present. It is the opinion of this author that the latter is the case. Meaningful relationships between visual function and reading efficiency do exist. Substantiating these relationships on the basis of group data may not be possible. For one thing, both vision and reading are highly complicated activities and neither one is fully understood. The multiplicity of tests available and lack of agreement as to any single index of reading function would certainly attest to the complexity of reading. The controversy regarding the most desirable method of teaching reading and the fact that some youngsters respond better to one approach than another would indicate that reading is certainly not a unitary function. It would seem illogical, therefore, for any particular cluster of visual functions to be optimum for all facets of reading. Even a superficial task analysis bears this out.

The term reading is utilized to encompass an enormous range of abilities. At one end of the spectrum, the ability to give the verbal sound of a two or three letter configuration represents successful reading for the beginner. At another level, the ability to read long pages and chapters for the purpose of deriving information and understanding with no need whatever to pronounce any of the words represents successful reading for a college student. These two activities present more differences than similarities, yet both are construed as reading and each would be appropriate for the circumstances.

The visual demands inherent in the two above cited reading activities are markedly different. The first grader learning to read is confronted with very large print, the word that he is asked to look at is generally well isolated on the page or blackboard, he is not asked to sustain visual attention for more than a few brief moments at a time, and his prime objective is not the conceptualization behind the word but rather the transformation of a visual symbol into a vocal symbol. The college student looks at considerably smaller print, is asked to sustain attention for long periods of time, and has as his prime objective the acquisition of the conceptual thought of the author and the factual content of the material rather than the sight-sound association that marks his entry into the world of reading. It is no wonder then, that no simple index of visual function could be expected to be equally appropriate for both activities (Flax, 1967).

Even if one were to limit the investigation of reading to word recognition and pronunciation skills, it would still be unlikely that a straightforward correlation with any particular visual functions would emerge since sophistication in this aspect of reading would be expected, by itself, to change the visual requirements of the activity. Decoding a word to enable a sight-sound transformation involves a series of probability determinations. Each letter of the printed word or each group of letters is capable of standing for a variety of sounds. Application of phonic rules reduces the number of possible sounds that might be made in response to a particular visual configuration but rarely reduces the number of choices to one. The English language is so erratic that there are often two or more plausible pronunciations of a particular visual configuration. With experience in reading, with development of language function in general, there is a gradual reduction in the number of available choices based upon the incorporation of other cues such as context and meaning. The neophyte reader, the language impoverished neophyte in particular, cannot draw upon these cues with the facility that the experienced reader can. Thus it becomes necessary for the neophyte to more closely scrutinize each aspect of the visual configuration in order to deal with the many more logical possibilities for pronunciation that he has to contend with than does the more experienced reader. The fact that reading involves encountering a myriad of different type styles further complicates the situation. In some instances individual letters or words bear but small shape relationship to one another among the various typographical forms encountered (Tinker, 1965). This places a further burden on the beginner or poorly experienced reader since the visual configuration of the sight-sound unit is continually varied. Here, again, experience plays an enormously important role in reducing the demand on precise inspection of shape configuration. Pre-set and anticipation are important determinants in word recognition.

The beginning reader therefore must pay considerably more attention to internal details of word structure than does the more sophisticated reader. Thus, any attempt to correlate visual functions that pertain to determination of internal details in word structure and reading proficiency might lead to spurious group data. The experienced reader may score poorly in those aspects of visual function that pertain to discernment of internal detail in words and yet not show any appreciable loss in reading proficiency. It is conceivable that a significant fall off in this aspect of visual function could have little or no impact upon reading provided that this function was intact at that stage in the individual's reading progress when it was most needed. Conversely, an aspect of visual function, such as maintenance of binocular stability, might be anticipated to have greater interference for the proficient reader than for the beginning reader. The beginning reader is not expected to sustain high level attention for long periods of time and thus even a relatively severe binocular problem might not influence reading at all at that stage, whereas in the better reader who attempts to read for long periods of time with small print with emphasis on comprehension, the same binocular instability might have a significant adverse effect.

Group investigations of vision and reading also suffer from the widely recognized difficulties in matching for associated factors which could conceivably influence the vision-reading relationship. This is frequently attempted but, unfortunately, is no small task. Even the most obvious matching, which is usually done, of intellectual capability has inherent weaknesses. Most often the matching is done on the basis of a group administered intelligence test. This, in itself, is a source of contamination since the group tests often show depressed scores due to either poor reading or visual deficiencies, although not necessarily in a predictable manner. Individual tests of intelligence, such as the WISC, often indicate sub-factor scattering that could play an important role in the relationship between vision and reading. Certainly, when one investigates that severely retarded group of children known as dyslexics, the discrepancies between verbal and performance scores on intelligence testing become important considerations (Rabinovitch, 1954; Silver, 1960; Kinsbourne, 1963). Merely specifying intelligence as a single number is hardly sufficient. The youngster whose intelligence is depressed by deficiency in language function is not apt to show visual factors as important contributors to reading deficiency whereas the youngster who scores low in intelligence due to depressed visuo-motor function is apt to show clinically modifiable visual factors which adversely influence reading success.

Further complication of group data comes about because of the extreme difficulty of matching groups in terms of home environment, drive to learn, personality factors, experiential background, motivation, general physical development, and personality style. The assumption that these factors will randomize in large samples may be justifiable but it also will tend to blunt the emergence of any meaningful relationship from large sample data. It may be that particular visual functions relate to reading only in particular subclassification groups. For instance, a youngster with high drive, good motivation, exposed to good teaching, who comes from a family that is reading oriented may be willing to tolerate a headache in order to continue to read whereas a youngster with a different personality make-up and background might withdraw from the reading activity at the first sign of discomfort. The same visual inefficiency which might have only minor influence on reading ability in the first child would have major influence on reading ability in the second child.

There is still another important consideration that is rarely given proper emphasis. This has to do with the adaptability available within the visual function itself. In cases of strabismus, suppression and suppression amblyopia are considered as adaptive mechanisms to permit more effective overall performance by elimination of diplopia or the confusion that would arise from two different images being projected directly ahead of the individual (Borish, 1949). These are not necessarily completely desirable adaptations but their function is recognized as a means of permitting increased efficiency. Unfortunately, the same concept has not been widely applied in the matter of relating vision and reading. The general rule of thumb is to order visual function starting with an arbitrary clinical model of vision and then correlating on the basis of degree of departure from this norm. There is failure to recognize that, at times, a larger departure from standard measures may actually permit better function in the reading activity than a lesser deviancy. For instance, the individual who totally suppresses the vision of one eye is less apt to have difficulty at reading than the individual who only partially suppresses vision of one eye. This confounds any attempts at correlational studies since the ordering of visual proficiency is not appropriate to the requirements of reading but is rather based upon arbitrary clinical standards. The concept of adaptive responses within the visual system to permit increased efficiency at a particular task certainly pertains to the fusional and accommodative-convergence systems and might even explain some refractive errors. Under these circumstances, correlation of reading scores against standard clinical norms becomes almost fruitless.

What, then, might be an approach to the problem of relating visual function to reading? One attempt might be to recognize the varying visual demands of reading at different stages in development of reading skill and to proceed accordingly. The visual demnds of learning to read are quite different from the visual demands of reading as a sustained activity once word recognition has been mastered. In the former situation, that of learning to read in the first place, end organ visual functions such as refractive error, visual acuity, fusion, convergence, and accommodative skill can be expected to play only a small role. The visual functions that are most important to learning to read have to do with visual form perception, visual-memory, the ability to appreciate directional differences on a visual basis, visuo-motor control, and eye-hand coordination. The fundamental visual problem inherent in learning to read is the ability to appreciate the shape and directional orientation of configurations and the ability to integrate these configurations with verbal data. The referential framework vital to provide directional stability to ocular images is derived from integration of visual stimuli with postural information available from vestibular, tactile, and proprioceptive cues.

Visual deficiencies that might interfere with learning to read in the first place are not apt to be identifiable on the basis of investigation into eye health, acuity, or standard refractive findings but are rather to be found in investigation of the development of vision and visuo-motor function with particular emphasis upon intactness of form perception ability and the interrelatedness of vision with other sensory-motor systems. Many individuals who suffer from severe early reading difficulty, characterized by an inability to learn word recognition skills, suffer from deficiencies in visual form perception and inappropriate integration of vision with other senses. Much of the symptomatology of dyslexia is explainable in terms of inadequate visual performance (Flax, 1968).

While visual form perception difficulties and deficiencies of adequate visual development may play a part in early reading failure or dyslexia, they are not apt to be major factors in reading inefficiency noted in the higher grades of elementary school. Visual perception deficits may make it impossible to learn to read but they play a decreasing role in reading efficiency once basic word recognition skills have been mastered. This should not be surprising in view of the fact that the level of proficiency of visual perception necessary to learn to read in the first place is achieved by most children who are six years of age. Thus, while a youngster in the upper grades of elementary school may still show a lag in visual perception abilities when rated against his chronological age, the base line necessary to begin to learn to read is only that of a normal six-year-old. Persistent visual perception deficits may interfere with conceptualization during reading but they generally tend to play a decreasing role in word recognition abilities as the youngster matures and as the reading level itself increases.

On the other hand, certain end organ functions can be expected to significantly interfere with efficiency at reading if proper account is taken of adaptations within the visual system. It becomes possible to make order out of chaos if it is recognized that sometimes lesser amounts of fusional instability or milder interferences in accommodative-convergence function can be more interfering to reading success than clinically greater problems. Unfortunately, the determination as to whether a particular clinical visual problem could be an interfering factor in the process of reading requires an individual judgment and must be based upon an in-depth clinical evaluation of the subject. Generalizations about the contribution of end organ vision problems to the process of reading are not highly predictive in individual cases. This is particularly true when the role of compensatory changes is properly assessed. Even on a group basis, some data would indicate that a worsening of certain clinical measurements may accompany an increase in reading proficiency (Kelley, 1957). At the risk of generalizing, however, it might be stated that end organ visional functions tend to play a much greater role in the maintenance of high efficiency reading than they do in the more rudimentary aspects of learning to read.

References

Bond, Guy L., and M. A. Tinker (1957). *Reading Difficulties: Their Diagnosis and Correction*. Appleton-Century-Crofts, New York, pp. 85-92.

Borish, Irvin M. (1949). *Clinical Refraction*. The Professional Press, Chicago, pp. 372-374.

Eames, Thomas H. (1959). Visual handicaps to reading. *J. of Education 141*, 1-35.

Flax, Nathan (1967). The development of vision and visual perception: Implications in learning disability. *Proceedings of 1967 International Convocation on Children and Young Adults with Learning Disabilities*. Home for Crippled Children, Pittsburgh, Pennsylvania, pp. 130-134.

Flax, Nathan (1968). Visual function in dyslexia. *Am. J. of Opt. 45*, 574-587.

Flom, Bernice C. (1963). The optometrist's role in the reading field. *Vision of Children*. Monroe J. Hirsch and Ralph E. Wick, eds. Chilton Books, Philadelphia.

Gates, Arthur I. (1947). *The Improvement of Reading*. The Macmillan Company, New York, pp. 84-94. Third edition.

Harris, Albert J. (1961). *How to Increase Reading Ability*. David McKay Co., Inc., New York, pp. 231-240. Fourth edition.

Huelsman, Charles B., Jr. (1958). Some recent research in visual problems in reading. *Am. J. of Opt. 35*, 559-564.

Kelley, Charles R. (1957). *Visual Screening and Child Development*. North Carolina State College, Department of Psychology, School of Education, Raleigh, North Carolina.

Kinsbourne, M., and E. R. Warrington (1963). Developmental factors in reading and writing backwardness. *Brit. J. Psychol. 54*, 145-156.

Rabinovitch, Ralph D., Arthur L. Drew, Russell N. DeJong, Winifred Ingram, and Lois A. Withey (1954). A research approach to reading retardation. *Neurol. and Psychiat. in Childhood, Res. Publ. Ass. Nerv. Ment. Dis. 34*, 363-396.

Silver, Archie A., and Rosa Hagin (1960). Specific reading disability: Delineation of the syndrome and relationship to cerebral dominance. *Compreh. Psychiat. 1*, 126-134.

Spache, George D. (1964). *Toward Better Reading*. Garrard Publishing Company, Champaign, Ill., pp. 2-3, 103-113, 301-303.

Tinker, Miles A. (1965). *Basis for Effective Reading*. University of Minnesota Press, Minneapolis, pp. 115-156.

William Feinbloom

Visual Rehabilitation of the Partially Blind, Today and the Future

One of the broad areas of science is the search for truths in the natural life processes to help man overcome his handicaps. Thus the visual sciences includes in its investigations and studies all aspects of visual function and malfunction and methods to overcome his handicaps. Visual rehabilitation of the partially blind is one of the subheadings relating to vision in the visual sciences. An understanding of the field of visual rehabilitation includes the studies of normal and abnormal physiologic optics function of the human eye as well as the study of normal and abnormal visual perception.

The application of these basic visual science studies to vision rehabilitation includes professional practice with partially blind patients. As will be indicated throughout this paper, advances in visual rehabilitation during the past forty years followed closely upon new developments or progress in the visual sciences.

In 1932 the author first reported the results of his experimental vision rehabilitation work with a patient, Mr. A., which laid the groundwork for such vision rehabilitation with optometric methods and procedures (Feinbloom, 1932A). At that time we already had the advantages of the work of Von Rohr (1912), who had designed and perfected a 1.8x full field telescopic spectacle. We also had a number of papers, reporting on the use of such systems by Scott (1911), Stall (1912), Laufer (1915), Erggelet (1916), Wolf (1918), Nordenson (1924), and Graedel and Stein (1924). These reported on experiences with the static use of the then existing form of telescopic spectacles. This reference to static use means that the patients had no mobility while using these rehabilitation devices either for distance seeing or for reading. The patient could not move about freely while wearing these rehabilitation aids, and so their use was limited to those tasks where the patient remained stationary or seated.

Additional work was reported in the literature by Metzger (1927), Meuger (1927, 1929), Levy (1928, 1929), and Bruner (1930). All of these cited further experiences with additional patients and various attempts to extend the use of the then existing optical systems.

The first effort to overcome the static status associated with the use of telescopic spectacles was reported by the author in 1932 (Feinbloom, 1932B). Here by the use of cylinder lenses instead of spherical lenses it was shown that it was possible to overcome the phenomenon of telescopic parallax, i.e., the perceptual interpretation that objects appear closer than they really are when seen magnified through a telescopic system. Even though this development was only a beginning in dealing with the many problems that confront this important field, it served to stimulate great interest in this field.

During the ensuing years additional developments and improvements were introduced by the author (Feinbloom, 1933A, 1933B, 1933C), all directed to increase the properties of the optical systems used in visual rehabilitation. These new designs were directed to include those properties that would be most useful to partially blind eyes rather than normal eyes. The basic system of the telescopic spectacle was expanded to give magnifications of 1.3x, 1.7x, 2.2x, and 3.0x and their placement in the field of view so as to permit mobility in those patients with normal peripheral fields, for walking, working, and even automobile driving. Also, the basic microscopic spectacle was developed as an air spaced doublet system with a magnification range from 2x to 20x.

In addition, special study was given to the design and engineering of the finished spectacle form to make them cosmetically acceptable. It is always important in rehabilitation work to have all devices or aids used resemble as near normal appearances as possible. These present rehabilitation spectacles are light in weight, mounted in very modern frames, and the lenses are held together with clear plastic mountings so as to take the "optical instrument" look out of them.

The original designs of the author's telescopic spectacles were made by the Kollmorgan Optical Company (Kollmorgan, 1934), and the latter developed designs including those in use today by Designs for Vision, Incorporated (1968).

To understand the status of the present day program of visual rehabilitation of the partially blind it is both necessary to know the size and the nature of the partially blind population. In addition we must know the problems of the partially blind, including the different levels of vision rehabilitation possible, the trained professional personnel available, and the existing optical and other aids available in this rehabilitation program.

Let us first define the "partially blind" as that portion of the "legally blind" population who have some measurable visual acuity and so are potentially amenable to vision rehabilitation.

The "legally blind" person is defined as one whose central visual acuity is 20/200 (10 percent) or less in the better eye after spectacle correction or if there is a field defect in which the widest diameter of the visual field subtends an angle of no greater than 20 degrees. Thus the definition of the "legally blind" includes all of our "partially blind." However the "partially blind" excludes all those who are "totally blind," and excludes those who have only light perception or have only movement perception or shadow vision. This definition does not suppose that the totally blind are visually in the same class as the "light perception" or "shadow" groups but for our discussion we include all three groups with the "totally blind."

Both definitions of "partially blind" and "legally blind" have as their upper limit of visual acuity 20/200. It is the lower limit that is different; the partially blind supposes some certain measurable visual acuity, e.g., 10/500 (2 percent), whereas the legally blind takes its lower limit as total blindness or a complete absence of visual acuity.

This distinction between partially blind and totally blind is very important, for while both groups need rehabilitation, the problems and methods of rehabilitation toward the two groups is radically different.

In the first instance the partially blind who have some measurable vision can conceivably be helped through optical aids; the second group, the totally blind, can often be helped, but only by other than optical aids. While the two groups have "blindness" in common, the care of the first group (partially blind) is the province of the optometrist as to visual rehabilitation whereas the care of the second group is not.

As with all such classifications there are individual instances that defy clearcut classification; these borderline cases on initial examination may be classified in the one group, and after some further investigation and study may be reclassified in the other.

In our definition of partially blind we used as a lower limit that of any measurable visual acuity. At present our existing measuring methods make 10/500 (2 percent) the lower limit of the likelihood of some visual rehabilitation being possible.

The true size of the partially blind population in the United States is not known. It has been estimated as up to 2 million persons exclusive of the totally blind, which again is only an estimate, at about four hundred thousand. Among the specific suggestions for further research noted later in this paper is one for the study of the size and composition of the partially blind population. It is worth noting that the New York Times, in announcing its large type edition for use by the partially blind, estimated that there were 6 million persons in the United States who could not read standard newsprint with their regular correction eyeglasses.

Existing data relative to the partially blind are fragmentary and result from data collected in "blind" centers, or in private practices, so are not meaningful in terms of the general population.

One of the greatest problems confronting the partially blind is lack of specific information as to what visual rehabilitation is possible for them and where such suitable evaluation and treatment can be obtained. Public education in this field has been limited due to the professional reticence of the practitioner even though most public media are interested in presenting new advances that offer some hope to the partially blind. A number of public education articles in national magazines (Abramson, 1956; Berg, 1959; Irwin, 1963) have helped supply such information, and have also helped kindle a spirit of interest among practitioners to undertake this work of rehabilitation. In addition, the Xerox Corporation has undertaken publication of a large type edition of *Reader's Digest,* and, as mentioned, the *New York Times* puts out a weekly summary of the news in a large type edition.

The problem of the partially blind person, whether of school age, work age, or senior citizen, involves the question of how to see well enough to regain mobility (if it has been lost), to go to and from his school, work, or home, and how to see well enough to read the blackboard and the textbook, to do the work on the job, either at the desk or at the machine, how to recognize persons in the street, and how to see to read a newspaper.

The degree or level of such rehabilitation possible for all groups varies, of course, with many factors, including the patient's age, cause of partial blindness, type and degree of vision remaining, length of disability, motivation, and adaptability.

The present program helps to visually rehabilitate, to some higher level, about 50 percent of the patients seen and cared for. Of the backlog of the partially blind, 2 million persons, no more than 1 percent are seen annually for evaluation and possible rehabilitation and these patients are about evenly divided between clinics and private practice. At present in the author's opinion the total number of adequately trained practitioners in this field is not over one hundred fifty men and women, including all those who serve in private practice, optometric clinics, or eye clinics at a hospital. By adequately trained we mean practitioners who have had suitable postgraduate training in visual rehabilitation and who have successfully completed at least twenty cases.

While fully twenty times this number have adequate undergraduate training and have treated an occasional patient or two, only the limited number mentioned above have had postgraduate training and have actually completed twenty cases.

At present, there are only very limited postgraduate facilities to insure that each graduate student will be able to care for twenty cases. Such a teaching program requires a clinic setting with emphasis on vision rehabilitation. After such training, the practitioner can carry on vision rehabilitation in his private practice very well. Here he can arrange to spend all the necessary time he needs on such cases both with the patient and his family. Frequently the family must also be instructed in the retraining necessary in vision rehabilitation. The time involved in carrying out a visual rehabilitation program varies from a minimum of four hour-long visits in a simple case to be a possible twenty hour-long visits with involved cases.

Specific Areas of Experimental Research, Studies and Developments Needed to Further the Program of Vision Rehabilitation of the Partially Blind

1. Statistical studies of the partially blind population relative to:
    a. Number of persons in the population
    b. Distribution of these persons with respect to:
        1. age
        2. sex
        3. geographical location
        4. degree of remaining vision
        5. reasons for loss of vision, primary cause, secondary cause, by location of pathology and environmental causes
2. Experimental research in physiological optics of partially blind eyes to determine their variation from the normal in the areas of:
    a. Light threshold
    b. Dark adaptation
    c. Intensity discrimination
    d. Relation between visual acuity and illumination
    e. Flicker frequency
    f. Changes in localization, fixation, pursuit, saccadic and tracking movements with varying loss of central or peripheral vision
    g. Retinal summation and inhibition effects
    h. Binocular visual function where partial vision exists in both eyes
    i. Telescopic parallax or the misjudgment of depth perception associated with telescopic magnification
3. Studies relating to professional facilities and staff required for visual rehabilitation of the partially blind population
    a. Present availability of professional facilities and staff—degree of training, experience, and geographical distribution

    b. Professional time required—minimum and maximum for cases of visual rehabilitation depending on age, length of time of existing disability, degree of involvement, motivation, and level of rehabilitation expected

    c. Projection of estimated need of number of practitioners based on population studies and time of treatment studies above

    d. Planning of adequate postgraduate study including clinic experience for training of professional staff

4. Optical development and engineering

    a. Investigation in further design of optical aids, such as telescopic and microscopic spectacles for further increase of visual acuity at a distance and near with increased fields of view

    b. Further investigation in design of binocular optical aids for near vision

    c. Design of optical aids to permit maximum mobility

    d. Design of optical aids to increase the field of vision of persons with reduced fields of vision as in retinitis pigmentosa and glaucoma

    e. Optical and physical methods for improving contrast of viewed objects

    f. Illumination required and best methods for obtaining such for the partially blind

    g. Improvement of the coordination of the reading material to eye movements of the partially blind, through auxiliary use of stationary and moving reading tables

5. Studies relating to methods of visual rehabilitation

    a. Development of visual rehabilitation code; however best to describe the loss of visual function, i.e., take an existing "visual profile inventory"; this is to include suitable test charts and other measures of visual function

    b. Psychological problems related to partial blindness

    c. Sociological problems relating to partial blindness

    d. Methods of motivating the partially blind toward rehabilitation

    e. Development of techniques of teaching motor coordination between eye movements relating to localization, fixation, pursuit, saccadic and tracking, and the acts of walking, reaching, hand manipulation, and reading

    f. Study of the visual acuity requirements and other cues needed in performing everyday vision tasks, as:

        1. recognizing a person as a known person at a distance of 10 feet

        2. recognizing the features of a person

        3. recognizing food on a plate while eating

        4. writing one's signature on a line

5. reading handwriting and printed material
6. reading textbook print
7. reading newsprint

6. Educational programs:
   a. Development of methods of attracting and interesting professionals in the field of visual rehabilitation
   b. Development of a public information program to acquaint the public of the present status and future developments in visual rehabilitation for the partially blind

Technical Developments to Date

The present day optical systems that are primarily used in visual rehabilitation are the bioptic and trioptic telescopic units for distance vision improvement, and the binocular reading systems for near work. Where the rehabilitation is not sufficient for near vision, e.g., for reading newsprint with the binocular reading systems, the microscopic spectacle is tried.

The distance unit in bioptic telescopic form at present has a maximum magnification of 3.0x, i.e., an existing visual acuity of, e.g., 10 percent will theoretically be increased to 30 percent. This increase in visual acuity is in the central area, while still allowing full use of the patient's own full peripheral field. In this way the person moves about as usual using his ordinary visual acuity of 10 percent with normal judgment of distance and normal field of view. When he wants increased acuity he lowers his head slightly and looks through the telescopic portion mounted above the visual axis of the Rx carriers in the spectacles. Young persons between the ages of six and twenty-one (the school years) learn to adjust to this type of supra-bifocal vision almost at once. In fact, with many of these cases we use the trioptic unit system, which embodies the bioptic unit described above and in addition has a microscopic spectacle mounted as an infra-bifocal in the lower part of the carrier. Thus the unit has three optical parts, first, a telescopic unit mounted above the visual axis, second, in the Rx carrier for full use of field, and third, in the microscopic unit used for reading vision and mounted interiorly as a regular bifocal segment.

Both of these systems, the bioptic and the trioptic are useful aids in the visual rehabilitation of many partially blind persons who require increased acuity with mobility, either outdoors, indoors, or in the school, home, office, or factory. As further developments occur in telescopic magnification, they too can be incorporated into the bioptic and trioptic systems.

Persons in their working years may find that their adjustment after loss of vision is more difficult than it is for the young. A great deal depends on the patient's own motivation or how well he can be motivated, and the level of rehabilitation that can be held out to him. The more readily he can maintain his present employment, the quicker is his acceptance of the rehabilitation procedure. Where he must settle for a reduction in job classification due to the more limited nature of the rehabilitation possible, his own acceptance becomes more difficult.

The use of the more recently developed 2.5x and 3.5x reading or working binoculars is a boon to those working people who suffer vision loss and who can be rehabilitated sufficiently to permit them to continue in their jobs. These binocular systems are telemicroscopes designed with working distances of from 14″ to 16″ or other if needed. Frequently they are used at the desk or work bench, with acuities increased for the particular job requirements, and then too with additional caps, can have their magnification further increased for the more difficult task of reading newsprint.

Persons past the working years, who represent probably one half of the partially blind population, are most interested in being able to read. This act of reading represents important independence on their part, as well as a desire to use their leisure time. Here the involvement of loss of vision may be more severe, and often we must utilize the higher magnifications possible in the doublet microscopic system. This doublet design insures a longer working distance and wider corrected field than can any single lens system, as well as a greater depth of focus. These optical characteristics are of importance when used for visual rehabilitation of a partially blind person.

Where the senior citizen understands the nature of the rehabilitation process that one proceeds from the simple to the more complex, from the possible to the more difficult, a high enough level of rehabilitation is obtained. The establishment of new motor coordination patterns between the paramacular area and proper eye movements is possible in many of these cases. Only after repeated trials can the practitioner as well as the patient be sure of success of the rehabilitation effort. As more research and studies are made of the many problems involved, more techniques will emerge to further increase the success of the rehabilitation process.

As was stated in the introduction of this paper, the field of visual rehabilitation of the partially blind is a part of the general field of visual science. As has been demonstrated again and again in scientific investigation, discoveries in one field lead to knowledge and advances in another. Thus the various theories of vision, of the photoreceptor process of the mechanism of visual acuity, of vernier acuity, depth perception, amblyopia, binocular vision, and myriad other basic visual problems are all involved in how the partially blind eye interprets and adjusts to the normal stimulus, and how even with limited optical aid he often adjusts far beyond the theoretical value of the optical aid.

References

Abramson, M. (1956). Escape from blindness. *Better Homes and Gardens.* March.

Berg, R. H. (1959). New ways to save. *Look.* January 6.

Bruner, A. B. (1930). Telescopic lenses as an aid to poor vision. *Amer. J. Ophth. 13,* 667.

Designs for Vision (1968). Don't close the door on the partially blind. New York, New York (brochure).

Erggelet, H. (1916). Ein neuer fernrohrlupenfuss. *Z. Ophthal. Opt. 4,* 149.

Feinbloom, W. (1932A). A case of vision rehabilitation. *Annual Optometric Year Book.* New York State Optometric Society.

Feinbloom, W. (1932B). The new subnormal vision lens. *Transactions of the Amer. Acad. of Optometry 11.*

Feinbloom, W. (1933A). A new wide angle telescopic spectacle. *Optometric Weekly 24,* 685.

Feinbloom, W. (1933B). Report of special communication on the Feinbloom new type subnormal vision lens. *Optometric Weekly 24,* 89.

Feinbloom, W. (1933C). Report of 500 cases of subnormal vision. *Transactions of the 12th Annual Meeting of the Amer. Acad. of Opt. 8,* 58.

Graedel, H. S., and J. C. Stein (1924). Telescopic spectacles and magnifiers as aids to poor vision. *Transactions of the A.M.A. Section on Ophthalmology 262.*

Irwin, T. (1963). Mom! Mom! I can see! *The Saturday Evening Post.* May 24.

Kollmorgen, E. O. (1934). Telescopic spectacles. *Optometric Weekly 24,* 1280.

Laufer, H. (1915). Telescopic spectacles. *Arch. Ophth. Chicago 89,* 401.

Levy, A. H. (1928). Telescopic spectacles. *Diopt. Bull. 30,* 439.

Levy, A. H. (1929). Telescopic spectacles. *Br. J. Ophth. 13,* 593.

Mayer, L. L. (1927). Visual results with telescopic spectacles. *Amer. J. Ophth. 4,* 256.

Mayer, L. L. (1929). Further visual results with telescopic glasses. *Arch. Ophth.* Chicago 2, 135.

Metzger, E. (1927). The Iggersheimer mirror spectacles. *Arch. Ophth. 118,* 487.

Nordenson, J. W. (1924). The spectacle magnifier of Gullstrand. *Acta Otolaryng.* (Stockholm) 5, 405.

Scott, K. (1911). Telescopic eyeglasses. *Ophthalmology 7,* 445.

Stall, K. L. (1912). Telescopic spectacles. *Lancet Clinic 108,* 120.

VonRohr, M. (1912). A 1.8x telescopic spectacle. U.S. Patent #1,028,281.

Wolf, H. (1918). Mannschaftsuntersuchunger mit Von Rohrschen Fernrohrbrillen (Zeiss). *Z. Augenheilk 40,* 235.

Herbert A. W. Schober

The Role of Exact Eye Correction in Industrial Hygiene

An impressive number of experimental, statistical, and theoretical research concerns a fundamental problem in industrial hygiene, correlations between visual performance, working success, and prevention of accidents. Many factors, such as age and sex of working people, sanitary conditions, illumination, etc., have been discussed as additive factors. Unfortunately, most of these papers are too old and do not fit with the present situation because most of the working conditions and the working processes have changed fundamentally during the last decade. This change touches nearly all factors. Full employment is connected with increase of working age and with the fact that physical constitution has to be normalized as much as possible. Automation and modern mechanical engineering, as well as material and working speed claim optimal visual performance. Because this trend is found in all industrialized countries, the national differences in the working systems play only a secondary role. Most of the results of relevant research may be generalized.

This situation required teamwork, which was activated by two academic institutions in Munich, the Institute for Ergonomy (H. Schmidtke and H. Schmale) and the Institute for Medical Optics (H. Schober and G. Wallmann, 1967). This work consisted in finding out exact correlations between the different visual qualities and the working success. The teamwork started in early 1964 and was extended over a period of nearly two years. In different German factories 9,468 male and female workers of all ages were examined. Fields of activity were preferred where good vision is expected, such as optical, electronical, graphical, textile, and watchmaker factories. From the side of the visual organ the most important factors of visual performance were tested at the beginning by screening with the Ortho-Rater. That concerns monocular and binocular visual acuity for distances of 25 feet (far vision) and for distances of 1 foot (near vision), accommodation, vertical and horizontal phoria, stereoscopic vision, and color defects. Though the papers of Imus (1950), and Gordon, Zeidner, Zagorski and Uhlaner (1954A, 1954B) prove that the Ortho-Rater produces the results with nearly the same accuracy as the clinical methods, all persons which were suspected to show defects in one of these qualities had to pass also a strong optometrical and ophthalmological inspection. That included objective and subjective control of the refractive power, of astigmatism, and of differences between monocular and binocular vision. For all these persons the accommodation range was controlled with the hand-optometer of Schober (1958); horizontal and vertical phoria and the ACA gradient were measured; defects in color vision were scrutinized with the anomaloscope of Nagel. If the subjects worked with spectacles the visual acuity was measured with and without their eyeglasses. In these cases differences between the visual acuity with full correction for far and near distances and the visual acuity with their spectacles were marked out.

This distribution of ages was slightly different between male and female workers; this distribution also depends on the kind of work. On the average, the distribution of the refractive power was the same in the examined group as in the extended statistical results of Betsch (1929). It was in range with the statistical reports given in other European countries and also with the United States results. Our distribution was nearly Gaussian because we could not detect cases of pathological myopia and other disturbing eye diseases between the examined persons. The peak of the refractive power was situated at $+0.25$ D. That means a low hyperopia.

Table 1 shows the distribution of binocular visual acuity.

Table 1. Distribution of visual acuity measured for 9,468 workers in West German factories without or with their usual eye correction during working process.

Visual Acuity	Near Vision	Far Vision
20/20 and better	4940 (52.18%)	5422 (58.27%)
Less than 20/20	4528 (47.82%)	4046 (41.73%)
Between 20/22 and 20/25	2969 (31.35%)	2878 (30.40%)
Between 20/29 and 20/40	860 (9.09%)	1307 (13.80%)
20/50 and worse	217 (2.29%)	343 (3.62%)

Table 2 shows the percentage of spectacle wearers.

Table 2. Percentage of eyeglass wearers in a representative group of 9.468 workers in West German factories.

	Number of Persons	Percentage
No eyeglass wearers	7100	75%
Near eyeglasswearers	1132	12.2%
Far eyeglasswearers	392	4.2%
Far and near eyeglass bifocal or trifocal wearers	844	8.6%

As shown in Table 1 nearly 42 percent of the subjects worked with a visual acuity of less than 20/20 for far distances and nearly 48 percent produced the same result for near distances. This outcome is alarming because Germany has a very extended education program for better vision propagated in journals by some associations such as the "Green Cross" and by radio and television. German ophthalmologists and opticians know the importance of good and exact visual correction. Because the public health systems take all charges for medical inspection and spectacles, it is difficult to understand why the visual correction is so bad. Some reasons may be found in comparing Table 1 to Table 3.

Table 3. Visual acuity of spectacle wearers in West German factories

a. Near Vision

Number of Spectacle Wearers	Visual Acuity	
	20/20 and higher	Less than 20/20
1,976	38%	62%

b. Far Vision

Number of Spectacle Wearers	Visual Acuity	
	20/20 and higher	Less than 20/20
1,236	32%	68%

The result for spectacle wearers is—as shown—still worse than the result for non-spectacle wearers. Sixty-eight percent of this group had a smaller visual acuity than 20/20 for far distances, and 62 percent for near distances. Even neglecting the fact that a certain percentage of spectacle wearers never get full visual acuity, we have to learn that also persons who know about the importance of visual correction frequently use wrong glasses. An attending questionnaire worked out by the Demoscopic Institute of Allensbach shows that many persons do not use their own spectacles, but spectacles of relatives, or neglect repeated medical and optometrical examinations. They use scratched eyeglasses, not exactly fitted frames, etc. On the one hand it is a general experience over the whole world that younger people are less careful than older ones. On the other hand the visual acuity drops remarkably with increasing age. Therefore the drop in acuity shown in Table 4 may be still more marked.

Table 4. Percentage of subjects with visual acuity less than 20/20 in different ages (all 9,468 workers in West German factories were examined with their usual spectacles if eyeglass wearers)

Age and Number of Persons	Far Vision	Near Vision
Less than 20 (1,329)	28%	15%
21-30 (3,193)	37%	32%
31-40 (1,857)	39%	35%
41-50 (1,497)	55%	53%
51-60 (1,203)	63%	68%
Over 60 (389)	65%	56%

For the comparison of visual acuity with working success Professor Schmidtke and his staff worked out a special criterion. All objective methods which describe the working effect do not allow for a comparison of the different kinds of work. Therefore a combined objective-subjective method was preferred. It consisted of interviews with the superiors which are then combined with the results of working effect within a period of not less than half a year. A special correlation factor was computed with characteristic samples. In this way the subjects were classified into four groups according to their working success. These groups are enumerated from 1 (optimal group) to 4.

Table 5 shows the percentage of spectacle wearers in the different groups and a characteristic increase of spectacle wearers from the optimal group to the worst group. This result, surprising at first, originates from the drop of the visual acuity as shown in Table 3. Because the amount of undercorrected persons is higher for the group of spectacle wearers than for non-spectacle wearers this result may change as correction is improved. There is no reason for any argument that spectacle wearers have less working success than persons with normal eyes. Indeed, all statistical calculations show without any doubt that not the use of spectacles but the drop of visual acuity produces also a drop of working success. This correlation has the same high significance in all samples which were selected for the exact computation. Details may be extracted from the book of Schmidtke and Schober (1967).

Table 5. Number and percentage of working success for 9,468 workers in West German factories.

| | Working Success | | | |
	Very high	High	Medium	Low
Number and Percentage	1028	4323	3305	812
	11.2%	44.9%	35.3%	8.6%
Number and Percentage of Eyeglass Wearers in These Groups	186	1062	860	260
	18%	24%	26%	32%

The probability for an exact correlation between visual acuity and working success is higher than 99 percent. If the reliability coefficients are introduced the correlation between working success and visual acuity for near distances gets 0.3535 (for far distances 0.3063). That means that variations of the visual acuity for near distances influence the variations of the working success by more than 12.5 percent. For distinct working places this correlation is still much higher.

If the same correlation is worked out for watchmaker factories, the variations of visual acuity influence the variation of working success by more than 56 percent. Also in the textile factories, in transistor factories, and in working places which require a good visual performance, the correlation probability is higher than 25 percent. Surprisingly a very small percentage was sometimes found for working places which need a very high visual performance. By analyzing this paradoxical effect we found that the workers at all these places had to pass a very strong medical inspection with special regard for visual acuity before getting their jobs.

With such a preselected sample the variations of visual acuity are too small to produce an influence on the working success. Therefore also this result is a remarkable proof for the correlation marked out above. No high differences were to be found among persons who wear bifocal, trifocal or separated far and near spectacles. For this purpose the samples have been too small. A much more extended statistical work is necessary to solve this question with similar accuracy.

Although there is such a strong correlation between visual acuity and working success, the difficulties increase remarkably if the same research concerns other visual qualities. Even for the correlation between heterophoria and working success it is impossible to adapt the same statistical method. For this purpose another computer program is unavoidable. Exophoria and esophoria have different influences on far vision and on near vision. Therefore such a simple correlation as for visual acuity is not to be found. Heterophoria must also influence the working success because some other research (Schober, 1960; Carapancea, 1958; Roser, 1952) proves that uncorrected or undercorrected heterophoria produces headaches, visual fatigue, and some other nervous troubles. By dividing the subjects according to their working success into only two groups, an above-normal and below-normal working success, Schmidtke and Schmale were able to show that the percentage of persons with heterophoria or anomalous ACA-gradient is much higher in the below-normal success group than in the above-normal group. That means that the probability for low. working success is higher for subjects who have heterophoria or an anomalous ACA-gradient than for normal subjects.

The difficulties in finding out a distinct correlation between visual qualities and working success increase still more if stereoscopic vision or color vision are included. The tests for stereoscopic vision are not as exact as the tests for visual acuity and heterophoria. Their usefulness has been frequently discussed in literature. However, it is also well known that car drivers and other persons who have lost an eye through an accident compensate for this disadvantage very quickly. Even crane operators with the same defect show no conspicuous drop of working success one or two years after such an accident. They use monocular factors such as size of the items, distribution of light and shadow, motion, and geometrical perspective for the compensation of their defect. Similar results exist for color vision. It is well known that the statistical distribution between trichromatic people and color defective people is not strikingly different in normal life and in factories, where good color vision is expected (paper mills, painters, color factories, textile factories, etc.). This is at least the case if color defective persons have not been excluded from such a job by special medical control. Ophthalmologists and psychologists are familiar with the camouflage of color deficiency in all these cases.

The exact correlation between visual acuity and working success allows for the composition of visual claim profiles for different working places, which are stated in distinct schedules for the Ortho-Rater or for other types of screening equipment. Schmidtke, Schober, Schmale, and Wallmann (1967) designed such schedules for 44 different working places in textile factories, watchmaking factories, mechanical and optical factories, and for the electrotechnical and graphical industries, as well as for distinct working places in engine factories. Examples are given in this book. These schedules concern the different success groups and may be very useful to personnel offices and hygiene staffs. Because these schedules are explained by exact statistical computations they may be more useful than some older ones which are designed only by experimental experiences.

Summary

Many experimental and theoretical papers concern the correlation between the visual qualities of the human eye and working success. Most of them do not fit with modern situations because nearly all factors which influence the results have changed during the last decade. A team of biophysicists, physiologists, and psychologists in Munich, Germany, using exact techniques, examined 9,468 male and female workers in different factories to find out the correlation between visual acuity, heterophoria, range of accommodation, stereopsis, color vision, and other visual qualities with the working success. The results were computed with the methods of mathematical statistics. Nearly 50 percent of the examined persons worked with a visual acuity of less than 20/20. This surprising result looked even worse when only spectacle wearers were considered: 68 percent of far glass wearers and 62 percent of near glass wearers had insufficient visual acuity. That demonstrates strikingly that the propaganda for the need of optimal vision is not adequate. A questionnaire proved that many workers wore the wrong spectacles, e.g., the spectacles of their parents, etc. A strong correlation is to be found between age and degree of visual acuity. There is also a strong correlation between visual acuity and working success (a probability of more than 99 percent). The variations of visual acuity explain more than 12.5 percent of the variations of the working success. For many types of employment the visual acuity is a decision factor.

The correlation of heterophoria and defects of accommodation and convergence with the working success was only indirectly proved. The percentage of orthophoria and normal ACA-gradient is higher in the groups with higher working success than in the groups with the lower working success. No proof could be given for the correlation between working success and stereopsis or color vision. The causes for this result are discussed. Forty-four visual profiles for the different working places were worked out.

References

Betsch, A. (1929). Uber die menschliche refraktionskurve. *Klin. Monatsblatter f. Augenheilkunde 82*, 365-379.

Carapancea, M. (1958). La neurose hypometropique. *Rev. Oto-Neuro-Ophthal. 30*, 321-324.

Gordon, D. A., Zeidner, J., Zagorski, H. J., and Uhlaner, J. E. (1954A). A psychometric evaluation of Ortho-Rater and wall-chart tests. *Amer. J. Ophthal. 37*, 699-705.

Gordon, D. A., Zeidner, J., Zagorski, H. J., and Uhlaner, J. E. (1954B). Visual acuity measurements by wall-charts and Ortho-Rater tests. *J. Appl. Psychol. 38*, 54-58.

Imus, H. A. (1950). Comparison of Ortho-Rater with clinical ophthalmic examinations. *Trans. Amer. Acad. Ophthal. and Otolar.*

Roser, J. (1952). *Untersuchungen über die Wirkung des Leuchtstoffrohrenlichtes auf Gesundheit und Leistung des Arbeitenden Menschen.* Arbeitsgemeinsch. Med. Verlag, Berlin.

Schmidtke, H., and H. Schmale (1961). *Sehanforderung und Berufseignung.* Verlag, Bern und Stuttgart.

Schmidtke, H., H. Schober, H. Schmale, and G. Wallman (1967). *Sehandorderunger bei der Arbeit.* A. W. Genter, Verlag, Stuttgart, West Germany.

Schober, H. (1960). Astenopische Beschwerden, ihre Uraschen und die Möglichkeiten zu ihrer Behebung. *Studium Generale Heidelberg 13.* Issue 9.

Schober, H. (1958). Ein Handoptometer fur den Klinischen Gebrauch. *Klin. Monatsblatter f. Augenheilkunde 132*, 246-248.

Richard L. Feinberg

The Role of NINDB
in Furthering Vision Science

Perhaps at the outset, I should explain what "program analysis" implies. This may answer the many queries I have received regarding my post as Program Analyst for Vision for the National Institute of Neurological Diseases and Blindness.

For the greatest advancement in research, there must be a continuing review of research development; a search for new and important opportunities; and means found for the rapid and effective development of new areas and techniques. Within the Office of the Director, National Institute of Neurological Diseases and Blindness (NINDB), the program analysis group has the responsibility for a review of the total research program. When it is evident that certain areas require increased efforts, symposia are organized or review committees are created to consider the promotion of research within the area of neurological disorders and blindness.

May I say, parenthetically, that I am not in grants management. The individual who says, as in the jest in *Science,* "While you're up, get me a grant," is apt to be disappointed.

In this paper I hope to review the growth of vision research and training supported by NINDB during its sixteen years of existence, the relationship of NINDB with other institutes at the National Institutes of Health, with outside agencies, and with the health professions. Such a large part of the United States' support for health research is derived from appropriations of the National Institutes of Health that these Public Health Service programs, and the principles that govern their administration, should be known to all segments of the scientific community.

A breakdown of the sources of support of medical and health-related research shows, as is true with all research and development in the United States, that federal programs have contributed the major share in supporting expansion in these fields. Of the estimated two billion dollars spent on health research during the fiscal year 1967, the federal government contributed two-thirds.

Today, seven federal agencies conduct or support medical research or fundamental research relating to the health fields. These are: Department of Health, Education and Welfare, Department of Defense, Atomic Energy Commission, Veteran's Administration, National Science Foundation, Department of Agriculture, and the National Aeronautics and Space Administration. The Department of Health, Education and Welfare is responsible for over 88 percent of the total federal expenditure for such research.

The National Institutes of Health was given more responsibility and independence under another reorganization step announced April 1, 1968, by Secretary Wilbur J. Cohen of the Department of Health, Education and Welfare. The expanded National Institutes of Health includes the Bureau of Health Manpower and the National Library of Medicine.

Within the new National Institutes of Health, educational activities now have an equal status with research functions, a goal long sought by NIH's director, Dr. James A. Shannon. Dr. Shannon continues as director of the expanded NIH.

About 85 percent of the HEW's appropriation for research in the health fields is invested in Public Health Service programs conducted and administered by the National Institutes of Health. As the Public Health Service's primary research arm, the National Institutes of Health has certain functional responsibilities. These include:

1. Broad support of research in non-federal, academic and other research institutions. Research grant applications are reviewed initially by study groups composed of nongovernment scientists. Their consensus is the determinant of the technical merit and relevance of specific projects. Their recommendations are forwarded to the Offices of Extramural Programs and, in turn, to the various Institute Advisory Councils which make the final decisions on grant applications. Once these hurdles have been crossed, they are ready for funding.

2. Training of specialized manpower to expand and enhance the nation's research and teaching capability.

3. Construction of research facilities so as to permit the further development of health-related research and training resources of the nation.

The first of the nine institutes to be established was the Cancer Institute authorized by Congress in 1937; NINDB was created thirteen years later in August, 1950.

Among its principal research directives, the National Institute of Neurological Diseases and Blindness was charged with research in disorders of vision. The Institute's eye research and training program represents one of its most rapidly growing programs, showing a 600 percent growth over the past decade. In fact, its growth rate has exceeded that of the Institute as a whole.

Several task forces have been established, and more will be, under the Subcommittee on Vision. These task forces review current research and recommend organized programs of directed research and development. Among the major task forces established are: Ocular Pharmacology and Toxicology, Glaucoma, Eye Pathology Banks, and Refractive Anomalies of the Eye. Others will be established as the need becomes manifest.

Under the Subcommittee's direction, a Workshop on Refractive Anomalies of the Eye was held in 1965. A report of the Institute-sponsored workshop on refractive problems has just been published and is about to be distributed. It contains papers by many of the world's leading authorities on the subject, both from the professions of optometry and ophthalmology.

In a preface to the report, Dr. Richard L. Masland, Director of NINDB, notes that refractive anomalies are among the most widespread of all health impairments, and that, paradoxically, there are few active research programs aimed at exploring their underlying mechanisms.

Primarily intended as a stimulus to investigators in the visual sciences to pursue new research possibilities, the document also contains much information of general interest. Papers are grouped into sections on the present state of knowledge, diagnostic and therapeutic techniques, promising avenues for research, clinical importance of refractive anomalies, and recommendations on areas of need.

I should like to state the recommendations of the Workshop.

1. It is recommended that there be developed accurate, reliable, and easily applied techniques for measuring the components of the refractive apparatus of the eye, including techniques suitable for infants under three and for clinical and anthropological field studies. Instrumentation should be developed with the aim that it be sufficiently simple and inexpensive to find widespread application in the studies recommended below.

2. Since refractive errors and presbyopia involve structural factors and interaction of biochemical systems in the various tissues of the eye, it is recommended that there be further studies in this direction and that these make maximum use of material obtained from eye banks, enucleated eyes, autopsies, and parallel animal studies.

3. It is recommended that there be intensified studies, including the experimental manipulation of environmental and physiological factors of animals, particularly primates.

4. It is recommended that there be utilization of carefully controlled

clinical material to assess the role of environmental, biochemical, embryological, pharmacological, neurological, and physical factors in the development and progression of refractive anomalies. As far as possible, these studies should be conducted using double-blind experimental designs.

5. Continued attention should be given to new procedures and wider utilization of known ones for the rehabilitation of patients afflicted with substantial refractive errors, including the surgical, optical, physiological, psychological, and social facets of the problem. Particular attention should be directed to sequelae of refractive errors which give rise to visual disability. Important areas appear to be high myopia with degenerative changes, marked hypermetropia associated with nystagmus, strabismus, and amblyopia, and refractive conditions causing abnormalities of binocular vision.

6. It is recommended that there be large-scale, long-term longitudinal studies of refractive errors and the components of the refractive apparatus, starting at an early age, to provide a description of the frequency, biological variations, and natural history of refractive abnormalities. These studies should be designed to facilitate correlation between known or suspected genetic, gestational, and postnatal factors and should include selected groups allowing maximum insight into genetic factors (e.g., twins, families with refractive errors) and into environmental factors (e.g., primitive societies and societies in the process of transition from the rural to the urban state).

It is our hope that these excellent recommendations will be carefully followed in extension of further NINDB-sponsored research.

Last year, the Institute established the Vision Information Center at the Harvard University Library. Its purposes are to hasten dissemination of scientific information and to aid in program analysis. Objectives of the Center are to define, identify, store, retrieve, and disseminate the literature of vision so that the information may be communicated more quickly and completely. The scientific community will be encouraged to make complete use of this facility. Integration of the activities of the Center with the national network of specialized information centers now being developed with the support of NINDB is underway.

An important element in the NINDB mandate is the training of scientists to do teaching and research. Indirect support in the form of training grants leaves the universities a free hand in conducting their training programs without interference from the federal government. These grants are awarded only to schools with established programs which meet certain standards, but the Institute occasionally offers financial help to aid a university in setting up a training program.

The Institute's training program also includes a variety of fellowships which provide stipends for individual scientists who need graduate training in research techniques and career development awards which give full-time support to scientists conducting basic laboratory work in universities or other research centers.

Of the total number of NINDB grantees, I am sure that it will interest you to know that one-half are non-M.D.s and possess degrees of Ph.D., O.D., D.O.S., Ed.D., D.Sc., Dr. Eng., and D.V.M.

This year the Institute allotted about $14 million for individual research projects on vision. In addition, there were 59 intramural projects, a number of collaborative and field projects, and some direct contracted studies. Twenty-one centers for specialized group interdisciplinary attack on specific problems of vision were also funded. Altogether for the fiscal year 1967, 23.3 million dollars have been expended on vision research.

The wisdom of these programs of research and training can be observed in the program that you have here this week. At least eight of the speakers on this program are, or have been, recipients of grants provided by NINDB. We are proud of the fact that two of the NINDB grantees have received Nobel prizes during the year 1968.

Tangible evidence of accomplishment may be found in the multidisciplinary areas of research which NINDB supports. Fundamental knowledge is still lacking in the understanding of causes of many blinding diseases and of refractive anomalies. Since not only the eye but also the nerve pathways in the brain are necessary for vision, basic neurological research is involved. A concerted attack on the eye disorders necessarily includes a consideration of the systemic disorders that may involve the eye, such as diabetes, diseases of the blood vessels, the diseases of the nerve and brain. Research is supported by NINDB in all of the major causes of blindness, including cataract, glaucoma, diabetic retinopathy, corneal scarring, uveitis, retinal detachment, tumors, and amblyopia.

Refractive anomalies occupy a special place since these produce the greatest number of victims, but it is also an area in which most of them can be helped with glasses or other prosthetic devices. The need for eye-glasses is probably the greatest single inheritable defect of the human race. Broadly based epidemiological genetic studies are underway to assist in solving the problems of refractive anomalies, and Institute research has been initiated relating to understanding the cause of amblyopia, together with the development of early diagnostic treatment methods. Amblyopia, rather than being a single disease, is a failure of learning how to see. Time is a vital factor because vision development ends by the age of six or seven years. After that, the amblyopic eye can no longer learn. Drs. Torsten N. Weisel and David H. Hubel of the Harvard Medical School have proved that pattern perception is the function of the cortical cells and occluded or unused (amblyopic) eyes quickly lose connection with these cells. Since it may take a year or two to correct the weak eye and to teach fusion, experts say that the child with amblyopia should be found by the time he is three or four years old. By the time many get their first real eye examinations the damage is irreparable. Each year, at least a hundred thousand children are passing the point at which they can be rescued.

The number of such victims has been made painfully clear by some recent surveys. One study of 60,000 airmen revealed that 3.2 percent of the Air Force personnel had virtually no sight in one eye. Another survey of 190,000 draftees showed amblyopia in more than 4 percent. The rate may be even higher among civilians. The challenge to the eye professions is clear. There can be no greater opportunity to preserve sight and avoid a lifelong, crippling handicap than in the case of amblyopia. Our Institute is particularly interested in encouraging the professions to examine the vision of children during their early years.

We have had some dramatic breakthroughs. The discovery of the cause of retrolental fibroplasia (RLF), which, in 1952, was causing blindness at the rate of 1,888 children a year, is one example. That year, the Institute helped initiate a coordinated study by clinics and hospitals throughout the country to examine the suspected relationship between RLF and the amount of oxygen given premature infants. In two years, the findings in these cooperating institutions offered convincing evidence that excessive amounts of oxygen was the cause of the blinding disease.

Scientists estimate that 10 percent of all infants under 2,500 gm. birthweight are afflicted with the respiratory distress syndrome, a condition affecting approximately 40,000 American-born premature infants annually. A combination of cardio-pulmonary abnormalities is believed to deprive these infants of an adequate tissue oxygen supply. Without added high concentration of oxygen it is estimated that at least 50 percent of such babies die.

The return to use of high oxygen therapy represents a change from rigid controls imposed in the 1950's following the discovery that overuse of oxygen causes blindness in infants. Nevertheless, increasing evidence demonstrates the need for high oxygen therapy to combat the life-threatening respiratory distress syndrome.

Other illustrations of the accomplishments of grantees of NINDB could be offered: investigations of the human macula (University of California, San Francisco Medical Center), adding to our understanding of the degenerative changes which take place in the eye as a consequence of aging; techniques of ultrasound in determining the dimensions of components of the eye and providing visualization of the posterior segment; the clinical usefulness of two new electrophysiologic methods; ERG and evoked occipitogram (EVOG) on infants and toddlers (in a study by Dr. J. Lawton Smith, Miami, and Thorne Shipley, Yale); a new method for freezing corneal tissues for keeping transplants alive (Dr. H. Kaufman, University of Florida), and many more. Time does not permit the extensive catalog.

The domain of vision sciences is not overlooked by NINDB; on the contrary, it is given every impetus through encouragement, advice, and funding. Our great concern is finding more researchers in the field capable of carrying on work against the challenging problems. Conferences like this may help impress upon our educators and scientific community the need for inspiring more young people to enter into this province. Robert Frost once defined wisdom as that quality which motivates man to act in spite of insufficient knowledge. Research provides the means for man to act with more knowledge. NINDB is a catalyst which makes such research possible.

Lawrence W. Macdonald

Visual Training—
Its Place in Visual Science

Visual training has been defined as the art and science of developing visual skills to achieve optimal visual performance and comfort. The term visual is used to connote function. The notion of visual function extends from consideration of the physiological processes involved on the one hand to how we apparently see the world about us on the other.

In a sense, visual training involves both of these aspects: the physiological correlates and the visual space world. The visual space world has been defined by Skeffington as the sum total of the visually related energies with which the organism is interacting at a given moment. It has also been called the volume of pertinence.

The value of visual training to visual science is that of applying and synthesizing the knowledge of visual science and applying this product to enhance the effectiveness of the visual processes of people. This in turn allows people to see more in less time.

Visual science in effect has been with us for a long time. The history of philosophy is largely a history of early visual science. Appearance was a major parameter in most philosophical attempts to explain natural science. As such, most philosophers have considered how and why things appear the way they do.

Heraclitus (about 470 B.C.) summed the situation by implying that the world is the way the world is. It is everchanging. The process of change is one. Man sees the world as he sees it. Each one of us views the process from different points of view, from different experience. It is only as we can continue to view the process of change that we can factor out the invariables in the situation. Through this process one can derive a notion of how the world exists. Through this process, one can attain wisdom.

On the other hand Parmenides (500–450 B.C.) stated that the world is motionless. Any apparent change is an illusion. The world does not change. It only seems to change. That which may be spoken of, or thought of, is what is. Parmenides believed there was no void, no empty space.

Empedocles (494–434 B.C.) also believed there was no empty space. Light, being finer, filters through fine "pupils" in space. These pupils were protected by fine coverings. Objects were perceived by having their light filter into the pores of the senses. In the case of vision, he claimed that the eye was made up of fire and water. It was surrounded by earth and air. Light being fine enters the pores and separates fire and water. Fire pores recognized white objects and water pores recognized black objects.

Since there was no empty space, there was no notion of light reflecting from the surface of objects and then entering the eye. It seemed as if models of the object filtered through the fine covering of what he called the pupil. With this theory there remained the problem of why one could not see in the dark.

Anaxagoras (500–428 B.C.) attempted to answer this question. He advanced the notion that sensations take place by opposite qualities. Like is not affected by like. Anaxagoras observed that the reflection from the anterior surface of the cornea imaged the terrain in the pupil of the eye. Since the pupil was dark, and the terrain was light he believed this allowed for the situation of opposites that allowed sensations to take place. This explained why man could not see in the dark.

Here we see visual science in operation. Observations were made. A theory is formed to explain the observations. The question is asked, does the theory account for all observations? When the answer is no the theory is modified to account for the inconsistencies.

Democritus (460–351 B.C.) introduced the notion of empty space. He claimed that the world was made up of atoms and empty space. He agreed with Empedocles and Anaxagoras that images passed through pores and into the senses. He believed that color did not exist in nature. He speculated that atoms had different shapes. White was represented by a smooth shape. Black was represented by a rough shape. He believed that color was represented by an arrangement of proportion and impulsions.

Democritus believed that vision was explained by reflection. The atoms were said not to strike the eye directly. Instead they imprinted the intervening air. This imprint was reflected by the moist eye.

These notions are similar to the corpuscular theory of Newton. As such they bear on contemporary thought in visual science.

The thinking of Democritus with no knowledge of refraction provided many problems. Some of them were: how does one get the great large image of the mountain inside the comparatively small eye? How can one explain afterimages? How can one explain mirror images? (Ronchi, 1957).

Ibn al Haitham (Alhazen) (about 1000) attempted to explain some of these problems with his theory of visual rays. He showed that the rays from the mountain could converge in a manner that would allow them to fit into the eye. The problem still remained as to why the rays would converge faster for anyone standing close to the mountain. It was not until 1500 when Maurolico of Messina advanced the notion that rays emanate in all directions from every point (Ronchi, 1957).

Alhazen was aware of the inverted image at the posterior eye. Since he realized that people do not see the world inverted, he concluded that the sensorium of the eye must be the crystalline lens (Ronchi, 1957). Held (1965) has shown the effect this notion had on the knowledge of the anatomy of the eye. The crystalline lens was placed at about the center of rotation of the eye.

It was not until 1600 when Kepler discovered the general laws of refraction that the sensorium was placed at the retina of the eye.

The science of optics was further advanced by the works of Kepler, Descartes, and Newton, to name a few. This work was mathematically oriented. It considered the figural aspects of the visual field only. Because it marked that first major advancement of science in the Western hemisphere since the days of Aristotle, it is called the "new science." Although this work marked an advancement in the area of physical science, it contributed little to our understanding of visual perception.

It remained for John Locke (1632–1704) and Bishop Berkeley (1685–1753) to make the next contributions to the area of visual perception. There was much in common between the theories of these two philosophers. Yet they had one critical and important difference. The difference was in the area of abstraction. Locke was a strong empiricist. He believed that all information came through the senses. He believed that information from the various sense modalities could be abstracted into a unity. Abstraction took place by finding that which was common to the sensory information. Locke claimed that we could abstract the notion of roundness through our sense of sight and touch. The common consistency between the visual inputs and the tactile inputs could be abstracted into an idea of roundness. This abstract idea could then be considered without referring back to the original sense information.

Locke's work was strongly influenced by the new science. His ideas on abstraction were an extension of the mathematical concepts of abstraction. His work in perception was fundamentally "figure" oriented.

On the other hand, Berkeley claimed there was no such thing as an abstract idea. One could have an association of ideas through use. One could develop general ideas in the sense that one idea could represent a whole family of ideas. However, he believed that tactile information was always tactile information, and visual information was always visual information and that the commonality was never abstracted.

He believed mathematically, for instance, that notions of general ideas may be formed, but that any such idea must always be referred to a particular problem: a particular line, a particular triangle, and the like. Any attempt to solve the equation for "n" number was purely speculative. He disagreed with Newton in the notion of absolute space and absolute time and claimed that we can think only of particular space and particular time.

Berkeley was the first to play down the emphasis on the part that figure plays in visual perception. Through his notion of association of ideas, he was the first to show that the interplay of all ideas was important in understanding how men see the world as they do.

In his "Essay Towards A New Theory of Vision," Berkeley did three things. He showed that one cannot explain visual perception on the basis of ray tracing alone. He developed his notion of what we now call the visual space volume including, to my knowledge, the first investigation into the area of size constancy. He talked of the problem of how man sees the world upright with the inverted retinal image, a problem which is apparently still being discussed today (Young, 1962).

Berkeley, in considering what he called the minimum visible, was the first to advance the notion of the interplay of figure and supporting ground. He used the term minimum visible very much as we now use the term "visual space volume" (the sum total of the visually related energies with which the organism is interacting at a given moment). At Berkeley's time conventional wisdom apparently used the term minimum visible as we use the term minimum separable. It was an expression of the minimum angle that could be subtended between two points and still have them seen as two points. Berkeley stated that this situation never exists. Berkeley indicated that never were just two points stimulated on the retina. The total retina was always involved in any visual situation. Therefore it was folly to talk in terms of minimum visible unless one considered the relation of the figure to the total field. Berkeley therefore insisted that minimum visible was the total visual field. Through his notion of association of pertinent ideas, this notion can be extended to the contemporary concept of the visual space volume.

The role of optometry and visual training in visual science has been to "kitchen test" the various theories of vision and visual function. As such, it has found support for some notions, found many inconsistencies in other notions, and developed some elaboration of knowledge in visual science to explain some of the apparent inconsistencies.

One of the notions of the new science was that of the need for conjugate focus of the eye and the object of regard in order to see it clearly. One of optometry's early contributions to the theory of visual science was to develop evidence which would lead to believing this notion is inadequate for explaining the part the eye plays in visual perception. This came about through the use of the retinoscope.

The retinoscope is an instrument used by optometrists to determine the physiological focus of the eye. In the early days it was used as a device for determining the absolute refractive error. One problem with this notion was that people could not wear the glasses that were prescribed by this technique alone.

In 1911 Andrew J. Cross published his observations of the use of the retinoscope at near point. Cross placed a nearpoint target on the retinoscope. Patients were asked to attend the target and see it clearly. According to the laws of refraction, in order to see the target clearly, the target should have been in conjugate focus with the eye. Cross observed that this was not the case. It was later called a lag of accommodation by Tait. Much discussion has been given to this lag of accommodation.

In 1924 Skeffington made another discovery with the retinoscope. In order to scope along the visual axis and thereby come up with an objective measure of the refractive error that was free from error caused by aberrations, a device called an Ambrewster was developed. This was a system that utilized mirrors, so that the patient could maintain fixation on the farpoint object while the optometrist scoped along the visual axis of the eye.

One of Skeffington's first patients was a three diopter hyperope. Skeffington asked the patient to read the chart while he observed with his retinoscope. The patient read the small type, indicating that the patient's eye was in conjugate focus with the chart. Yet, the retinoscope revealed that conjugate focus was not, in fact, present. The retinoscope showed three diopters of hyperopia. This can be approximated every day in an optometric office without the Ambrewster device. Have a hyperope read the fine letters of a test card placed at farpoint. When the patient is seeing the letters clearly, theoretically the eye is in conjugate focus with the chart. Yet, the retinoscope will reveal a "with" motion (Skeffington, 1931).

It was another observation with the retinoscope which led optometry to consider the role of problem solving in visual function. In 1946 while investigating how visual skills develop in children with Arnold Gesell, G. N. Getman reported a startling observation. The purpose of the study was to investigate the accommodative function of the eye of children. The subject was seated on the nurse's lap, the optometrist was in a position to use his retinoscope. In order to maintain the attention of the child, a series of pictures and line drawings were projected on a screen. The startling observation was that the child's focus often changed with the slide. Focus varied in appearance from hyperopia to myopia. Transient astigmatism also manifested itself (Gesell, 1949; Getman, 1958).

Prior to this time, it had been thought that the accommodation would vary only in response to changes in the external energy situations. For example, if the target was moved closer, the accommodation would change. The observation of Getman indicated that it was also responding with changes in internal energy. Upon further investigation, it was found that it was related to problem solving.

Retinoscope observations led to optometry's first contribution to visual science. This contribution was essentially negative in that it demonstrated that there were more parameters in visual optics than had previously been considered. We could not account for these observations with the prevailing theory.

Similar contributions came from visual training. Optometric visual training dates to the early 1900's (Gesell, 1949). The Arneson squint corrector developed during that time is still a valuable adjunct to any visual training office. Arneson was one of the first to bring in the notion of periphery into visual training (Arneson, 1934).

Peckham (1926) used visual training to enhance binocular function and reduce asthenoptic symptoms. At that time, the fundamental tenet of visual training was to build greater ranges into the system. It was thought that this would make the patient less susceptible to fatigue and the symptoms of discomfort. It was also thought that it would allow the patient with crossed eyes to straighten his eyes, and the like. Optometry was still oriented to the "New Science" of the seventeenth century. It was steeped in the notion of the physical eye, amplitudes of accommodation and convergence, and the relationship between accommodation and convergence.

In the early days, visual training was based on the notion of muscle strength. In theory, if a patient showed a muscle imbalance of exophoria, for example, the optometric practitioner would attempt to build adductive muscle strength to allow the patient to compensate for and overcome this imbalance. The clinicians found this did not work. Patients were trained to have adductive capacity of 100 prism diopters and more. Yet their symptoms prevailed.

Evidently there was more than amplitude involved. Later, Lancaster proved this by showing that the weakest muscle of the eye is 50 to 100 times stronger than it needs to be to perform the physiological act of moving the eyeball in its socket (Lancaster, 1937). All muscles are too strong to start with.

In the 1920's, Skeffington developed what I call his cushion concept. He postulated that the body and the visual system needs ranges in order to function effectively. He advanced the notion that the ideal eye was not emmetropic and orthophoric as had been previously thought. He claimed that the ideal eye would be slightly hyperopic and exophoric. He believed that hyperopia operated to protect the integrity of convergence.

He further elaborated this theory to account for various refractive and extraocular muscle problems. He wrote as follows:

It becomes obvious, therefore, if these things are true, that the refractive error is not necessarily a matter of the physical characteristics of the eye ball itself, but may be a matter of association of energy expenditure to the functions which control vision. It must be fully realized that the eye is merely a sensory end organ. It might very properly be called a collection of dendrites, whose functions are to receive the stimulation of light and carry that stimulation back to the brain where it is interpreted.

Quite properly it may be said that the eyes do no seeing, but merely record the varying intensity of light and that the brain does all the seeing which is done. Most certain it is that the eye is far from resembling a camera but is really the prototype of that much newer apparatus of tele-vision...

It is a truthful, accurate and scientific statement to say that the eye sees nothing, but that the brain does all seeing. The refractive error then is merely the measure of the end result of the amount of innervation which varying forces and demands have caused to be utilized in accommodation, outside of the habitual linking of the accommodative-adductive relationship. (Skeffington, 1931)

Through the work of Skeffington and associates visual training took on a new purpose. Visual training techniques were designed to build greater freedoms and more accuracy of eye movements along with greater flexibility of accommodation. For the first time visual training embraced the notion of prevention and rehabilitation.

With the demand for freedom and accuracy of visual function came increased awareness of the relationship of the viewer and the viewed. It was logical extension to move from here into areas of perception of visual space. For the first time, attention was paid to philosophy and techniques for developing visual skills. Along with this came the notion of the involvement of the total organism in the act of visual perception. Crow and Fuog reflected the thinking of the day in their chapters on visual training published through the Optometric Extension Program (Crow and Fuog, 1937). This thinking led to the work of G. N. Getman on visual development in children, some of which has been previously cited. It has been the merger of the notions of visual training and the studies of the development of visual abilities in children that have led to present day visual training.

Visual observation made in the practice of visual training is still raising many questions in visual science. Some of these questions relate to fundamental concepts. It has been cited that optometry has long been involved with the inconsistencies of visual science as they relate to the accommodative mechanism. Present thinking among some optometrists would hold that the purpose of accommodation is not to get a clear round focus on the retina. Indeed, we would hold that one of the purposes of accommodation is that of an equilibrating mechanism (Macdonald, 1963). Simply stated, the theory would suggest that by increasing the convexity of the crystalline lens, the energy input to the eye would be spread over a greater area, thus decreasing the intensity per any given area (Macdonald, 1962, 1965). This would serve to balance the external and the internal energy in the system.

There is presently coming into vogue the notion that the crystalline lens may not be wholly responsible for the effects of accommodation. This notion was first suggested by the work of Ludlam on photokeratoscopy (Ludlam, 1967). It is supported by observations in visual training. In a monocular situation, when accommodating to overcome the effects of a minus lens, some patients will report an apparent displacement of the field of view to the right or left. Since the situation is monocular, it would seem that the impact of any accommodative convergence relationship would be greatly reduced, if not eliminated. There would be no reason for the apparent lateral shift of the total field of view if the mechanism for accommodation was the ciliary mechanism only. Indeed, this observation may shed some insight on the nature of the accommodative convergence relationship itself, a phenomenon which has never been clearly understood. These observations are further discussed elsewhere (Macdonald, 1968).

Optometry through visual training has been working in the area of the organization of visual space and its relationship to performance since the late 1930's. I think it is generally possible to support the views of both John Locke and Bishop Berkeley at the same time. Yet there are some problems that require consideration.

I believe that most optometric practitioners doing visual training would basically agree with the following tenets:

1. A child is born. Characteristically the child is born with the standard complement of parts that will be needed to succeed in the episode of life.

2. The initial development is triggered by various reflex systems (as opposed to John Locke who did not believe in innate ideas).

3. Through the interaction of experience, one learns to organize his visual space world.

4. We can support both Locke and Berkeley simultaneously when we

attempt to develop awareness of differences (Berkeley) while at the same time trying to derive similarities (Locke). The process of abstraction might be defined as the deriving of similarities out of the differences without losing the identity of the differences.

5. It is only as we can expand our visual space volume to embrace the similarities and differences simultaneously that we can grow in our ability to process information and enhance our ability to perform in space.

6. Visual deviancies represent a skew in this information processing. They represent an adaptation on the part of the individual to force freedom into his visual space volume in response to particular demands imposed either externally or internally. He cared enough to try. He did what he had to do to succeed.

The differences between Locke and Berkeley should cause optometry to reexamine some of its purposes in visual training. We should not let the challenge offered by these two great thinkers go unanswered. We must try to be sure that we are not building patients in our own image (shades of *Brave New World*). I believe the ultimate purpose of visual training is to develop individual identity and virtuosity simultaneously.

Through visual training we teach and encourage patients to abstract. For example, in techniques to develop size constancy, we stress that things should *look* to the patient the way the patient *knows* them to be. That this is possible supports the notions of Locke.

There are those individuals who say that they see distant objects as small as would be predicted by the angle subtended by the object at the nodal point of the eye. These same patients will often state that they know how large the object is by their tactile sense. However, they state that they do not see it this way. This performance would support the notion of Berkeley. Are we doing wrong by our patients to teach them to abstract?

Berkeley did not consider this problem. His intent was to show that people see magnitude less in terms of their visual ideas and more in terms of their tactile ideas. It would seem from his work on size constancy that, when asked to give a judgment on the magnitude of a distant object, one would respond in terms of his tactile sense. Berkeley's purpose was to emphasize the difference between the ideas derived from the two sense modalities and to emphasize that one did not need the doctrine of abstract ideas to explain the result.

But what of the individual who seems to be too aware of the differences and gives both answers simultaneously? It would not seem that we should suggest to the patient that he suppress the visual and go on the basis of the tactile information alone. This would not resolve the problem. The alternative is to teach the patient to abstract.

In support of this notion is the observation that most patients with the seeming problem of lack of size constancy have difficulty with space-time and number concepts. It would seem desirable to develop the skill of size constancy to develop proficiency in the areas of space-time and number concepts.

Also, in support of the notion of abstraction is the observation that there seem to be at least two areas of performance which depend on the process of abstraction. These are related to the sense of vision and hearing and involve the process of fusion.

Fusion seems fundamentally to be a process of abstracting and unifying the commonality between the inputs to the right and left eyes and/or the right and left ears. Energy patterns or forms that are common to both inputs go together. Forms that are different are perceived as different.

Mechanisms can be developed to explain fusion. My first attempt to build a model to explain fusion was based on sound (Macdonald, 1961). A later attempt was based on the physiology of the visual system (Macdonald, 1964). A brief summary of this notion is as follows. It is based on the initial hypothesis that fusion is a process and not a place in the brain. The process of fusion involves the total organism.

The visual fields are divided into a right and left half by the distribution of the nerve fibers that allow for the semi-decussation at the chiasm. Energy from the visual field is transduced at the retina and travels along the optic nerve to the lateral geniculate bodies. Energy traveling through the uncrossed fibers finds its way to layers 2, 3, and 5 of the lateral geniculates, while energy traveling through the crossed visual fibers finds its way to layers 1, 4, and 6 of the lateral geniculates (Van Buren, 1963).

In any event, the energy from the right visual field finds its way to the left lateral geniculate while the energy from the left visual field finds its way to the right lateral geniculate.

The theory postulates a sweep of energy that arises from the brain stem and finds its way to the external geniculates, perhaps through the anterior pillar of the fornix. Arden and Soderberg (1961) have suggested that activity from the reticulum formation may well be responsible for the establishing of the resting activity of the lateral geniculates.

It is postulated that the energy from the brain stem would sweep through the geniculates and on through the optic radiations to area 17. This would allow for fusion of energy that was common to the two eyes from the right and left visual field. It would also allow for the picking up and carrying along with it any information of differences that may be contained in the energy distribution. It might be helpful to think of the layers of the lateral geniculates as a series of sieves, each with a separate flavor. As water is forced through the sieves it will pick up the qualities of the different flavors. Assuming each flavor maintains its identity, like flavors would fuse. Unlike flavors would remain separate. Such a theory would serve to explain how it is possible for a cheiroscope to work.

The sweep would carry the information contained in the energy patterns through the optic radiation to visual cortex. Yet, we still do not have complete fusion. We are still left with the "seam problem" as described by Gordon Walls (1948). The "seam problem" may be resolved by interconnections to the right and left visual cortex through the splenium of the corpus callosum. This has been reviewed by Geschwind (1962).

The cerebellum appears to have an afferent-efferent loop to the striate cortex as well as to the midbrain (Snider, 1958). The frequency of the cerebellum is the highest of all of the brain centers being in the order of 300 cycles. It is also interesting that it seems to be fundamentally frequency modulated.

When the cerebullum sends its signals to area 17, it would seem that the signals are of a similar pattern but opposite in sign. This is postulated because of the right left phasic relationship of the antigravity system. This cerebullar activity would set up a rapid oscillatory effect between the right and left striate cortex. The visual information being contained in a frequency that is very slow when compared to that of the cerebellum would tend to piggy-back on the cerebellar frequency. The rapid oscillation that is set up by action of the cerebellum would tend to insure the complete integration of the visual field.

There may be those who wonder how I can avoid talking about the action of the extraocular muscles when discussing the process of binocularity. It is my belief that the action of the extraocular muscles are the result of the organization of the visual energies. The action of the extraocular muscles is not causal. The extraocular muscles take orders. They do not give orders. Deckert suggests that a retinal image in itself is neither the necessary nor sufficient condition for the development of pursuit eye movements. Instead he suggests that the necessary prerequisite is the development of an appropriate cerebral image (Deckert, 1964).

Thus we can postulate a mechanism for abstraction as it relates to fusion of energy from the two eyes. Perhaps we can resolve the apparent differences between Locke and Berkeley in the history of visual science by thinking of abstraction as a verb rather than a noun. For it would seem that both fusion and abstraction are processes.

Perhaps we could extend the science of vision by considering the word vision as a verb rather than a noun. Indeed this is why the Committee on Orthoptics and Visual Training of the American Optometric Association has continued to advise the use of the term visual training rather than vision training. The term visual connotes process.

References

Arden, G. B., and V. Soderberg (1961). The transfer of optic information through the lateral geniculate body of the rabbit. *Sensory Communication.* Massachusetts Institute of Technology Press, pp. 536-7.

Arneson, T. J. (1934). *The New Optometry.* Arneson Clinic, Minneapolis, Minnesota. Second edition.

Cross, A. J. (1911). *Dynamic Skiametry.* A. J. Cross Company. p. 111.

Crow, G., and H. L. Fuog (1937). Fundamental principles of visual training. *Optometric Extension Program.*

Deckert, G. H. (1964). Pursuit eye movements in the absence of a moving visual stimulus. *Science 143* (3611), 1192-1193.

Geschwind, N. (1962). The anatomy of acquired disorders of reading. *Reading Disability.* Johns Hopkins Press, pp. 115-129.

Gesell, A., et al. (1949). *Vision, Its Development in Infant and Child.* Paul Hoeber, pp. 172-185.

Getman, G. N. (1958). Book retinoscope, Parts I and II, Developmental vision. *Optometric Extension Program 2* (10, 11), 65-75.

Held, R. (1965). Object and effigy. *Structure in Art and Science.* Gyorgy Kepes, ed. Brazillier, New York, p. 44.

Lancaster, W. B. (1937). Physiology of disturbances of ocular motility. *Arch. Ophth. 17,* 983-993.

Ludlam, W. H. (1967). Human experimentation and research on refractive state. *Synopsis of Refractive State of the Eye.* American Academy of Optometry.

Macdonald, L. W. (1961). The significance of binocularity. St. Louis Visual Training Conference, Washington, D.C. Transcript by Caryl Croisant, Morro Bay, California.

Macdonald, L. W. (1962). A theory for interpreting book retinoscope observations. *Opt. Weekly 53*(29), 1446.

Macdonald, L. W. (1963). Accommodative rock, its purpose and value. Visual training. *Optometric Extension Program 2* (1,2), 2-15.

Macdonald, L. W. (1964). On the physiology of the binocular system. Visual training. *Optometric Extension Program 2* (7, 8), 45-58.

Macdonald, L. W. (1965). The dispersion factor of a plus lens. *West Coast Visual Training Conference,* San Jose, California. Transcript by Caryl Croisant, Morro Bay, California.

Macdonald, L. W. (1968). Accommodation. *Eastern Seaboard Visual Training Conference,* Morro Bay, California.

Peckham, R. M. (1926). *The Modern Treatment of Binocular Imbalances.* Shuron Optical. Second edition.

Ronchi, V. (1957). *Optics.* New York University Press. Pp. 25, 26, 28, 30, 39.

Skeffington, A. M. (1931). *Differential Diagnosis in Ocular Examination.* Wilton Publishers. Pp. 16-17, 49-52.

Snider, R. S. (1958). The cerebellum. *Scientific American 199* (2), 84-90.

Van Buren, J. M. (1963). *The Retinal Ganglion Cell Layer*. Charles C Thomas.

Walls, G. L. (1948). Is vision ever binocular? *Optical Journal and Review of Optometry LXXXV* (15), 33.

Young, J. Z. (1962). Why do we have two brains? *Interhemispheric Relations and Cerebral Dominance*. V. B. Mountcastle, ed. Johns Hopkins Press. Pp. 20-21.